New Developments in Liquid Crystals

Edited by
Georgiy V. Tkachenko

New Developments in Liquid Crystals
Edited by Georgiy V. Tkachenko

Republished by InTech
Janeza Trdine 9, 51000 Rijeka, Croatia

Technical Editor Teodora Smiljanic

Additional hard copies can be obtained from orders@intechopen.com

New Developments in Liquid Crystals, Edited by Georgiy V. Tkachenko
p. cm.
ISBN 978-953-307-015-5

For EU product safety concerns:
IN TECH d.o.o., Prolaz Marije Krucifikse Kozulić 3, 51000 Rijeka, Croatia,
info@intechopen.com or visit our website at intechopen.com.

Meet the editor

Georgiy Tkachenko obtained his MSc in Lasers and Optoelectronic Techniques form the Kharkiv National University of Radioelectronics (Ukraine) in 2007. He carried out his PhD studies at the same university working on tunable optical filters based on porous silicon multilayer films infiltrated with nematic liquid crystals. The degree in Optics and Laser Physics was granted to him in 2012. While working at the University of Bordeaux (France), he has realized a series of quantitative experimental studies on optical radiation forces controlled by the interplay between the chirality of matter and the chirality of optical fields. Georgiy joined the group in September 2014 as a research fellow working on optical trapping and rotation of microscopic particles in viscous environments. Georgiy left the group in May 2017 to take up a position at OIST.

Preface

Liquid crystal technology is a subject of many advanced areas of science and engineering. It is commonly associated with liquid crystal displays applied in calculators, watches, mobile phones, digital cameras, monitors etc. But nowadays liquid crystals find more and more use in photonics, telecommunications, medicine and other fields. Accordingly, first half of this book is dedicated to fundamental properties of liquid crystals, while Chapters 7-12 give a picture of recent trends in the sphere of liquid crystal displays and light modulators.

Development of tunable photonic devices is very promising field of use for liquid crystals. Chapters 1 and 2 are focused on one- and two-dimensional photonic crystals whose optical properties are tuned by means of thermal, electrical or optical influence upon the infiltrated liquid crystals. Special attention is paid to infiltration efficiency and the molecule equilibrium organization within hollows of the host material. Chapter 3 presents the study of electrically tunable magneto-optical effects in magnetophotonic crystals filled with nematic liquid crystals.

Chapter 4 demonstrates the use of a liquid crystal spatial light modulator for simulation of atmospheric turbulence. This technique can be applied for design and testing of high-precision telescopes, adaptive optical and laser communication systems.

Chapter 5 describes a technique based on thermochromic liquid crystal films to obtain two-dimensional thermal images produced by ultrasound physiotherapy equipment.

Chapter 6 proposes simple and accurate optical methods for determining the nonlinear refractive coefficient, the nonlinear absorption and the rotational viscosity coefficient in the dye-doped nematic liquid crystal.

Of course, the book gives consideration to numerous issues of up-to-date liquid crystal displays. Chapter 7 suggests a polarizer-free display using dye-doped liquid crystal gels whose physical mechanism is mainly the combination of both light scattering and absorption. Potential applications are paper-like flexible displays, electrically tunable light shutters and decorative displays. Chapter 8 is focused on the fundamentals of an active matrix liquid crystal display, namely the operation description, the driving methods and circuitry and the analog circuits design by using polycrystalline silicon thin-film transistors. Chapter 9 presents a 10-bit liquid crystal display column driver consisting of piecewise linear digital-to-analog converters. Chapter 10 offers the optimization of anisotropic conductive film curing process. This study can provide an important support to optimize the curing process for various packaging applications, such as the chip-on-glass packaging for liquid crystal displays. Chapter 11 introduces some light emitting diode backlight driving systems and discusses their advantages over conventional cold cathode fluorescent lamps as applied to liquid crystal display panels. Chapter 12 presents the characteristics of a liquid crystal holographic memory to generate binary patterns and describes an optically reconfigurable gate array with a liquid crystal - spatial light modulator.

The goal of this book is to show the increasing importance of liquid crystals in industrial and scientific applications and inspire future research and engineering ideas in students, young researchers and practitioners.

Editor

Georgiy V. Tkachenko
Kharkov National University of Radio Electronics
Lab. „Photonics"
E-mail: tgogy@mail.ru
Ukraine

Contents

1

Nematic Liquid Crystal Confined in Electrochemically Etched Porous Silicon: Optical Characterization and Applications in Photonics

Georgiy V. Tkachenko[1], Volodymyr Tkachenko[2], Giancarlo Abbate[2], Luca De Stefano[3], Ilaria Rea[3] and Igor A. Sukhoivanov[4]
[1]Kharkov National University of Radio Electronics,
[2]CNR-INFM Lab Coherentia, Università di Napoli Federico II,
[3]Istituto per la Microelettronica e Microsistemi (CNR-IMM),
[4]Universidad de Guanajuato,
[1]Ukraine
[2,3]Italy
[4]Mexico

1. Introduction

Liquid crystals (LC) confined in curved geometries have been a fundamental challenge for more than a century, starting from the study of supra-micrometre nematic droplets suspended in an isotropic medium (Lehmann, 1904). In the mid-1980s, a new period began with this topic stimulated by the discovery of various composite materials suitable for electro-optic and thermo-optic applications in controllable light scattering windows, flat-panel displays, holography, optical networking, and computing. In these materials LC molecules are confined within polymer or porous networks, therefore a competition arises between surface ordering and disordering effects on formation of stable director configurations and configurational transitions, critical temperatures of mesogenic phase transitions, orientational and hydro-dynamics and other properties. So far the behaviour of mesogens enclosed in different porous matrixes such as Nuclepore polymer membrane, Anopore aluminium oxide membrane, Vycor glass, and others with pores of different size and shape have been investigated by means of various experimental techniques: specific heat calorimetry, nuclear magnetic resonance, dielectric spectroscopy, polarization microscopy, dynamic light scattering etc.; for a review see (Crawford & Žumer, 1996).
Another host material namely electrochemically etched porous silicon (PSi) (Canham, 1997) has appeared to be promising for tunable and switchable optoelectronic devices due to the simplicity of fabrication, flexibility in wavelength design and compatibility with silicon microelectronic technology. Since PSi film is only several microns thick, most of the above mentioned techniques exhibit difficulties in its characterisation. Analysis of PSi-LC composite is complicated by the anisotropic nature of both the PSi matrix and the infiltrated LC. Nevertheless, an advanced technique developed for characterization of thin films,

Source: New Developments in Liquid Crystals, Book edited by: Georgiy V. Tkachenko,
ISBN 978-953-307-015-5, pp. 234, November 2009, I-Tech, Vienna, Austria

namely variable angle spectroscopic ellipsometry, can give crucial information on the amount of the infiltrated LC and preferential director orientation inside the pores (Marino et al., 2007).
The great advantage of PSi technology is an opportunity to fabricate multilayer structures with customized porosity and thickness of each layer. These structures are used in photonic devices such as Bragg reflectors (Pavesi & Dubos, 1997), optical microcavities (Weiss & Fauchet, 2003; Ouyang et al., 2005; Weiss et al., 2005; De Stefano et al., 2007), and even nonperiodic sequences, such as quasi-crystals (Moretti et al., 2006). Despite the many experimental and theoretical studies of the PSi-based 1-D photonic bandgap structures, rigorous simulations of their tuning properties, when filled with LC, have been scarce. Actually two rough approximations were usually done: first, the pores were considered filled with LC completely; second, the spatial distribution of the LC director was not taken into account or the simplest uniform axial configuration (the director oriented along the pore axis) was assumed. Whereas more complicated director configuration in the Si macropores with a diameter more than 1 micron were observed (Leonard et al., 2000; Haurylau et al., 2006], for the pores with a diameter less than 150 nm there is a lack of experimental information on the nematic director configuration.
Both electrical and thermal tuning has been achieved in the PSi-LC photonic bandgap microcavity realized on a silicon wafer (Weiss et al., 2005). Applying an electric field along the pore channels, the electrical reorientation of the LC with positive dielectric anisotropy was obtained. This fact indicates the non-axial orientation of the LC director inside the pores without field. However, more detailed study of the orientational properties of LC molecules in pores and their influence on the PSi-LC microcavity spectrum is still needed.
In the present chapter we are focusing on properties of a nematic LC confined in porous silicon with random pore distribution to be used in 1-D photonic devices. Section 2 gives a brief excursus into the physics of porous silicon and describes the methods applied for fabrication of monolayer, multilayer and free-standing PSi films; techniques for oxidation of the samples and infiltration with liquid crystals are also described. In the Section 3 we present the results of an ellipsometric study of refractive indices and birefringence of PSi and porous silica ($PSiO_2$) monolayers infiltrated with the commonly used nematic liquid crystal mixture E7. The effective ordinary and extraordinary refractive indices of the confined LC are derived from the experimental data using the effective medium approximation (EMA) model for the anisotropic composite. The temperature dependence of the refractive indices is compared with that in a bulk. Section 4 is dedicated to theoretical and experimental study of a free-standing PSi microcavity placed in a glass cell and infiltrated with E7. We present temperature dependence of the microcavity spectral characteristics and rigorous simulation of the LC effect on the spectral tuning. The distribution of the LC director within pores is simulated using the Frank's free energy approach. From the comparison between experimental spectra and the results of numerical calculations, we obtain the LC volume fraction in the composite, information on the LC director configuration inside the pores, and a rough estimate of the anchoring strength of LC molecules at the pore walls. In Section 5 we analyze the effect of an electric field on the director configuration of LC confined in pores and the PSi-LC microcavity spectrum.

2. Sample preparation and infiltration with liquid crystals

Porous silicon was discovered in the fifties trying to electropolish silicon in hydrofluoric acid (Uhlir, 1956, Turner, 1958). For low current densities, respectively high electrolyte concentrations, silicon is not electropolished but pores are formed. Increasing the current density over a threshold value which decreases with the electrolyte concentration, results in electropolishing. In the beginning of the nineties visible luminescence in PSi at room temperature was discovered (Canham, 1990; Lehmann & Gösele, 1991). The possibility to produce optoelectronic devices based on PSi started enormous research activity. Meanwhile many applications for porous silicon are still developing. Most of these applications are based on the morphology of PSi.

PSi is formed by electrochemical etch of crystalline silicon wafers in HF-based solutions (Canham, 1997). The wafer is the anode; it is placed in back-side contact on an aluminum plate while the front side is sealed with an O-ring and exposed to the anodizing electrolyte. The cathode is made of platinum or any HF-resistant and conductive material. When a potential is applied to silicon in an aqueous environment, an electric current is induced to flow through the system.

The only important charge transfer reaction in the silicon/HF system is the reduction of water with the subsequent liberation of hydrogen gas. It is only under anodic polarizations that silicon dissolution occurs. The exact dissolution chemistries of silicon are still under debate, although it is generally accepted that holes h^+ are required in the initial steps for both electropolishing and pore formation. The dissolution mechanism can be expressed in the simplified reaction:

$$Si + 6F^- + 2H^+ + 2h^+ \rightarrow SiF_6^{2-} + H_2 \uparrow . \quad (1)$$

Heating and illumination increase the hole/electron pair generation in the substrate and affect the dissolution process. *P*-type silicon has an excess of holes, so it can be etched in a dark, whereas in the case of *n*-type silicon holes are minority carriers and illumination is generally required.

Due to quantum restrictions in thin Si walls, the electrochemical dissolution occurs only at the pore tips (Lehmann & Gösele, 1991). Thus pores are growing deep into the substrate according to the orientation of its crystal planes. Electrochemical etch of commonly used Si wafers with <100> orientation leads to the formation of columnar pores oriented normally to the plane of the silicon substrate (Canham, 1997).

PSi shows a great variety of morphologies dependent on the doping type and level in the silicon substrate and the electrochemical etching parameters. Usually for a given substrate and electrolyte, only one type of porous structure can be obtained. Guidelines of the International Union of Pure and Applied Chemistry define the ranges of pore sizes (Canham, 1997): pores characterized by a diameter less than 2 nm define microporous silicon; for sizes in the range 2-50 nm the PSi is mesoporous; pores having diameters more than 50 nm are typical for macroporous silicon. Highly doped *p*- or *n*-type silicon anodized in aqueous HF solution usually forms mesopores with sizes from 20 nm to 50 nm. In the case of lightly doped *p*-type silicon, the pore size distribution is normally found in the range 1 - 5 nm. The electrochemical etching of *n*-type substrates in the dark (for light doping) or with back-side illumination (for moderate doping) results in the formation of a macroporous material with radii in the micrometer range (Leonard et al., 1999).

The most important parameter of the PSi is the porosity, defined as the fraction of void within the porous layer. The porosity depends directly on the anodisation conditions and can be measured by means of gravimetric, profilometric or ellipsometric analysis of PSi monolayers. We have used here the ellipsometric technique, which is discussed in the next section.

In the present work porous structures were fabricated using highly doped p+-silicon wafers with 0.01 Ω/cm resistivity and 400 μm thickness. The anodisation was performed in the HF-based solution (50 wt% HF : ethanol = 3:7) in the dark at room temperature. Electrochemical etch is quasi linear only for a short period of time (Canham, 1997), thus the pulses of electrical current were applied with duration of 0.2 s alternated with pauses of 5 s. The PSi structures formed in above-listed conditions were studied by means of the scanning electron microscope with a field emission gun (SEM-FEG Gemini 300, Carl Zeiss). As may be seen in Fig. 1 (a), the pores etched have average diameter of 50 nm, which corresponds to the threshold value between mesopores and macropores. Such pores are suitable for tunable PSi-based devices for near-infrared wavelength range: the pore size is large enough for infiltrating with different soft materials and at the same time it is much smaller than the wavelength, so scattering can be neglected and EMA can be used for electrodynamics simulation.

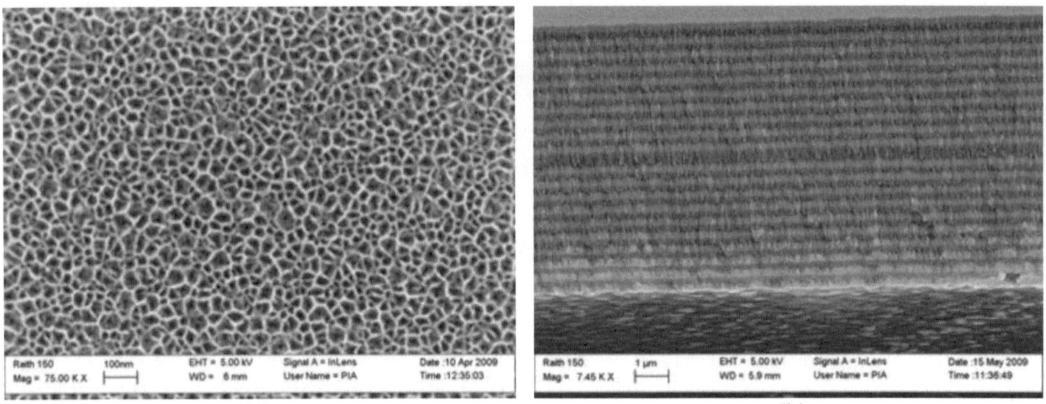

(a) (b)

Fig. 1. SEM-FEG image of a multilayer PSi: a) top-view, b) lateral view

For the given substrate, electrolyte and environment conditions (temperature and illumination), the porosity and etch rate are uniquely defined by the current density applied during the anodisation process (Canham, 1997). Thickness of the porous layer depends on the duration of the etch process monotonically. By changing the current applied and the duration of the etching it is possible to fabricate the multilayer PSi structures with desired porosity and thickness of each layer. Lateral view SEM-FEG image of a multilayer PSI is shown in Fig 1(b). It represents a couple of periodical structures with alternating layers of high and low porosity.

The multilayer PSi samples studied in this work were made free-standing by means of the Turner "lift-off" technique (Turner, 1958): at the end of etch process the current density was ramped to 800 mA/cm^2. That current density corresponds to the electropolishing mode and results in the PSi film detachment from the substrate. Free-standing films are thin and extremely fragile (Ghulinyan et al., 2003), however, this disadvantage is eliminated as soon as the film is placed inside a solid glass cell. But then the free-standing PSi film has some advantages as compared with the one on a wafer. A cell containing the PSi film can be filled with LC easily by the capillarity effect. Pores pass through the entire PSi film and are open from both ends, so very high volume fractions of LC inside pores can be obtained. Furthermore, free-standing PSi-based optical devices can operate both in transmission and reflection modes, while on-wafer designs work only in reflection.

Fresh PSi is characterized by a high chemical reactivity. If stored in ambient air, the texture becomes partially oxidized, which leads to the change of its optical properties. To eliminate the problem of aging, the thermal oxidation of the structure is used. Usually, this treatment is also applied to PSi-based devices to utilize the visible range (Pirasteh et al., 2006). The oxidation reduces or completely removes silicon from the skeleton substituting it with SiO_2, which grows isotropically also into the pores. According to (Pirasteh et al., 2006), volume fraction of SiO_2 grown into silicon walls is twice as much as that grown into the void. PSi is generally oxidized in a pure O_2 atmosphere by a two step thermal treatment. The first oxidation step, defined as pre-oxidation, at low temperature (300 - 400 °C) is required in order to selectively oxidize the silicon backbonds thus assuring an easy propagation of the oxidant into the silicon structure. Moreover, the pre-oxidation reduces the damage of the material texture in the second oxidation step at higher temperatures between 800 and 900 °C (Pap et al., 2005).

Infiltration of the PSi and $PSiO_2$ samples with liquid crystal mixture E7 is performed under vacuum at the temperature above the clearing point of LC. The sample and LC are placed on the processor-controlled heater within a special vacuum chamber. The temperature is increased to 80°C and the chamber is evacuated to the pressure of about 1 Torr. At these conditions air and moisture are removed from the pores and liquid crystal. Then sample surface is completely covered by the LC drop. Liquid crystal fills the pores through capillary forces, low viscosity in isotropic phase facilitates infiltration. After about 5 hours, the pressure is rised to the atmospheric value, which pushes LC into the pores, providing excellent infiltration. After that the temperature is decreased slowly below the clearing point to promote LC transition to the nematic phase with equilibrium configuration of director.

3. Ellipsometric study of liquid crystals confined in porous Si and SiO_2

Standard spectroscopic ellipsometry is based on the measurement of two physical quantities: the relative amplitude change, Ψ, and the relative phase change, Δ, suffered by incident light when reflected (or transmitted) by a layered structure. These two parameters are linked to the reflection (or transmission) coefficients, which are themselves related to the optical response of the surface: the Ψ's and Δ's spectra depend on the refractive indices of the layers, on their thickness and, in the case of anisotropic films, on the orientation of their optical axis (of course, they depend on any physical parameters that affect the optical behaviour of the material, for instance on temperature).

For anisotropic media, generalised spectroscopic ellipsometry (GSE) is applied requiring the adoption of the 2×2 Jones matrix formalism. In this way it is possible to generalise the Ψ and Δ parameters to the case when a change occurs in the light polarisation. The six GSE parameters are linked to the Jones matrices of reflected ($\mathbf{J}^r$) or transmitted ($\mathbf{J}^t$) beam through the following equations:

$$\tan\Psi \cdot e^{i\Delta} = \frac{J_{pp}}{J_{ss}}; \quad \tan\Psi_{sp} \cdot e^{i\Delta_{sp}} = \frac{J_{sp}}{J_{ss}}; \quad \tan\Psi_{ps} \cdot e^{i\Delta_{ps}} = \frac{J_{ps}}{J_{pp}} \tag{2}$$

where $\mathbf{J} = \mathbf{J}^r, \mathbf{J}^t$; $\mathbf{J}^r = \begin{pmatrix} r_{pp} & r_{sp} \\ r_{ps} & r_{ss} \end{pmatrix}$; $\mathbf{J}^t = \begin{pmatrix} t_{pp} & t_{sp} \\ t_{ps} & t_{ss} \end{pmatrix}$ with r_{pp}, r_{ss}, r_{ps}, r_{sp} (t_{pp}, t_{ss}, t_{ps}, t_{sp}) representing the reflection (transmission) coefficients for *p*-, *s*-, and cross-polarisations, respectively.

Obtaining quantitative information on the wanted parameters of a sample requires setting of an optical model of the latter and a fit procedure matching the data generated by that model with the experimental ones. The mean squared error value, which is a measure of the difference between experimental and model data, is the indicator of the goodness of the employed model, as usual in GSE measurements.

Due to inherent inhomogeneity at mesoscale in the layer plane of porous silicon, the optical model is widely applied adopting the Bruggeman Effective Medium Approximation (EMA) (Spanier & Herman, 2000). As an electrochemically etched PSi contains anisotropically aligned nanowires and pores, the isotropic EMA has to be extended to take into account a form birefringence.

The pore shape effect can be examined by using adjustable depolarization factors L_i for ellipsoidal voids (*i* denotes projections along the main ellipsoid axes). For a <100> oriented Si wafer the depolarization factors fulfill the relations: $L_x=L_y=L_{xy}$ and $L_z+2L_{xy}=1$ (z axis is along the normal to the wafer). The components ε_i of the effective dielectric permittivity tensor of the PSi film are calculated using the generalized Bruggeman formula for two media (Spanier & Herman, 2000):

$$\frac{\varepsilon_{Si} - \varepsilon_i}{\varepsilon_i + L_i(\varepsilon_{Si} - \varepsilon_i)} p_{Si} + \frac{\varepsilon_{air} - \varepsilon_i}{\varepsilon_i + L_i(\varepsilon_{Si} - \varepsilon_i)} (1 - p_{Si}) = 0 \quad (3)$$

where i=xy, z; ε_{Si}, ε_{air} - dielectric permittivities of silicon and air, respectively; p_{Si}– volume fraction of silicon. For a completely oxidized sample, the volume fraction of silicon has to be substituted with that of silica (p_{Silica}).

Two samples of the PSi and $PSiO_2$ layers were fabricated, both on the doped silicon substrate, and characterized by the J. A. Woollam Co. Inc. Variable Angle Spectroscopic Ellipsometer (VASE®). In order to get a large amount of data, the GSE parameters were measured as a function of both angle of incidence and wavelength, in the spectral range from 500 to 1700 nm (the details of data acquisition are described elsewhere (Marino et al, 2007)). All data were simultaneously analysed using WVASE32® software (Woollam et al, 1999) to determine the sample refractive index, thickness, and volume fractions of the constituents.

The samples were initially studied before the LC infiltration in order to find the parameters of the empty porous matrix. After that the E7 liquid crystal was infiltrated and the surface of the porous layers was cleaned from the LC by a flow of pure nitrogen. Temperature of the filled samples was adjusted and stabilized by a processor-controlled hot-stage from CaLCTec S.R.l. which provided temperature stability and accuracy of 0.1 °C.

The empty samples were modelled using the equation (3). The layer thickness, volume fraction of the porous matrix material and depolarization factors were found by fit as follows: h=874±4nm, p_{Si}=15.7±0.1%, L_{xy}=0.356±0.002, and L_z=0.288±0.002 for the PSi sample; and h=423.5±4nm, p_{Silica}=37.0±0.2%, L_{xy}=0.37±0.02, and L_z=0.26±0.02 for the $PSiO_2$ one. The obtained refractive indices n_{xy} and n_z of the anisotropic layers are shown in Fig. 2(a) and 2(b) for PSi and $PSiO_2$ samples, respectively.

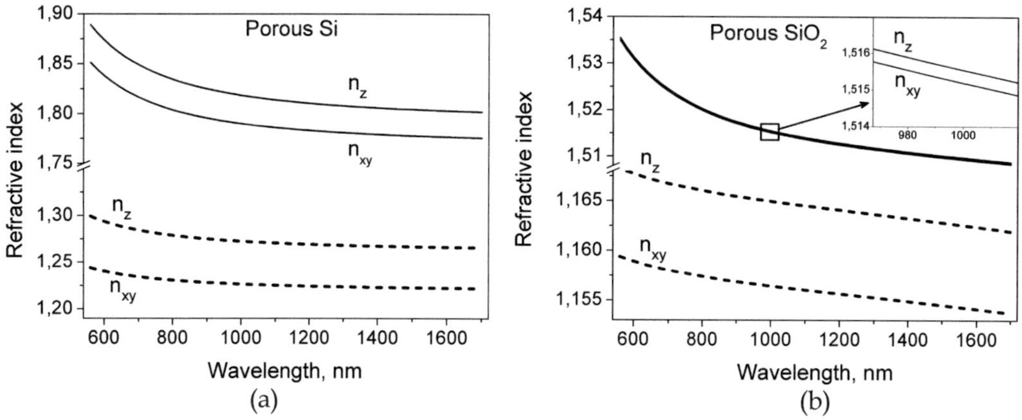

Fig. 2. Refractive indices of PSi (a) and $PSiO_2$ (b) measured for the empty samples (dashed lines) and simulated for the samples filled with E7 at 60°C (solid lines)

One can see that the form birefringence of the PSi matrix is much higher than that of the $PSiO_2$ one (0.05 versus 0.009 at wavelength of 600nm) although the corresponding refractive indices of the samples are quite close. This is explained by much less contrast between refractive index of the pore wall and that of air for the $PSiO_2$ sample as compared with the PSi one.

We have simulated the refractive indices of the porous layers filled with E7 in isotropic phase, using the equation (3) and model parameters for empty samples just substituting the refractive index of void with that of LC (Fig. 2). The dispersion curve of E7 at 60°C was extrapolated from the data of Ref. (Li et al, 2005).

Birefringence of the filled samples decreases to 0.03 and 0.0004 for PSi and $PSiO_2$ samples, respectively. We note that the birefringence of the filled PSi sample is still high while that of the $PSiO_2$ sample can be neglected because of very low refractive index contrast of composite constituents.

Let us remind that the refractive indices of LC have to be found by the fit procedure. Thus the model parameters for the empty porous matrix must be fixed for the filled sample. Otherwise, the strong correlation between the matrix parameters and those of LC gives rise to uncontrollable error in LC refractive indices. Form birefringence of the filled PSi sample is quite high, so the anisotropies of both the porous silicon matrix and the nematic LC contribute to the effective refractive indices of the composite material. As a result, these contributions can not be distinguished and it is impossible to get actual values of refractive indices of the nematic liquid crystal confined in PSi sample.

Since form birefringence of the oxidized sample filled with LC is negligible, we are able to obtain the refractive indices of E7 confined in porous silica. In general LC fills the porous matrix partially (Marino et al, 2007), so we used the Bruggeman formula for three media (Spanier & Herman, 2000): silica, air (in un-filled pores) and LC:

$$\frac{\varepsilon_{Silica}-\varepsilon_i}{\varepsilon_{Silica}+2\varepsilon_i}p_{Silica}+\frac{\varepsilon_{air}-\varepsilon_i}{\varepsilon_{air}+2\varepsilon_i}p_{air}+\frac{\varepsilon_{LC,i}-\varepsilon_i}{\varepsilon_{LC,i}+2\varepsilon_i}p_{LC}=0\,, \tag{4}$$

where p_{air}, p_{LC} – volume fractions of air and liquid crystal in the medium, respectively; $\varepsilon_{LC,i}$– effective dielectric permittivity of LC inside the pores; other notations are as in (3). Here the LC birefringence is presented by different values $\varepsilon_{LC,xy}$ and $\varepsilon_{LC,z}$.

Using the equation (4) in the optical model we obtained porosity of 62±0.2% which is in a good accordance with the silica volume fraction found above by means of (3). The percentage of LC is 58.4±0.2%, which corresponds to about 94% of pore volume.

The obtained values of E7 indices are shown in Fig. 3 for different temperatures and for two wavelengths.

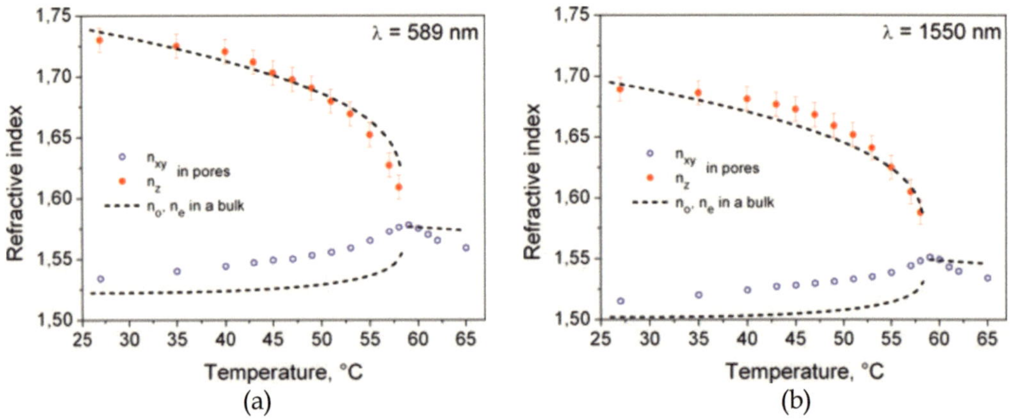

Fig. 3. Refractive index of E7 confined in porous silica (circles) for polarizations parallel (n_{xy}) and perpendicular (n_z) to the sample plane as compared with ordinary (n_o) and extraordinary (n_e) indices of E7 in a bulk (dashed lines) for wavelength of 589nm (a) and 1550nm (b)

For comparison the temperature dependences of the ordinary (n_o) and extraordinary (n_e) indices of E7 in a bulk are presented as extrapolated from the data of Ref. (Li et al, 2005). The clearing point of E7 in pores equals that in a bulk within the error of measurements. Below the clearing point the temperature dependences of n_{xy} and n_z were repeatable while heating or cooling. But if LC becomes isotropic we could not repeat the previous data by cooling and found essential decrease of the refractive index. It is explained by the LC leakage from the pores in isotropic phase.

We have found that n_z of the nematic E7 confined in $PSiO_2$ sample approaches extraordinary refractive index in a bulk, however, the error bars of n_z and n_e measurements are both rather high. The n_{xy} data increase with temperature that means the LC molecules are oriented preferably along the pore axis. At the same time, these data exceed the ordinary refractive index in a bulk. So we conclude that a significant part of the molecules are tilted around the pore axis.

4. Free-standing porous silicon microcavity containing nematic liquid crystal

4.1 Experimental

The design for fabrication represents a quarter-wave microcavity layer with high porosity (etched with current density J = 125 mA/cm²) sandwiched between two 11-period distributed Bragg reflectors (DBR). Alternating high and low (J = 25 mA/cm²) porosity layers form each DBR. We were aimed to obtain the microcavity having resonance within the O-band of telecom optical fibers which utilize wavelengths from 1260 to 1360 nm. Since about 300 nm red shift of the microcavity resonant wavelength is expected after the LC infiltration (Weiss & Fauchet, 2003), we have designed our structure to have resonance at 1040 nm without LC.

The applied lift-off technique allows the use of the same Si wafer for fabrication of several free-standing structures without disassembling the anodisation container. The first sample, being etched directly from the wafer surface, is always cracked, possibly because of strong mechanical strains. This sample serves as a sacrificial layer leaving the tips of pores on the Si surface after electropolishing (Ouyang et al., 2005). The next sample is free of the thin layer of nanoporous Si on the top and is usually intact. It has open pores which can be easily filled with a liquid crystal. Treatment of the fresh PSi sample with KOH/ethanol solution for 15 minutes scours the pores (Moretti et al., 2006) and slightly improves LC infiltration. The obtained PSi film was placed between two parallel glass plates divided by two spacers (12 μm thick stripes of mylar). After LC infiltration all chinks between the plates were sealed up with an ultraviolet adhesive glue NOA68. Our experimental sample is shown in Fig. 4.

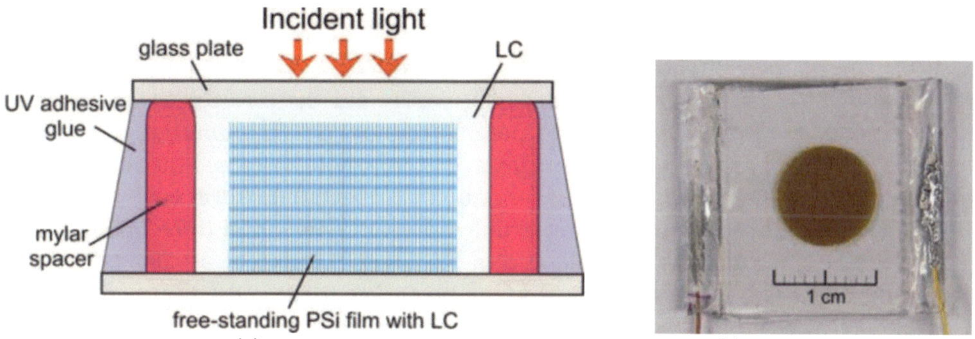

(a) (b)

Fig. 4. Microcavity sample: a) scheme, b) top-view picture

Spectral characteristics of the experimental sample were measured using the ellipsometer VASE®. The monochromator of the ellipsometer has the operating waverange from 270 to 1700nm and spectral resolution up to 2 nm. Moreover, we used the fitting facility of the WVASE32® software to specify the parameters of the multilayered PSi structure and to measure the effective refractive index of LC confined in it. During the measurements the PSi-LC filter was contained in a processor-controlled hot-stage, which was mounted on the sample holder of the VASE®.

4.2 Simulation of optical properties of a PSi film with LC

To calculate spectra of a multilayer PSi structure one has to define the refractive indices and thicknesses of all layers. When the sample is not filled with LC, the Bruggeman formula (4) for two media, namely silicon and air, is applied. Infiltration with LC complicates simulation of the layer refractive index. First of all, the effective refractive index n_{LC} of liquid crystal in pores depends on the spatial distribution of the local director **L**. In the nematic phase, the LC director field depends on the interplay between molecular surface anchoring, elastic energies of the liquid crystal, and forces caused by external fields, if any. Here the director field of the nematic mixture E7 confined in cylindrical pores is calculated using the Frank's free energy approach (Crawford et al, 1992; Tkachenko et al, 2008). If there is no external influence, the free energy F of the confined nematic is given by (Crawford et al, 1992):

$$F = \frac{1}{2} \int_{vol} \{ K_{11} \cdot (\mathrm{div}(\mathbf{L}))^2 + K_{22} \cdot (\mathbf{L} \cdot \mathrm{curl}(\mathbf{L}))^2 + K_{33} \cdot (\mathbf{L} \times \mathrm{curl}(\mathbf{L}))^2 - K_{24} \cdot \mathrm{div}(\mathbf{L} \times \mathrm{curl}(\mathbf{L}) + \mathbf{L} \cdot \mathrm{div}(\mathbf{L})) \} dV + \frac{1}{2} \int_{surf} \left(W \cdot \sin^2 \Omega \right) dS, \tag{5}$$

where K_{11}, K_{22} and K_{33} are the specific elastic constants for the bulk deformations: splay, twist and bend, respectively; K_{24} is the saddle-splay surface elastic constant; $\mathbf{L} = \cos(\Omega(r)) \cdot \mathbf{z} + \sin(\Omega(r)) \cdot \mathbf{r}$ is the nematic director, which represents the local direction of the LC optical axis; W is the molecular anchoring strength; Ω is the angle between **L** and the pore axis.

For cylindrical cavities there are five stable non-twisted configurations of the nematic director field depending on the anchoring conditions: uniform axial (UA), planar radial, planar polar, escaped radial (ER) and escaped radial with point defects. According to (Leonard et al., 2000; Haurylau et al., 2006) E7 exhibits ER configuration in the pores larger than 1 μm in diameter. This configuration occurs when LC molecules are anchored homeotropically, i.e. perpendicular to the pore walls, and the director **L** escapes into the third dimension toward the axis of the pore. While for the UA configuration the local director is oriented parallel to the pore axis in the whole volume. In our case of the mesoporous silicon there is a lack of information on the LC director field configuration, however the ellipsoid of the effective refractive index of LC is found to be uniaxial with its long axes oriented normally to the plane of a PSi layer (Marino et al, 2007). This can be explained by ER or UA configurations of LC director. We assume the ER configuration for our samples because it matches better the experimental spectra as will be shown in Section 4.3. So the molecular system in a pore has axial symmetry and the equation (5) depends only on the radial coordinate r.

To simulate the LC director field $\Omega(r)$, we used a well-known method of minimization of the free energy (Crawford et al, 1992). In the case of ER configuration, the expression (5) can be set in the form:

$$F = \pi h K_{11} \int_0^R F_0 dr + \pi R W , \tag{6}$$

$$F_0 = \frac{\sin^2 \Omega}{r} + r \cdot (\Omega')^2 \left(\cos^2 \Omega + k \cdot \sin^2 \Omega \right) - 2 \sigma \Omega' \sin \Omega \cos \Omega , \tag{7}$$

where Ω' is the first derivative of $\Omega(r)$; $k = K_{33}/K_{11}$; R is the pore radius; $\sigma = RW/K_{11} + K_{24}/K_{11} - 1$ is a dimensionless surface parameter. Minimization of the free energy F_0 gives the second order differential equation:

$$\Omega'' \left(\cos^2 \Omega + k \sin^2 \Omega \right) + (\Omega')^2 \sin \Omega \cos \Omega \, (k-1) + \frac{\Omega'}{r} \left(\cos^2 \Omega + k \sin^2 \Omega \right) - \frac{1}{r^2} \sin \Omega \cos \Omega = 0 , \tag{8}$$

which can be solved numerically using the following boundary conditions:

$$\Omega\big|_{r=0} = 0; \quad \Omega'\big|_{r=R} = \frac{\sigma \sin\Omega_R \cos\Omega_R}{R\cdot\left(\cos^2\Omega_R + k\cdot\sin^2\Omega_R\right)}, \quad \text{where } \Omega_R = \Omega\big|_{r=R}. \tag{9}$$

When the director field $\Omega(r)$ of LC inside the silicon pore is found, the effective dielectric permittivity and refractive index of the nematic LC in a cylindrical pore is calculated by the formula (Tkachenko et al, 2008):

$$\varepsilon_{LC} = n_{LC}^2 = \frac{1}{R^2}\int_0^R\left[2\varepsilon_o + \left(\varepsilon_e - \varepsilon_o\right)\sin^2\Omega(r)\right] r\,dr, \tag{10}$$

where ε_o, ε_e - ordinary and extraordinary components of the LC dielectric permittivity at near infrared range.
The effective dielectric permittivity of each layer of the PSi film for in-plane polarization (i=xy) is calculated by means of the equation (4) replacing "Silica" by "Si" in the subscripts. We assume percentage of the pore volume filled by LC (p'_{LC}) to be independent on porosity. This value is defined as:

$$p'_{LC} = p_{LC}/\left(1 - p_{Si}\right). \tag{11}$$

The LC volume fraction in PSi can not be measured directly. So we determine it by fit of the simulated spectra of the microcavity to the experimental data when the sample is heated over the clearing point of the LC. For the fit, the p'_{LC} value is the only variable parameter. Finally, the well-known transfer matrix method (TMM) (Born & Wolf, 1980) is applied to calculate the theoretical spectra of the PSi-LC multilayer structure.

4.3 Results and discussion

The experimental transmittance spectrum of the empty free-standing PSi microcavity is shown by solid dots in Fig. 5(a). The spectrum has the central stop-band in the wave range of 950-1150 nm with several side lobes and the second harmonic band-gap centered at 540 nm which are the features of the Bragg reflectors. A sharp resonant peak is observed at the wavelength of 1038 nm inside the central stop-band due to the presence of the microcavity layer. Light absorption by silicon causes a steep decrease of the transmittance in the visible.
We used the WVASE32® software to analyze this spectrum. The optical model contained all the layers of the filter described above including the glass substrates. The PSi film was described as a periodic lattice of EMA layers. The parameters of the PSi layers were obtained by the fit procedure matching the data generated by the model with the experimental ones (Fig. 5(a)). The fit gave void fraction of 79.8% and 72.3% for the layers with high and low porosity, respectively.
The spectra of the structure filled with E7 are presented in Fig. 5(b) for temperature values of 27°C and 65°C. The lower temperature value corresponds to the nematic phase while the higher one corresponds to the deep (8 degree higher than the clearing point) isotropic phase of the E7 liquid crystal. As may be seen, the microcavity peak is red-shifted by more than 300 nm as compared with the empty structure. This shift comes from the increase of the effective refractive indices of PSi layers. After LC infiltration the microcavity Q-factor falls to 37 from the value of 58 exhibited in the case of the empty filter, since the refractive index contrast decreases and the Bragg condition is not held for all the layers at the same wavelength.

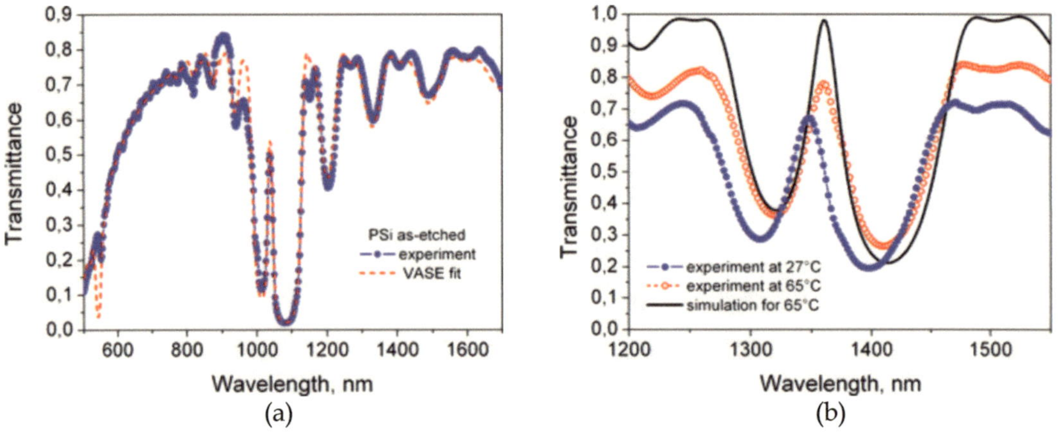

Fig. 5. a) Transmittance spectra of the empty PSi microcavity: experimental (solid dots) and fitted by WVASE32® (dashed line). b) Spectra of the sample filled with E7: measured at 65°C (solid dots) and 27°C (circles) and calculated for 65°C (solid line)

Optical properties of the free-standing film with LC were simulated using the numerical method developed as described above. The main advantage of this method, as compared with WVASE32®, is the ability to simulate orientational properties of LC.

In Fig. 5(b) the solid line shows the simulated spectrum which can be considered in a good accordance with the experimental curve if one takes into account the simplifications of the model used for calculations. Our model neglects energy losses due to reflection at the first interface, absorption in silicon, losses in the LC layer above the PSi film, and scattering at the interfaces of the multilayer structure. That is why the maximum transmittance of the experimental curve is about 20% lower than the calculated one. What is important for our study, the simulated position of the resonance peak and the stop-band shape match the experimental data.

We note that the finite size glass substrates (with thickness of 1 mm) should give rise to a great number of very sharp peaks in the simulated spectra of the multilayer PSi structure. These peaks correspond to the modes of the glass layers and appear at the spectral curves with a period of 0.3 nm and amplitude up to 0.3 in transmittance. However, these peaks were not resolved in experiments due to the low spectral resolution of the used monochromator. To avoid the presence of these un-necessary peaks in the simulations, the glass substrates were considered infinite.

In the simulation, we used the dispersion curve of the n_{LC} for E7 in the isotropic state, derived from literature (Li et al, 2005). The best fit of the simulated spectrum to the experimental one corresponds to $p'_{LC} = 0.843$. The fraction of the LC in the composite is equal to 67.3% and 60.9% for the high and low porosity layers, respectively.

Heating the sample gives rise to a red shift of the spectrum (Fig. 5(b)), due to the increase with temperature of the effective refractive index of the LC confined in the pores. The detailed temperature dependence of the resonance wavelength shift is plotted in Fig. 6(a). This thermal tuning is due to the temperature dependence of n_{LC}. It is nonlinear in the nematic phase with the highest slope near the clearing point of E7. The red shift of the resonance wavelength proves that possible director configuration is either UA or ER. Above the clearing point the resonance position is almost independent of temperature. For our sample, thermal tuning of the resonance wavelength was entirely reversible within the range of 13 nm.

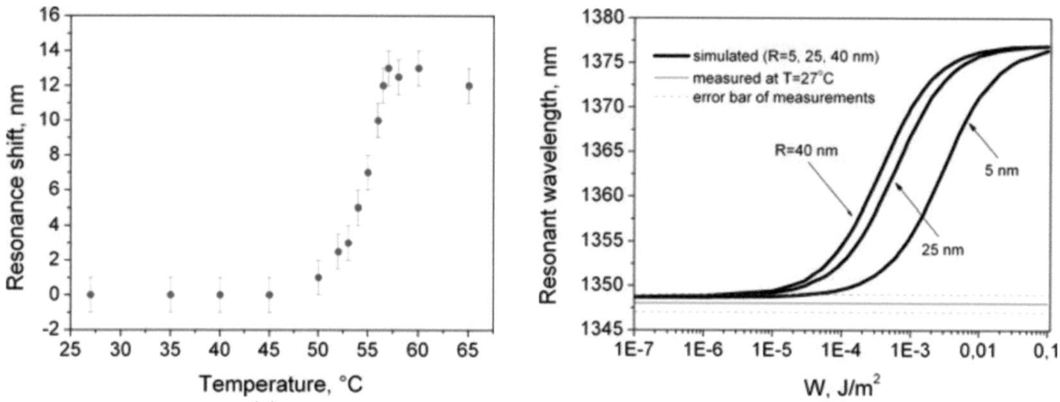

(a) (b)

Fig. 6. a) Temperature dependence of the microcavity resonance shift. b) Resonant wavelength of the microcavity versus the surface anchoring strength W

The tuning range is defined by the difference between the n_{LC} values at the room temperature and at the clearing point. The latter value is well known (Li et al., 2005), while to predict the former one we have to simulate the effective refractive index of E7 taking into account the nematic director configuration inside the pores. Comparing the spectra simulated for the different director fields with the experimental one helps to define the actual LC director configuration in the investigated PSi film.

The director field in the ER configuration was calculated using the Frank's free energy approach. We have used the following constants for E7 (Crawford et al, 1992; Leonard et al., 2000; Tkachenko et al, 2008]: K_{11} = 11.1 pN; K_{33} = 17.1 pN; K_{24} = 28.6 pN and the dispersion curves for ordinary and extraordinary indices from (Abbate et al., 2007). According to (Crawford et al, 1992; Leonard et al., 2000), the surface anchoring strength W for the E7 in supramicrometer silicon pores is estimated to be 10^{-5} J/m^2. Because the magnitude of W in mesopores is unknown, we took it variable in our computations. The dependence of the microcavity resonance wavelength on the molecular anchoring strength is shown in Fig. 6(b). The curves computed for different values of the pore radius are shown by the thick solid lines. The value of 25 nm is the averaged pore radius, while the values of 5 and 40 nm are the minimum and maximum pore radii occurred in our experimental PSi films. Experimental position of the resonance peak at 27°C and the error bar of the measurements are presented by the horizontal thin solid and dashed lines, respectively. As may be seen from the figure, the simulated curves approach the experimental resonant wavelength for W<10^{-6} J/m^2. Moreover, in this case the calculated resonance position does not depend on the pore radius. Thus, we take W = 10^{-6} J/m^2 and R=25 nm in simulations of $\Omega(r)$ for the ER director configuration.

Finally, we have performed the simulations of the spectra using the n_{LC} value given by equation (10) for the ER configuration and n_{LC} = n_o for the UA configuration. The LC fraction inside the pore volume was taken equal to 84.3% as found above. The calculated spectra both for the ER and UA cases are shown in Fig. 7(a) together with the spectrum measured at 27°C.

As may be seen, the ER-curve is much more similar to the experimental spectrum (the values of the resonance wavelength match very well). Consequently, the actual LC configuration in silicon mesopores is not UA but it is close to ER.

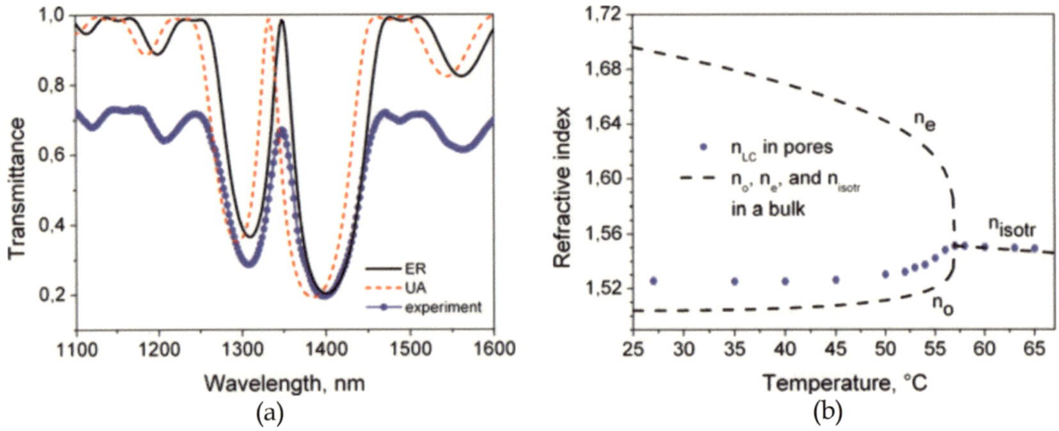

Fig. 7. a) Calculated spectra of the PSi-LC structure for ER (solid line) and UA (dashed line) director configurations of the LC. Experimental spectrum (solid dots) at 27°C is given for comparison. b) Effective refractive index of E7 in pores fitted by WVASE32® (solid dots) and values of n_o, n_e and n_{isotr} in a bulk (Li et al., 2005) (dashed lines)

Simulation of the spectra by the WVASE32® confirms this statement. Unlike the abovementioned numerical method, WVASE32® does not compute the effective refractive index of the LC in the pores but finds it from the fit of the generated and experimental spectra. The optical model of the structure implied the layer thicknesses, porosity values and the fraction of the LC as specified above, while two parameters of the Cauchy formula for the LC refractive index were varied during the fit procedure. Fig. 7(b) shows the temperature dependence of n_{LC} at 1300 nm in comparison with the refractive indices of E7 in the bulk (Li et al., 2005) in the nematic (n_o, n_e) and isotropic (n_{isotr}) phases. For the UA configuration of the LC director n_{LC} would be equal to n_o (at normal incidence of the light). As it is evident from Fig. 7(b), in our case n_{LC} is significantly larger than n_o. This fact is in accordance with the results obtained for E7 confined in the porous silica monolayer (see Section 3).

5. Electrical reorientation of LC molecules inside cylindrical pore: theoretical approach

Fig. 8 shows the model of a cylindrical pore filled with a liquid crystal under the influence of an electric field. In the ER configuration the LC director field has axial symmetry, so it is described by only one parameter, namely the angle Ω between LC director and the pore axis. At the pore edges, transparent electrodes are connected to a voltage supply to produce the electric field.

The director field configuration of the LC inside a pore depends on its elastic properties, the strength and preferred orientation of molecular surface anchoring, and the electrostatic forces caused by the applied electric field. The free energy of a confined nematic is given by (Crawford et al., 1992):

$$\widetilde{F} = F - \frac{1}{2} \int_{vol} (\mathbf{D} \cdot \mathbf{E}) dV. \tag{12}$$

where **E**, **D** are the electric field strength and displacement vectors, respectively. In the case of ER configuration, the expression (5) can be set in the form:

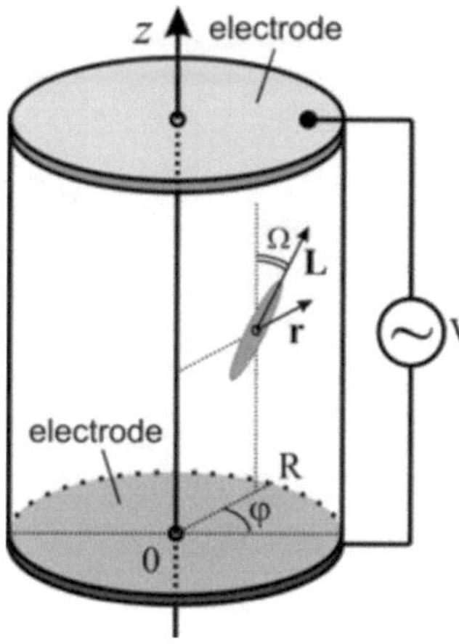

Fig. 8. Liquid crystal molecule inside a cylindrical pore

$$\widetilde{F} = \pi h K_{11} \int_0^R \left(F_0 - \frac{r\,\varepsilon_v \varepsilon_\perp \varepsilon_\parallel}{K_{11}(\varepsilon_\perp + \Delta\varepsilon \sin^2\Omega)} E^2 \right) dr + \pi R W\,, \tag{13}$$

where ε_v is the permittivity of vacuum; $\varepsilon_\perp$, $\varepsilon_\parallel$ are the components of the LC permittivity normal and parallel to **L**; $\Delta\varepsilon = \varepsilon_\parallel - \varepsilon_\perp$; E – the electric field component parallel to the pore axis. It is important to distinguish $\varepsilon_\perp$, $\varepsilon_\parallel$, low frequency permittivities, from ε_o and ε_e. Minimization of $\widetilde{F}$ gives the second order differential equation:

$$\Omega''\left(\cos^2\Omega + k\sin^2\Omega\right) + \left(\Omega'\right)^2 \sin\Omega\cos\Omega\,(k-1) + \frac{\Omega'}{r}\left(\cos^2\Omega + k\sin^2\Omega\right)$$
$$-\frac{1}{r^2}\sin\Omega\cos\Omega - \frac{\varepsilon_v\,\varepsilon_\perp\,\varepsilon_\parallel\,E^2\Delta\varepsilon\sin\Omega\cos\Omega}{K_{11}\left(\varepsilon_\perp + \Delta\varepsilon\sin^2\Omega\right)^2} = 0\,. \tag{14}$$

The equation (14) is solved numerically using the boundary conditions (9), where Ω_R is a function of E.

For simulation of nematic E7 director field within a cylindrical pore we used the following constants: R = 10, 25, 75, and 150 nm; $W = 10^{-6}$, 10^{-5}, $5{\cdot}10^{-5}$, 10^{-4}, and $5{\cdot}10^{-4}$ J/m^2; $\varepsilon_\parallel = 19.0$; $\varepsilon_\perp = 5.2$ (Crawford et al., 1992; Leonard et al., 2000). The simulated director field for different values of the electric field E at $W = 10^{-5}$ J/m^2 and R=75 nm is shown in Fig. 9.

The director is axially aligned at the pore axis (r = 0) and rotates as a function of radius to a certain angle Ω_R at the pore wall ($r = R$). In the case of zero electric field the director distribution agrees with that simulated in (Leonard et al., 2000). The LC molecules reorient toward the pore axis direction with E increasing. Above the critical field value E_{UA} which is about 3.8 V/μm for the used pore parameters, the LC molecular configuration becomes uniform axial.

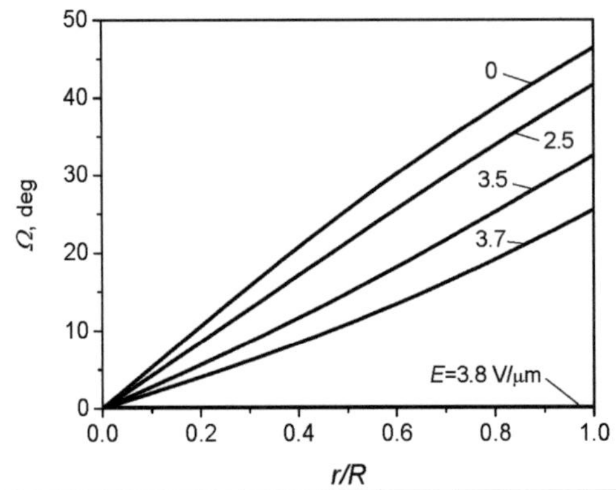

Fig. 9. Nematic director distribution in a pore for E=0, 2.5, 3.5, 3.7, and 3.8 V/μm; R =75 nm

The calculated n_{LC} versus E at different surface anchoring strength W are shown in Fig. 10(a). While electric field increases, the effective index tends to the minimum value of 1.501, which corresponds to the case of the uniform axial configuration. Furthermore, the higher is the surface anchoring strength W, the wider the range of refractive index tuning and higher the corresponding E_{UA} value.

The value of n_{LC} versus the applied electric field at different pore radius R is shown in Fig. 10(b). Reduction of the average pore radius causes insignificant decrease of the tuning range of refractive index. At the same time, E_{UA} value promptly grows. Therefore, the use of PSi with wider pores is required for devices operating at lower voltages. Because multilayer microcavities usually have an overall thickness of about 10 micron, a pore radius above 75 nm has to be chosen for the applied voltage to be less than 40 V, in the case of weak anchoring (W=10^{-5} J·m^{-2}). For stronger anchoring, the pores should be larger. However, it is noteworth remembering that excessive increase of the pore size is restricted by the growth of light scattering and violation of the Bruggeman approximation. On the other hand, these restrictions do not hold anymore when the pores are distributed periodically, as in 2-D photonic crystals. Hence, strong anchoring conditions can be used to increase Ω_R and the tuning range of such devices.

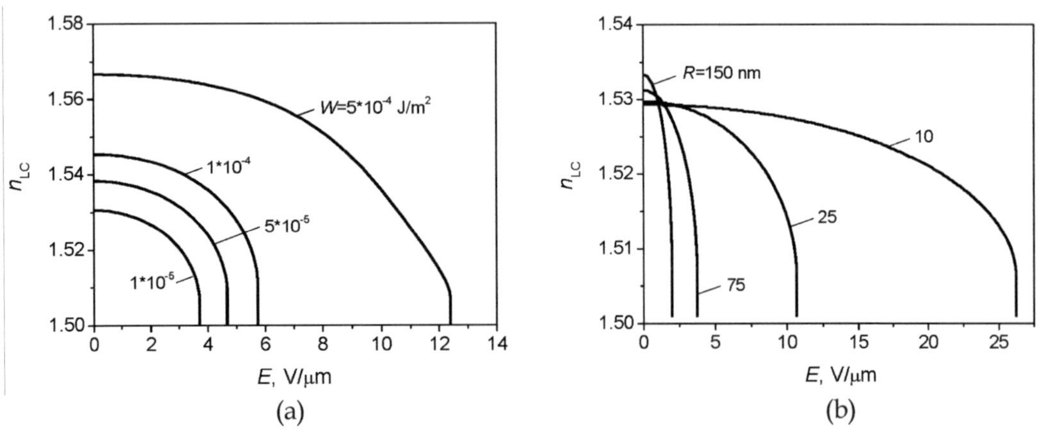

Fig. 10. Effective refractive index of the pore volume filled with E7 versus electric field: a) for $W = 10^{-5}, 5\cdot10^{-5}, 10^{-4}, 5\cdot10^{-4}$ J/m^2; R=75 nm; b) for R = 10, 25, 75, 150 nm; W=10^{-5} J/m^2

While electric field is applied to a multilayer PSi filled with the liquid crystal the value of n_{LC} goes down causing the decrease of the effective refractive index of each porous layer. As an example, we simulated spectra of the multilayer structure containing a microcavity sandwiched between two PSi distributed Bragg reflectors with alternating layers of 50% and 80% porosity, filled with E7 and tuned by the external electric field (Tkachenko et al., 2008). Shift of the microcavity resonance versus electric field is shown in Fig. 11.

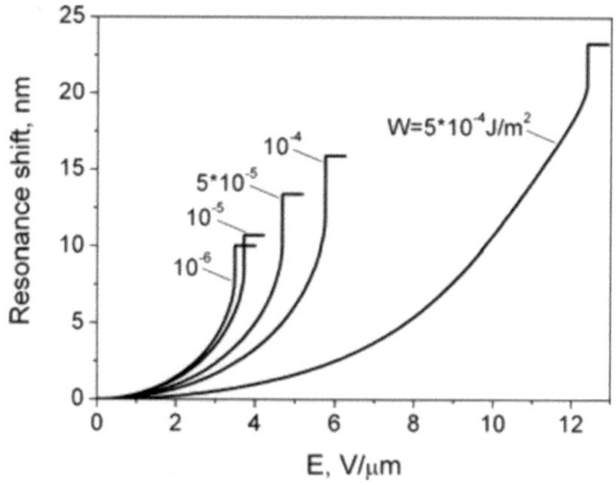

Fig. 11. Blue shift of the resonance versus electric field for $W = 10^{-6} \div 5 \cdot 10^{-4}$ J/m^2; R=75 nm

As may be seen, the electrical tuning range of the microcavity resonance varies from 10 nm up to 23 nm for weak and strong surface anchoring conditions, respectively. The electric field required for the maximum shift in the case of weak anchoring is about 3.5 V/μm, while for the strong anchoring it rises to 12.4 V/μm.

6. Conclusion

We have investigated properties of the nematic liquid crystal mixture E7 confined in thin porous films fabricated by electrochemical etch of silicon wafers. The use of spectroscopic ellipsometry is proposed for deriving information about volume fraction, effective ordinary and extraordinary refractive indices and preferred director orientation of the confined nematic.

An empty porous silicon film has rather high birefringence and after infiltration with the isotropic liquid crystal the birefringence of the resultant composite is still significant. Anisotropy of the porous silicon matrix hinders the measurements of the refractive indices of the nematic liquid crystal confined in pores. However, ellipsometry was successful in characterizing E7, in the completely oxidized sample. Relatively small form birefringence of porous silica decreases by a factor of 20 when E7 is infiltrated into the pores, because of the low refractive index contrast between the liquid crystal and silica. Thus, we considered the porous host as isotropic and derived the refractive indices of the anisotropic liquid-crystalline guest.

The free-standing mesoporous silicon microcavity infiltrated with E7 was designed and studied. Transmission spectra of the device were measured at different temperatures using the spectroscopic ellipsometer. Heating the nematic in pores results in the continuous red shift of the peak in the range of 13 nm. The Frank's free energy approach with assumption of escape radial configuration was applied for simulation of orientational properties of the nematic confined in silicon mesopores. The proposed method allows reliable calculation of the range of thermal tuning of interference filters based on porous silicon with liquid crystals.

We have simulated the reorientation of the local director of a nematic liquid crystal confined inside a silicon pore under external electric field influence. On the base of this simulation the maximum tuning range for porous silicon microcavity infiltrated with E7 was obtained for different values of the surface anchoring strength and pore radius. It was found that for strong anchoring a wider range of electrical tuning can be obtained than for weak anchoring, but a higher electric field is required.

Basically, devices with thermal tuning are much slower than electrically tuned ones. In this connection, an alternative and attractive idea would be to produce a local heating of liquid crystals in porous silicon by laser beam illumination, for the realization of a fast all-optical modulator, which is the subject of our future work.

7. Acknowledgements

The authors would like to thank Lucia Rotiroti, Edoardo De Tommasi and Principia Dardano from Istituto per la Microelettronica e Microsistemi (CNR-IMM, Naples, Italy) for their help with fabrication of samples, and Ivo Rendina, head of the Institute, for helpful discussion and financial support.

8. References

Abbate, G.; Tkachenko, V.; Marino, A.; Vita, F.; Giocondo, M.; Mazzulla, A. & De Stefano, L. (2007). Optical characterization of liquid crystals by combined ellipsometry and half-leaky-guided-mode spectroscopy in the visible-near infrared range. *Journal of Applied Physics.* Vol. 101, No. 7, April 2007, 073105-073105-9, ISSN: 0021-8979

Born, M. & Wolf, E. (1980). *Principle of Optics,* 6th edn., Pergamon Press, ISBN: 0080264816, Oxford

Canham, L. (1990). Silicon quantum wire array fabrication by electrochemical and chemical dissolution of wafers. *Applied Physics Letters.* Vol. 57, No. 10, September 1990, 1046-1048, ISSN: 0003-6951

Canham, L. (1997). *Properties of Porous Silicon,* Inspec/IEE, ISBN: 0852969325, London

Crawford, G.; Allender, D. & Doane, J. (1992). Surface elastic and molecular-anchoring properties of nematic liquid crystals confined to cylindrical cavities. *Physical Review A,* Vol. 45, No. 12, June 1992, 8693-8708, ISSN: 1050-2947

Crawford, G. & Žumer, S. (1996). *Liquid Crystals in Complex Geometries,* Taylor & Francis Ltd, ISBN: 0-7484-0464-3

De Stefano, L.; Rea, I.; Rendina, I.; Rotiroti, L.; Rossi, M. & D'Auria, S. (2006). Resonant cavity enhanced optical microsensor for molecular interactions based on porous silicon. *Physica Status Solidi A.* Vol. 203, No. 5, 886-891, ISSN: 0031-8965

Ghulinyan, M.; Oton, C.; Bonetti, G.; Gaburro, Z. & Pavesi, L. (2003). Free-standing porous silicon single and multiple optical cavities. *Journal of Applied Physics.* Vol. 93, No. 12, June 2003, 9724-9729, ISSN: 0021-8979

Haurylau, M.; Anderson, S.; Marshall, K. & Fauchet, P. (2006). Electrical modulation of silicon-based two-dimensional photonic bandgap structures. *Applied Physics Letters,* Vol. 88, February 2006, 061103-1 - 061103-3, ISSN: 0003-6951

Lehmann, O. (1904). *Flüssige Kristalle*, W. Engelmann, Leipzig

Lehmann, V & Gösele, U. (1991). Porous silicon formation: a quantum wire effect, *Applied Physics Letters.* Vol. 58, No. 8, February 1991, 856, ISSN: 0003-6951

Leonard, S.; Mondia, J.; van Driel, H.; Toader, O.; John, S.; Busch, K.; Bimer, A.; Gosele, U. & Lehmann, V. (2000). Tunable two-dimensional photonic crystals using liquid-crystal infiltration. *Physical Review B*, Vol. 61, No. 4, January 2000, R2389 - R2392, ISSN: 0163-1829

Leonard, S.; van Driel, H.; Busch, K; John, S.; Birner, A.; Li, A.; Muller, F.; Gosele, U. & Lehmann, V. (1999). Attenuation of optical transmission within the band gap of thin two-dimensional macroporous silicon photonic crystals. *Applied Physics Letters.* Vol. 75, No. 20, November 1999, 3063-3065, ISSN: 0003-6951

Li, J.; Wu, S.; Brugioni, S.; Meticci, R. & Faetti, S. (2005). Infrared refractive indices of liquid crystals. *Journal of Applied Physics.* Vol. 97, No. 7, March 2005, 073501-1 - 073501-5, ISSN: 0021-8979

Marino, A.; Abbate, G.; Tkachenko, V.; Rea, I.; De Stefano, L. & Giocondo, M. (2007). Ellipsometric study of liquid crystal infiltrated porous silicon. *Molecular Crystals and Liquid Crystals.* Vol. 465, April 2007, 359-370, ISSN: 1542-1406

Moretti, L.; Rea, I.; Rotiroti, L.; Rendina, I.; Abbate, G.; Marino, A. & De Stefano, L. (2006). Photonic band gaps analysis of Thue-Morse multilayers made of porous silicon. *Optics Express,* Vol. 14, No. 13, June 2006, 6264 – 6272, ISSN: 1094-4087

Ouyang, H.; Christophersen, M.; Viard, R.; Miller, B. & Fauchet, P. (2005). Macroporous Silicon Microcavities for Macromolecule Detection *Advanced Functional Materials.* Vol. 15, No. 11, 1851-1859, ISSN: 1616-301X

Pap, A.; Kordás, K.; Tóth, G.; Levoska, J.; Uusimäki A.; Vähäkangas, J.; Leppävuori, S. & George, T. (2005). Thermal oxidation of porous silicon: study on structure. *Applied Physics Letters.* Vol. 86, No. 4, January 2005, 041501-1 – 041501-3, ISSN: 0003-6951

Pavesi, L. & Dubos, P. (1997). Random porous silicon multilayers: application to distributed Bragg reflectors and interferential Fabry-Perot filters. *Semiconductor Science and Technology.* Vol. 12, No. 5, May 1997, 570-575, ISSN: 0268-1242

Pirasteh, P.; Charrier, J.; Soltani, A.; Haesaert, S.; Haji, L.; Godon, C. & Errien, N. (2006). The effect of oxidation on physical properties of porous silicon layers for optical applications *Applied Surface Science.* Vol. 253, No. 4, December 2006, 1999–2002, ISSN: 0169-4332

Spanier, J. & Herman, I. (2000). Use of hybrid phenomenological and statistical effective-medium theories of dielectric functions to model the infrared reflectance of porous SiC films. *Physical Review B.* Vol. 61, No. 15, April 2000, 10437-10450, ISSN: 1098-0121

Tkachenko, V.; Dyomin, A.; Tkachenko, G.; Abbate, G. & Sukhoivanov, I. (2008). Electrical reorientation of liquid crystal molecules inside cylindrical pores for photonic device applications. *Journal of Optics A: Pure and Applied Optics.* Vol. 10, No. 5, May 2008, 055301-1 – 055301-6, ISSN: 1464-4258

Turner, D. (1958). Electropolishing silicon in hydrofluoric acid solutions. *Journal of the Electrochemical Society.* Vol. 105, No. 7, July 1958, 402-408, ISSN: 0013-4651

Uhlir, A. (1956). Electrolytic shaping of germanium and silicon. *The Bell System Technical Journal.* Vol. 35, March 1956, 333-347, ISSN: 1089-7089

Weiss, S. & Fauchet, P. (2003). Electrically tunable porous silicon active mirrors. *Physica Status Solidi A*. Vol. 197, No. 2, May 2003, 556-560, ISSN: 0031-8965

Weiss, S.; Ouyang, H.; Zhang, J. & Fauchet, P. (2005). Electrical and thermal modulation of silicon photonic bandgap microcavities containing liquid crystals. *Optics Express*. Vol. 13, No. 4, February 2005, 1090-1097, ISSN: 1094-4087

Woollam, J.; Johs, B; Herzinger, C.; Hilfiker, J.; Synowicki, R. & Bungay, C (1999). Overview of Variable Angle Spectroscopic Ellipsometry (VASE), Part I: Basic Theory and Typical Applications, Proceedings of Optical Metrology, CR72, pp. 3-28, ISBN: 0819432350, Denver, Colorado, July 1999, SPIE, Bellingham, Washington

2

Liquid Crystals into Planar Photonic Crystals

Rolando Ferrini
Laboratoire d'Optoélectronique des Matériaux Moléculaires (LOMM), Ecole Polytechnique Fédérale de Lausanne (EPFL)
Switzerland

1. Introduction

In the last decade, great effort has been devoted to the study of photonic crystals (PhCs), which are a new class of artificial materials that consist of a periodic arrangement of dielectric or metallic elements in one, two or three dimensions (see Figs. 1-2). The periodicity of these dielectric structures affects the properties of photons in the same way as the periodic potential affects the properties of electrons in semiconductor crystals. Consequently, light propagation along particular directions is forbidden within large energy bands known as photonic bandgaps. Due to such unique properties, PhCs have been proposed as a promising platform for the fabrication of miniaturized optical devices whose potential has been demonstrated both theoretically and experimentally in several applied and fundamental fields such as integrated optics and quantum optics (Busch et al., 2004; Lourtioz et al., 2005).

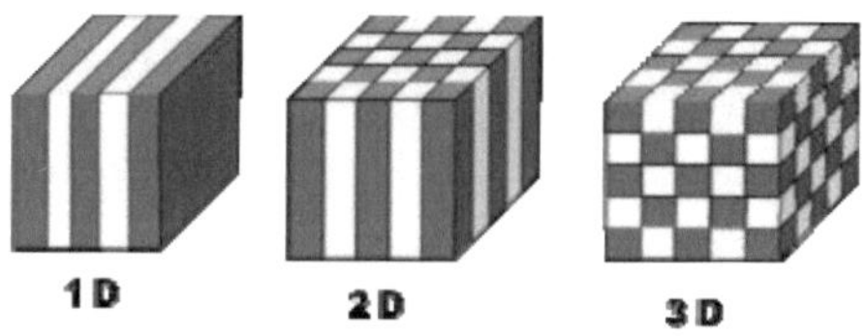

Fig. 1. Sketch of one- (1D), two- (2D) and three-dimensional (3D) photonic crystals (Joannopolous et al., 2008).

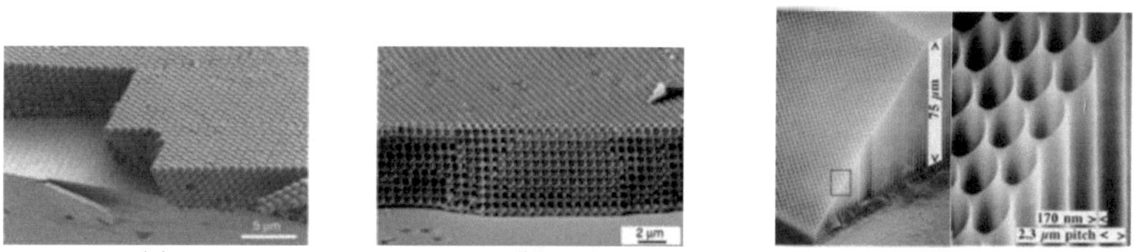

(a) (b) (c)

Fig. 2. Scanning electron microscopy images of three- [(a) opal and (b) inverse opal layers] and two-dimensional [(c) patterned and micromachined layer of macroporous silicon] photonic crystals [(a)-(b) Vlasov et al., 2001; (c) Grüning et al., 1995].

In particular, planar PhCs consisting of a periodic lattice of air holes etched through a high refractive index dielectric matrix (in general, a semiconductor-based vertical step-index

Source: New Developments in Liquid Crystals, Book edited by: Georgiy V. Tkachenko,
ISBN 978-953-307-015-5, pp. 234, November 2009, I-Tech, Vienna, Austria

waveguide providing the vertical light confinement: see Fig. 3) have been intensively studied as artificial materials that offer the possibility to control light propagation on the wavelength scale. For instance, PhC-based optical cavities with high quality factors have been proposed for the demonstration of cavity quantum electro-dynamic effects such as the control of spontaneous emission or the fabrication of single photon sources. Moreover, PhC devices have been studied as building blocks in wavelength division multiplexing applications for integrated optics, where the information is coded into light signals that are treated by either active or passive PhC components such as lasers, filters, waveguides, bends and multiplexers.

Fig. 3. Scanning electron microscopy images of (a) InP-based substrate-like and (b) GaAs-based membrane-like planar photonic crystals [(a) Ferrini et al., 2002a; (b) Sugimoto et al., 2004].

Nowadays, due to this extensive research effort, the conception and fabrication of such photonic structures have gained a complete maturity leading to the realization of the first real applications. PhC devices are routinely fabricated and their optical properties may be optimized at the design stage by modifying the size and/or the position of the air holes either inside or at the boundaries of the device (Song et al., 2005). Nevertheless, PhC-based structures are often lacking in versatility and tunability: On one hand, there are still a few factors that limit the use of PhCs in real devices, such as fabrication imperfections, losses and temperature sensitivity (Ferrini et al., 2003a-b; Wild et al., 2004). On the other hand, as a fundamental requirement for any practical application, the possibility should be guaranteed to adjust the optical properties of the fabricated components by external means. Therefore, the research has focused on the possibility of increasing the device functionalities either by correcting (after fabrication: trimming) or by controlling (on demand: tuning) the optical properties of the PhC in order *i*) to compensate either the temperature sensitivity or the imperfections of the PhC itself (Wild et al., 2004), *ii*) to create reconfigurable devices for integrated optics (Busch et al., 2004; Lourtioz et al., 2005), *iii*) to fabricate bio-chemical sensors (Barthelemy et al., 2007), and *iv*) to conceive new optical functions (Mingaleev et al., 2004). It is worth highlighting that this innovative and emerging research domain may have a huge potential for technological breakthroughs in various application fields such as integrated optics, quantum optics, detection and sensing.

The optical properties of PhCs can be modified by changing the optical length of the PhC structure. This can be achieved either by adjusting the geometrical parameters that define the PhC lattice, *e.g.* the lattice period (*i.e.* the filling factor f) (Joannopolous et al., 2008), or by modifying the refractive indexes of the PhC components. In the first case a mechanical stress may be applied to the PhC slab (Wong et al., 2004), whilst, in the second approach, it is possible to act either on the high index or on the low index component.

As for the high index (*e.g.* the semiconductor) medium, the standard tuning techniques used in classical optoelectronic devices can be applied to modify its refractive index by temperature, by optical pumping or by applying an external magnetic or electric field. For instance, in InP-, GaAs-, or Si-based PhCs, since the semiconductor thermo-optical coefficients are in the order of several 10^{-4}/°C, the thermo-optical effect has been largely used to modulate the optical response (Takeda & Yoshino, 2003; Wild et al., 2004; Tinker & Lee, 2005). Moreover, fast modulation rates in the order of ps down to fs have been obtained by both resonant (*i.e.* by means of carrier injection) and non-resonant (*i.e.* by exploiting the Kerr effect) optical pumping (Haché & Bourgeois, 2000; Leonard et al., 2002; Baba et al., 2003; Ndi et al., 2005; Raineri et al., 2005; Britsow at al., 2006; Hu et al., 2006; Ndi et al., 2006; Teo et al., 2006; Hu et al., 2007; Tanabe et al., 2007).

When either magnetic or ferro-electric or electro-optic non-linear materials are used to fabricate PhC devices, external magnetic or electric fields can be applied, respectively, to adjust the optical response (Kee et al., 2000; Lyubchanskii et al., 2003; Scrymgeour et al., 2003; Belotelov & Zvezdin, 2005).

Finally, other external means like mechanical probes or micro-electromechanical actuators (Märki et al., 2006; Kanamori et al., 2007) have been explored to tune the optical properties of the high index matrix and thus the PhC optical performances.

In the case of semiconductor-based planar PhCs, the classical tuning techniques described above are not always suitable for all the envisaged optics and photonics applications. For example, although the potential of temperature tuning, carrier injection and optical pumping has been fully demonstrated even in real devices, they still present a few intrinsic limits. On one hand, in planar PhC circuits, the large thermal conductivity of the semiconductor matrix makes difficult the temperature tuning of a single device on a chip (Wild et al., 2004). On the other hand, the optical pumping methods require high power optical sources that cannot always be easily integrated in an optical circuit (Teo et al., 2006) and can often change only locally the PhC response.

Therefore, in the last few years, the PhC community has considered with greater and greater attention the alternative approach consisting in the replacement of the low index component (*i.e.* the air) with a different material that can offer further degrees of tunability. Namely, if the refractive index of the infiltrated material can be modified by an external factor (*e.g.* electrical or magnetic fields, light irradiation, temperature, etc.), one can easily adjust and modulate the optical properties of the infiltrated PhC, thus enabling the development of tunable PhC devices for different application fields. Among the numerous techniques that have been proposed, infiltrating the air pores of a PhC with a synthetic organic material that has a tunable refractive index (*e.g.* liquid crystals, polymers, liquids, and liquid dispersions of colloidal quantum dots) has proved to be one of the most promising approaches for both trimming and tuning (Busch & John, 1999; Yoshino et al., 1999b; Leonard et al., 2000; Kubo et al., 2002; Gottardo et al., 2003; Mertens et al., 2003; Schuller et al., 2003; Maune et al., 2004; Mingaleev et al., 2004; Weiss et al., 2005a; Erickson et al., 2006; Ferrini et al., 2006; Haurylau et al., 2006a-b; Intonti et al., 2006; Martz et al., 2006; Tomljenovic-Hanic et al., 2006; van der Heijden et al., 2006a; Barthelemy et al., 2007; Smith et al., 2007; Tay et al., 2007). The effect of infiltrating a PhC with a low refractive index material is qualitatively shown in Fig. 4, where the transmission through a ΓM-oriented triangular lattice of holes (see Sect. 2.1, Fig. 5) etched in a dielectric matrix is simulated for the transverse electric (TE) polarization and for different refractive index values inside the holes. Once the PhC is infiltrated, due to the reduced refractive index contrast between the dielectric matrix and the holes, the photonic bandgap shrinks and the band edges red-shift. In particular, since for energies located in the low- (dielectric) and high-energy (air) transmission bands the electric field of the propagating Bloch modes is concentrated in the dielectric and in the holes, respectively (Busch et al., 2004; Lourtioz et al., 2005; Joannopolous et al., 2008), the energy shift of the air band edge is one order of magnitude larger than the corresponding shift of the dielectric band edge.

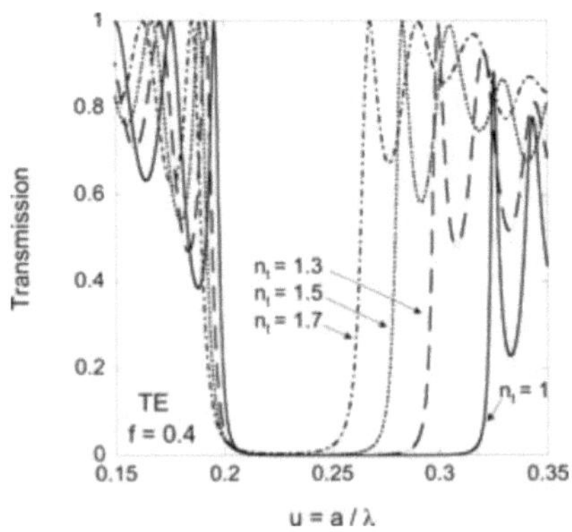

Fig. 4. Simulated transmission spectra through a ΓM-oriented triangular lattice of holes (air filling factor $f = 0.4$) etched in a dielectric matrix ($n = 3.3$) for the transverse electric (TE) polarization (see Fig. 5) and for different refractive indexes inside the holes: *i.e.* $n_t = 1$ (empty holes), 1.3, 1.5, and 1.7 (El-Kallassi, 2009).

The choice of organic molecules to fill the air holes of planar PhCs stems naturally from the following properties of organic materials:

- they offer a large variety of interesting optical properties both in the visible and in the near infrared spectral regions that often depend on the molecular organization;
- there exist a wide palette of different molecular organizations depending on the molecule interactions. The complex thermodynamic properties of such molecular mixtures enable one to easily modify the molecular order (*e.g.* by changing the temperature, by applying an electric field, by irradiating at certain wavelengths, etc.) and, thus, the optical properties of the material itself;
- the complex interaction between organic molecules and inorganic surfaces, such as the semiconductor native oxide layers, strongly affects the wetting properties of the organic blends. Therefore, controlling the surface quality allows one to control the molecule organization and the optical properties of the material itself.

Therefore, research efforts have been focused on the functionalization of PhC devices by trimming/tuning the optical properties of these organic materials. In particular, the potential of PhC infiltration with nematic liquid crystals (LCs) has been largely demonstrated for one-, two- and three-dimensional (1D, 2D and 3D) PhCs. Therefore, besides their classical fields of application, LCs are also having a strong impact in the PhC field (Busch & John, 1999; Yoshino et al., 1999b; Leonard et al., 2000; Kang et al., 2001; Shimoda et al., 2001; Kubo et al., 2002; Mertens et al., 2002; Gottardo et al., 2003; Mertens et al., 2003; Schuller et al., 2003; Weiss & Fauchet, 2003; Busch et al., 2004; Du et al., 2004; Kubo et al., 2004; Martz et al., 2004; Maune et al., 2004; Kosmidou et al., 2005; Lourtioz et al., 2005; Maune et al., 2005; Weiss et al., 2005a-b; Ferrini et al., 2006; Haurylau et al., 2006b; Martz et al., 2006).

The molecule shape and their long-range orientational order are at the origin of the optical anisotropy of a nematic LC, which is a birefringent medium with an uniaxial symmetry. When interacting with the nematic LC, light experiences the ordinary refractive index (n_o) or the extraordinary refractive index (n_e) if the electric field is polarized perpendicularly or parallel to the axis of the LC molecules, respectively. The molecular order of a LC can be easily modulated by means of external perturbations, thus allowing one to adjust their dielectric constant (*i.e.* the refractive index) on demand. Consequently, the optical response of a PhC infiltrated with nematic LCs can be either trimmed or tuned by applying an external electric field which modifies the orientation of the molecules with respect to the polarization direction of a light beam propagating through it. Moreover, when the temperature is increased above the nematic-isotropic phase transition temperature (clearing point), the molecular order is destroyed and the LC is in its isotropic phase: its optical properties are thus characterized by an isotropic refractive index (n_i) that is independent of the molecule orientation. Finally, when nematic LCs are mixed with photosensitive molecules like photochromes (see below), the latter phase transition can be photoinduced isothermally by light irradiation at certain wavelengths. All these tuning mechanisms are reversible and they have been exploited to tune the optical properties of PhC structures infiltrated with nematic LCs (Alagappan et al., 2006; Halevi et al., 2006; Haurylau et al., 2006a-b; Martz et al., 2006; Anderson et al., 2007; Reyes et al., 2008).

In this chapter, we will show how the infiltration of LCs in the air holes of III-V semiconductor-based planar PhCs can be used to both trim and tune the optical properties of PhC devices by controlling the orientation of the LCs inside the nanometer-size holes by means of a temperature, electrical or optical trigger.

In Section 2, we will illustrate the main optical properties of the InP-based substrate-like planar PhCs consisting of a hexagonal array of air holes (hole diameter = 200 - 400 nm; air filling factor f = 0.40-0.50) that were mainly used for our studies. Moreover, the choice of the nematic LCs [*i.e.* LC-K15 (5CB)] will be briefly discussed and their optical properties will be summarized (Martz et al., 2004; Martz et al., 2005).

In Section 3, we will focus on the infiltration of LC molecules in cylindrical holes with a diameter of a few 100's nm by means of a vacuum chamber specifically designed to empty the gaseous content of the holes, to clean and, when necessary, to chemically activate the device surface (Martz et al., 2004; Martz et al., 2006). The reliability and reproducibility of the infiltration process are assessed by the optical characterization of the infiltrated structures. Moreover, we note that, in order to exploit the tunable optical properties of nematic LCs for the functionalization of PhC devices, it is important to know both the filling efficiency and the orientation of the molecules inside holes. This latter property eventually depends both on the physics of the infiltration process of an organic viscous liquid in nanometer-size holes and on the surface aligning properties of the structured semiconductor matrix. In the last two decades, the LC community has made great efforts to understand the organization of LC molecules confined in nanometer-size cylinders. Both theoretical and experimental studies have been performed to quantify the influence of the surface anchoring and of the hole size on the LC molecular organization (Crawford et al., 1991a-b; Marroum et al., 1995; Burylov, 1997). With regard to this issue, we will show how optical measurements as a function of temperature and polarization can yield information on both the hole filling and the molecule orientation (Martz et al., 2006; Ferrini et al., 2006).

In Section 4, we will show how the optical response of PhC devices infiltrated with nematic LCs can be tuned by temperature, electric field and optical irradiation (Ferrini et al., 2006;

El-Kallassi et al., 2007). In particular, we observe that, in spite of a large amount of research on LC infiltrated PhCs, little has been done on their optical tuning (Maune et al., 2005), even though all-optical switching plays a very important role in the optical communication field (Asakawa et al., 2006) and, as we have briefly discussed above, several other approaches have already been explored to optically tune planar PhCs (Ndi et al., 2005; Raineri et al., 2005; Teo et al., 2006; Tanabe et al., 2007). Here, we will illustrate how it is possible to optically tune the response of planar PhC devices by infiltration with a photo-responsive LC blend doped with azobenzene photochromic molecules (Legge & Mitchell, 1992; Sung et al., 2002; Ikeda, 2003).

2. Photonic crystals and liquid crystals

2.1 InP-based planar photonic crystals

Planar PhCs consisting of a triangular lattice of air holes were etched by chemically assisted ion beam etching (CAIBE) through a nominally undoped InP/(Ga,In)(As,P)/InP vertical waveguide grown by metalorganic vapour phase epitaxy on a (100) *n*-doped InP substrate: see Fig. 4a (Ferrini et al., 2002b; Ferrini et al.; 2004; Mulot et al., 2004). In the wavelength range of interest (*i.e.* $\lambda \approx 1.5$ μm) the waveguide is single-mode for both the transverse electric (TE) and magnetic (TM) polarization directions (*i.e.* electric field perpendicular and parallel to the hole axis, respectively: see Fig. 5a) and the effective refractive indexes n_{eff} for both the TE and the TM guided modes varies linearly with temperature (Wild et al., 2004; Ferrini et al., 2004; Mulot et al., 2004; Martz et al., 2005; Ferrini et al., 2006; El-Kallassi et al., 2007).

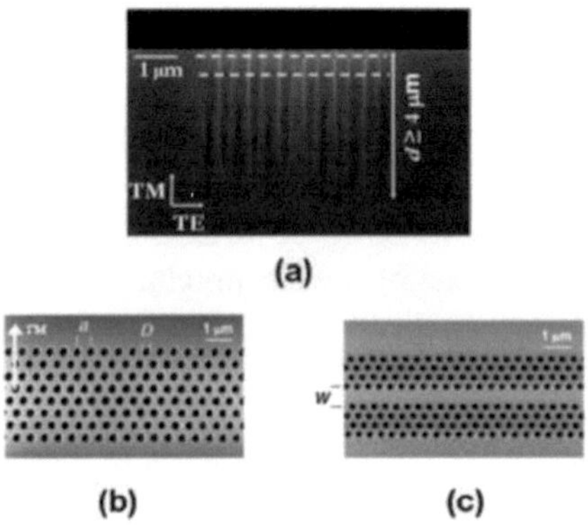

Fig. 5. Scanning electron microscopy images (El-Kallassi et al., 2007): (a) Cut view of a PhC etched through a InP/(Ga,In)(As,P)/InP planar waveguide [the GaInAsP core layer is sketched (dashed lines)]: the hole depth is d ≥ 4 μm. The white arrows indicate the orientation of the electric field for the transverse electric (TE) and magnetic (TM) polarization directions; (b) Top view of a 8-rows thick ΓM-oriented PhC slab (*a* = lattice period; D = hole diameter); (c) Top view of a Fabry-Perot cavity between two 4-rows thick ΓM -oriented PhC mirrors (*W* = cavity width).

A scanning electron microscopy cut view of the obtained PhCs appears in Fig. 4a: the hole depth *d* is of the order of 4 μm. Two types of structures were fabricated (see Figs. 5b-c): *i*) 8 rows-thick ΓM-oriented PhC slabs; and *ii*) Fabry-Pérot (FP) cavities consisting of two 4-rows thick ΓM-oriented PhC mirrors separated by a spacer *W*.

The internal light source (ILS) technique was used to optically characterize the infiltrated PhC structures (Ferrini et al., 2002b). To this aim, two GaAsInP quantum wells (QWs) were embedded in the core layer to act as a built-in light probe for optical measurements. The superimposition of the two QW emission peaks yields a 100 nm-wide photoluminescence (PL) spectrum centered at λ = 1.5 μm. The PL excited inside the QWs is used as a built-in probe beam. Part of the PL signal propagates parallel to the surface as a guided mode and interacts with the PhC structure before escaping from a cleaved edge and being collected and spectrally analyzed. The absolute PhC transmission is obtained normalizing the spectrum measured after transmission through the PhC slab with respect to a spectrum collected in a non-patterned region of the sample. Since ILS measurements on a single PhC structure yield the transmission spectrum only in a narrow spectral interval (~ 100 nm), the PhCs were replicated with different periods a (*i.e.* a = 300-600 nm with Δa = 20 nm) while keeping constant the intended air filling factor f = 0.40-0.50 (lithographic tuning). Thus, due to the scaling property of PhCs (Joannopolous et al., 2008), the absolute transmission through the PhC structures was measured as a function of the normalized photon energy $u = a/\lambda$ (energy reduced units). The real filling factor f was then measured by fitting the transmission spectra of the empty structures (Ferrini et al., 2002b). We obtained an average value $f = 0.43 \pm 0.01$ that corresponds to a hole diameter D = 200 – 400 nm.

2.2 Nematic liquid crystals

Commercial nematic LCs by Merck® were used to infiltrate the fabricated PhC structures. While preliminary experiments were performed with LC-E7, a blend of a cyanoterphenyl and several cyanobiphenyl molecules (see Fig. 6a), for the infiltration of the InP-based planar PhCs we chose LC-K15 that consists of only one type of molecules [4-cyano-4'-pentylbiphenyl (5CB): see Fig. 6b]. This latter property guarantees the stability of LC-K15 with respect to the infiltration procedure, during which the LC blend is heated at low pressure (see Sect. 3). This was verified by means of differential scanning calorimetry measurements, where the LC response was analysed before and after heating the molecule blends at low pressure, *i.e.* in similar conditions as during the in-filling process. While a change in the blend composition was detected in the LC-E7 blend, for the LC-K15 molecules the location of the peaks in the heat capacity signals (see Fig. 7a), which correspond to the nematic-to-isotropic phase transition (Mansare et al., 2002), does not change after the treatment, thus showing that the physical-chemical properties of LC-K15, as well as its optical properties (*i.e.* the refractive index), remain stable after the infiltration.

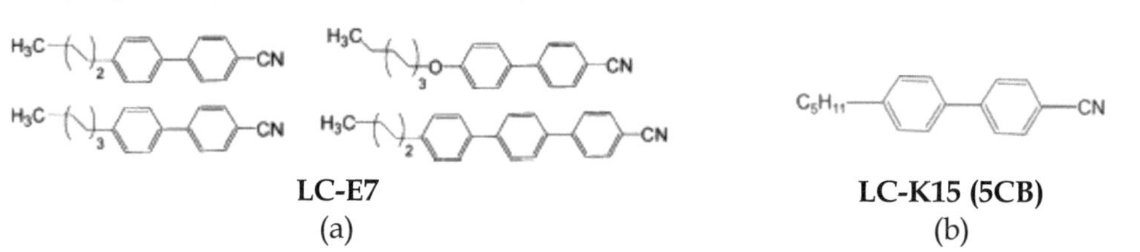

Fig. 6. Chemical structure of the used nematic liquid crystals (by Merck®): (a) LC-E7 blend [one cyanoterphenyl and three cyanobiphenyl molecules] and (b) LC-K15 [4-cyano-4'-pentylbiphenyl (5CB)]. All molecules present an aromatic structure with polar cyano headgroups and apolar aliphatic tails [(a) Martz et al., 2004; (b) El-Kallassi et al., 2007].

In addition, LC-K15 has lower nematic-to-isotropic and nematic-to-polycrystalline phase transition temperatures than LC-E7: *i.e.* $T_{c1} = T_{NI}$(LC-K15) = 35 °C (clearing point) (see also Fig. 7a) and $T_{c2} = T_{KN}$(LC-K15) = 23 °C (melting point), respectively (Mansare et al., 2002). This makes LC-K15 more suitable than LC-E7 for infiltration and temperature tuning experiments (see Sect. 4).
Finally, in Fig. 7b, the ordinary n_o, extraordinary n_e, and isotropic n_i refractive index values of LC-K15 are reported as a function of temperature for a wavelength λ = 1.5 μm (Li & Wu, 2004). In particular, at T = 24 °C, n_o = 1.501 and n_e = 1.656, while, at T = 37 °C, n_i = 1.548 (Brugioni et al., 2006).

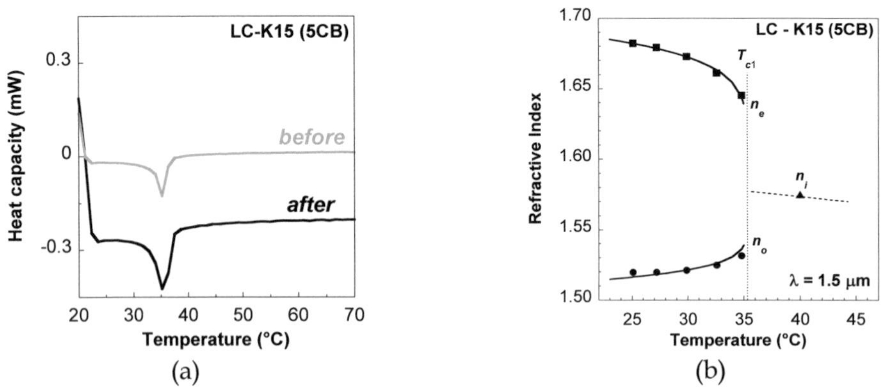

Fig. 7. Nematic liquid crystals LC-K15 (by Merck®): (a) Differential scanning calorimetry measurements on LC-K15 before (grey line) and after (black line) heating the LC at low pressure, *i.e.* in similar conditions as during the PhC infiltration process. (b) Ordinary n_o, extraordinary n_e, and isotropic n_i refractive index values of LC-K15 as a function of temperature. The nematic-to-isotropic phase transition temperature [$T_{c1} = T_{NI}$ = 35 °C (clearing point)] is reported. [(a) Martz et al., 2005; (b) Mansare et al., 2002; Li & Wu, 2004].

3. Infiltration of nematic liquid crystals in planar photonic crystals

3.1 Infiltration experiments

Using nematic LCs to fill InP-based planar PhCs, *i.e.* cylindro-conical holes with a diameter ranging from 200 up to 400 nm, an average depth of 4 μm and a closed bottom (see Fig. 5) is indeed far from being a trivial problem. For instance, when infiltrating nanoporous materials (*e.g.* nanometer-size tubes) with viscous liquids, the gravitational forces are several orders of magnitude smaller than the capillary forces, which eventually drive the infiltration process (Busch et al., 2004). The latter forces are proportional to the capillary diameter, the surface tension and the contact angle of the liquid on the tube surface (van der Heijden et al., 2006a-b). In particular, if this angle is lower than 90°, the capillary forces assist the infiltration, otherwise, for angles larger than 90°, the liquid is expelled from the tube. Therefore, in order to achieve a good infiltration efficiency (see Sect. 3.2), it is important to guarantee a LC wettability on the capillary surface as large as possible, *i.e.* the smallest possible contact angle. The liquid wettability on a solid surface depends on the chemical composition, the geometry and an eventual functionalization of the surface (Martz et al., 2004). In the case of III-V semiconductor-based (*e.g.* InP- or GaAs-based) planar PhCs, the sample surface tends to be hydrophobic, *i.e.* with a water contact angle larger than 90° (see

Fig. 8a). Nevertheless, chemical and/or plasma treatments, as well as chemical surfactants, can be used to increase the surface energy and thus to improve the wettability of such materials (D. Myers, 1991). For example, in Fig. 8a-b, the contact angle images of a water droplet on the surface of a non-patterned region of an InP-based PhC before and after a chemical attack with hydrofluoric (HF) acid are shown, respectively. The HF treatment removes the residual SiO_2 over-layer that is usually deposited on the InP surface for the fabrication of planar PhCs (Mulot et al., 2004) and contributes to strongly increase the water wettability on the sample, thus yielding a hydrophilic surface. In particular, LC contact angle images taken on the same surfaces show that, if LC-K15 presents already a good wettability (see Fig. 8c), the HF treatment yields a contact angle close to 0°, *i.e.* an almost perfect wettability.

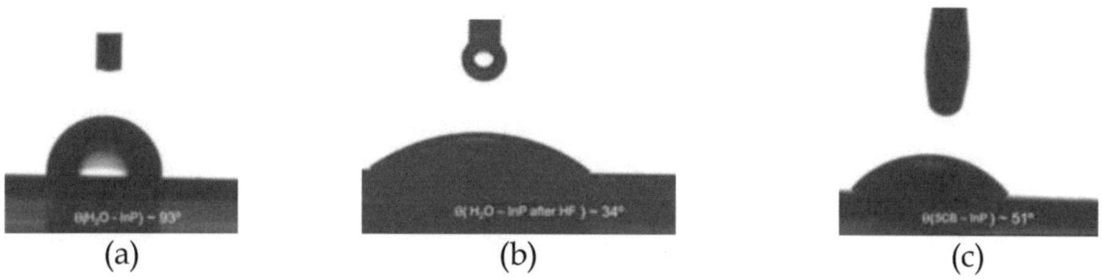

Fig. 8. Contact angle images of (a)-(b) water and (c) LC-K15 (5CB) liquid crystal droplets on the surface of a non-patterned region of an InP-based planar photonic crystal before and after a chemical treatment with hydrofluoric (HF) acid, respectively. For the LC-K15 (5CB) droplet, the HF treatment yields a contact angle close to 0°, *i.e.* an almost perfect wettability, so that no droplets can form on the surface (El-Kallassi, 2009).

Fig. 9. Deposition of self-assembled monolayers of (a) a *p*-(dimethylamino)-benzoic acid (AmBA) on a doped GaAs surface: contact angle images of LC-E7 liquid crystal droplets on (b) an unmodified and (c) a functionalized surface. (Martz et al., 2004)

Moreover, if necessary, it is possible to functionalize the semiconductor surface with molecular self-assembled monolayers (SAMs), which have been demonstrated to be very important for the surface modification of several optoelectronics devices and are widely used as molecular orientation layers for LCs (Martz et al., 2004 and references therein). In particular, we demonstrated the possibility of using surface-adsorbed monolayers on oxidized GaAs surfaces to improve the LC wettability. The surface of GaAs is unstable in an atmosphere containing oxygen and water: after exposure to air, a native oxide layer is usually formed. This latter layer can be used as a binding layer for the deposition of carboxylated derivatives using a vapor growth method under high vacuum on doped plasma-activated GaAs surfaces. Due to the amphiphilic nature of LCs, as orientation layers, we chose para-substituted benzoic acids that exploit the affinity to the highly polar cyano headgroups and to the aromatic structure of LC-E7 (see Fig. 6a). In Figs. 9b-c, the contact angle images of LC-E7 droplets on an unmodified GaAs wafer and on a GaAs surface functionalized with a *p*-(dimethylamino)-benzoic acid (AmBA: see Fig. 9a) are shown, respectively. This surface treatment strongly increases the LC-E7 wettability on the GaAs surface and static contact angles as small as 9° could be achieved.

In addition to the LC wettability, other parameters must be taken into account when infiltrating planar PhCs. If, in the case of 3D or 2D membrane-like PhCs, the samples can be partially immersed in a LC bath and, while LC molecules penetrate into the nanopores, the air is naturally evacuated, in substrate-based planar PhCs the air holes are accessible only from the top and the gas evacuation may be difficult. Indeed, it has been demonstrated both theoretically and experimentally that a low ambient pressure is beneficial to the infiltration process since it minimizes the effects of gas trapping in the LCs and contributes to drastically reduce the air volume that remains eventually encapsulated at the hole bottom (Leonard et al., 2000; Schuller et al., 2003; van der Heijden et al., 2006a-b). The residual air content is finally degassed through the LC column (Nielsen et al., 2005; van der Heijden et al., 2006a-b). Infiltration times of several minutes up to few hours were adopted in order to guarantee a complete gas evacuation and thus a perfect hole infiltration.

A vacuum chamber with controllable temperature and pressure was specifically designed for planar PhC infiltration experiments (see Fig. 10). The chamber is equipped with three sample holders and a small crucible containing the LCs that are heated by means of three resistances and a convection-heating ring, respectively; several thermocouples connected to the holders and to the crucible enable the separate control of the LC and the sample temperature. A pumping system allows the chamber pressure to be reduced down to 10^{-6} mbar and a vacuum gauge is used to control the vacuum level. A gas inlet (*i.e.* argon or nitrogen) is utilized to eventually release the chamber to the atmospheric pressure.

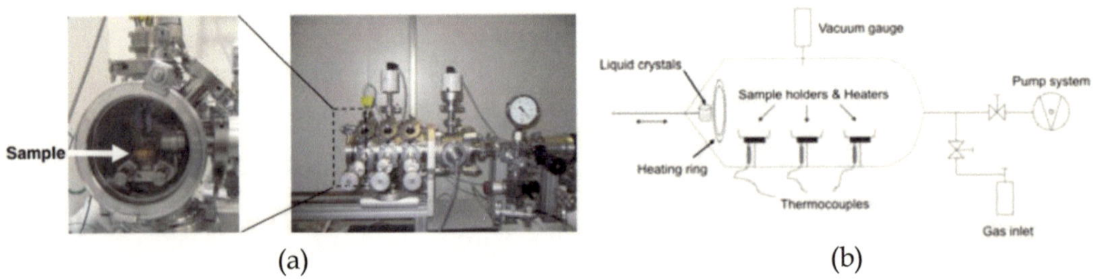

Fig. 10. (a) Images and (b) sketch of the vacuum chamber designed for the infiltration of planar photonic crystals (El-Kallassi, 2009).

The best temperature and pressure conditions were investigated to optimize the infiltration process and to improve its reliability and reproducibility. The chamber pressure is reduced to 10^{-5} mbar, which is enough to guarantee a small residual air content at the hole bottom (see above). The temperature is also a fundamental parameter: during the infiltration procedure, both the LCs and the samples are constantly kept at a temperature higher than the clearing point (*i.e.* 40 °C). This reduces the LC viscosity, thus favouring the pore infiltration, and avoids wall contamination by water residuals. The filling process itself can be divided into three steps: (*i*) *Evacuation*: The vacuum chamber is pumped to empty the gaseous content of the holes and to degas the LCs inside the crucible, which, to this aim, are heated at 30°C. This is indeed an essential step: if LCs are not properly degassed, trapped gases may combine to generate bubbles inside the holes, thus limiting the infiltration efficiency. (*ii*) *Contact*: The LCs are left in contact with the samples for several minutes up to few hours (see above). (*iii*) *Infiltration*: Finally, the vacuum chamber is instantaneously brought to the atmospheric pressure and the sample is slowly cooled down to the ambient temperature. This is an important step for the infiltration process: the pressure differential between the chamber and the holes introduces a supplementary force that assists the hole filling.

A systematic characterization of the infiltration procedure was carried out by means of optical measurements. Several infiltration experiments were performed on the same PhC samples under the same temperature and pressure conditions. Before each infiltration the samples were heated at a temperature much higher than the clearing point (see Sect. 2.2) to evaporate the LC molecules, *i.e.* to empty completely the holes. In order to exclude the presence of LC residuals in the holes, after heating, the samples were cooled down to room temperature, where transmission spectra through the emptied PhC structures were measured and compared with the reference spectra taken before infiltration. The same results were found in terms of the optical properties of the PhCs after infiltration, thus testifying to the great reproducibility and reliability of the in-filling technique (J. Martz et al., 2005).

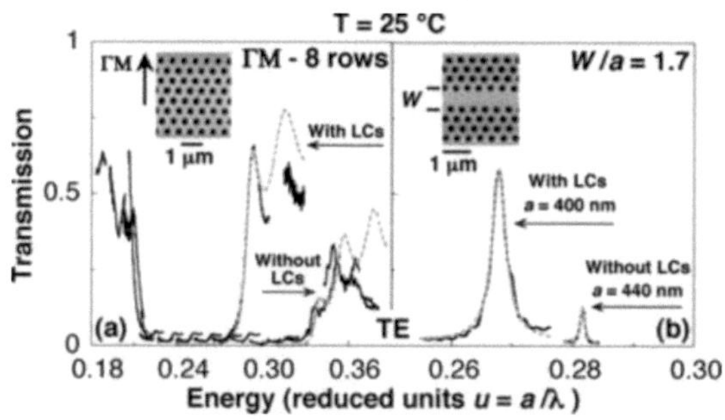

Fig. 11. Measured TE transmission spectra through (a) 8-rows thick ΓM-oriented PhC slabs (see the inset) and (b) a Fabry-Pérot (FP) cavity between two 4-rows thick ΓM-oriented PhC mirrors (see the inset) with and without LCs (black lines). *W* is the cavity width and *a* the PhC lattice period. The best-fit curves of the experimental spectra are shown (grey dotted lines) (Martz et al., 2006).

The TE polarized transmission spectra at room temperature through ΓM-oriented PhC slabs (see Fig. 5) without and with LCs are shown in Fig. 11a (black lines). A well-defined stopgap appears in the empty sample. Once the PhC is infiltrated, the photonic bandgap shrinks and the band edges red-shift (see Sect. 1). Moreover, the air band transmission increases, thus indicating a strong reduction of the PhC losses after infiltration. In low refractive index contrast planar PhCs perforating the vertical waveguide induces coupling of the guided wave to radiation modes into the claddings, thus resulting in out-of-plane losses (Ferrini et al., 2003a-b). When infiltrating PhCs with a material whose refractive index is higher than that of the air, the out-of-plane diffraction at the waveguide-hole interface is reduced. In order to confirm this, the spectra were fitted by means of a two-dimensional finite difference time domain (2D-FDTD) model assuming n_{eff} as the matrix index and n_{hole} as the hole refractive index (Qiu et al., 2002). Out-of-plane losses were included into the 2D model by adding a phenomenological imaginary dielectric constant ε'' in the holes (Ferrini et al., 2003a-b). The measured values $f = 0.42$ and 0.46 were assumed that correspond to the energy u (*i.e.* to the PhC period a in the lithographic tuning approach: see Sect. 2.1) where the fitted band edges are located in the spectra without and with LCs, respectively. The independent fit parameters were ε″ and n_{hole}: the calculated spectra are shown in Fig. 11a (grey dotted lines). The values n_{hole} = 1 and 1.59, and ε″ = 0.045 and 0.030 were found for the PhCs without and with LCs, respectively, thus confirming the increase of n_{hole} and the consequent decrease of losses after infiltration.

Transmission spectra at room temperature through the FP cavities (see Fig. 5) without and with LCs are shown in Fig. 11b. Due to the increase of n_{hole}, the FP resonance red-shifts with respect to the empty sample. The FP peaks were fitted using an Airy's formula (Ferrini et al., 2002b): the calculated curves are shown in Fig. 11b (grey dotted lines). Due to the reduced refractive index contrast, after infiltration the mirror transmission increases from 0.013 to 0.090 while its reflectivity decreases from 0.96 to 0.887. Therefore, in agreement with the results obtained for simple slabs, the loss coefficient L = 1 - R - T decreases from 0.017 to 0.013. Due to the change of both the mirror optical properties and the cavity losses (Mulot et al., 2004; Martz et al., 2005), the peak transmission increases and the cavity quality factor decreases from 310 to 100.

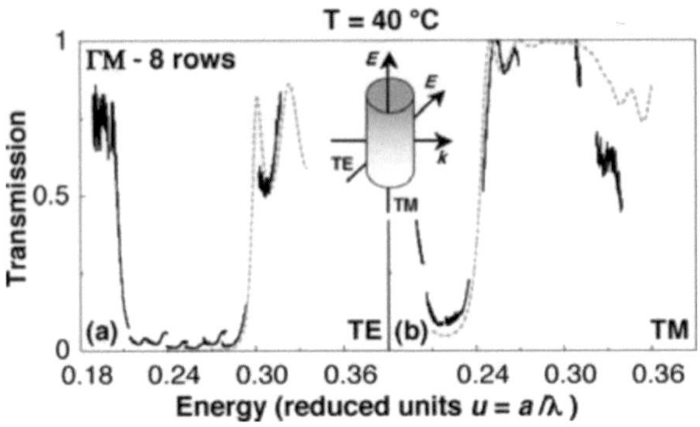

Fig. 12. Experimental (black lines) and calculated (grey dotted lines) (a) TE and (b) TM transmission spectra through infiltrated 8-rows thick ΓM-oriented PhC slabs (see Fig. 11) at T = 40 °C. The orientation of the E-field with respect to the hole axis and the light propagation direction (vector k) are sketched in the inset for both the TE and TM polarizations. (Martz et al., 2006).

3.2 Infiltration efficiency

The infiltration efficiency of substrate-like planar PhCs is a critical parameter, whose measurement is not easy (Schuller et al., 2003). While using standard scanning electron or atomic force microscopy is not straightforward, optical measurements have proved to be very useful in the characterization of infiltrated planar PhCs (Martz et al., 2005; Martz et al., 2006). The average infiltration efficiency η can be accurately determined by measuring the TE and TM transmission spectra (see black lines in Figs. 12a-b, respectively) through infiltrated PhC slabs at a temperature T = 40 °C > T_{c1}. Due to the birefringence of nematic LCs, the component of their dielectric constant along the electric field (and thus the effective LC refractive index n_{LC}) depends on the molecule orientation. However, when working at a temperature above the clearing point, LCs are in an isotropic phase and the dielectric constant is a scalar, independent from the molecule orientation and the field distribution, *i.e.* $n_{LC} = n_i$. As in Sect. 3.1, the TE and TM air band transmission edges were fitted by means of a 2D-FDTD model with f = 0.46 and 0.48, respectively, and ε″ and n_{hole} as free fitting parameters: the calculated spectra are shown in Fig. 12 (grey dotted lines). The fit yielded the value n_{hole} = 1.50 for both polarizations, thus confirming that at T = 40 °C the infiltrated LCs are in their isotropic phase. The same result is obtained by the 2D-FDTD fit of the resonance energy in the transmission spectrum through the FP cavities. The infiltration efficiency is defined as the percentage of the PhC slab that is filled, *i.e.* $\eta = (n_{hole} - 1) / (n_{LC} - 1)$. Since $n_{LC} = n_i = 1.575$ (see Fig. 7 in Sect. 2.2), the value $\eta = 0.87$ is obtained. We remark that, whilst the same value is obtained for several hole diameters, η does not contain any local information on the partial or complete filling of individual holes.

3.3 Optical characterization of the molecule orientation

Once the infiltration efficiency is known, in order to identify the better mechanisms to tune the optical properties of the infiltrated PhCs, it is important to understand how the nematic LC molecules organize inside the PhC holes. The configuration of the LC director field inside cylindrical capillaries can be indeed numerically obtained by minimizing the total energy of the LCs in contact with the cylinder surface. The molecule anchoring to the lateral surfaces, as well as, for nanometer-size tubes, the balance between the elastic and the surface forces, but also the LC properties (*e.g.* the LC elastic modulus), the cavity geometry, and the sidewall surface (*e.g.* its physical-chemical properties) primarily determine the exact molecule orientation (Burylov, 1991; Crawford et al., 1991a-c; Crawford & Zumer, 1996). Several equilibrium configurations are possible: On one hand, if the anchoring is homeotropic, according to the cylinder diameter and the anchoring force, the planar radial, planar polar or escaped radial configurations can be obtained (see Fig. 13a-c). On the other hand, in the case of a circular anchoring, the planar circular, the circular planar polar or the escaped circular configurations are privileged (see Fig. 13d-f).

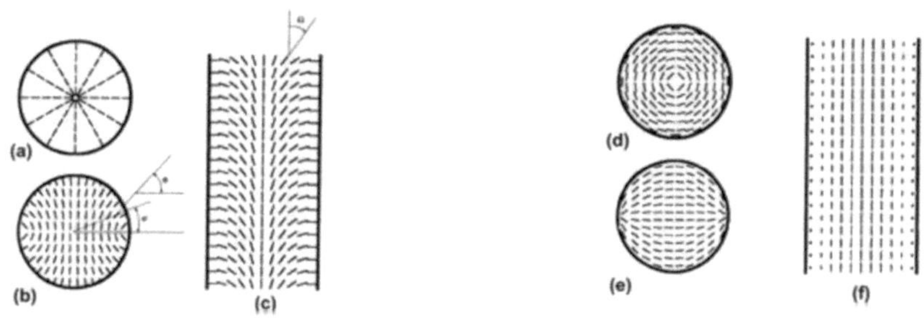

Fig. 13. Examples of liquid crystals director configurations in a cylindrical capillary. (a)-(c) Homeotropic anchoring: (a) planar radial; (b) planar polar; (c) escaped radial. (d)-(f) Circular anchoring: (d) planar circular; (e) circular planar polar; (f) escaped circular (Burylov, 1991).

In infiltrated planar PhCs, the comparison of optical measurements for different polarizations can provide information on the average molecule orientation. We note that all the possible configurations listed above can be grouped into three families: *i*) completely perpendicular (*i.e.* planar), *ii*) completely parallel, or *iii*) escaped (*i.e.* with no dominant orientation): see the inset of Fig. 14. On one hand, for the TM polarization the *E*-field vector is parallel to the hole axis, so that for both LC configurations n_{LC} depends on only one component of the LC dielectric tensor, *i.e.* $n_{LC} = n_e$ and $n_{LC} = n_o$ for the parallel and perpendicular configuration, respectively. On the other hand, for the TE polarization the *E*-field vector lies in the waveguide plane (*i.e.* it is perpendicular to the hole axis). While for the parallel configuration n_{LC} still depends on only one component of the LC dielectric tensor (*i.e.* $n_{LC} = n_o$), for the perpendicular configuration the average n_{LC} value depends on all the components of the LC dielectric tensor weighted by the local *E*-field. Therefore, it is only for the TM polarization that n_{hole} can be calculated for both LC orientations without any information on the exact director configuration: if $\eta = 0.87$, $n_{hole} = 1.593$ and 1.449 for the parallel and perpendicular configuration, respectively.

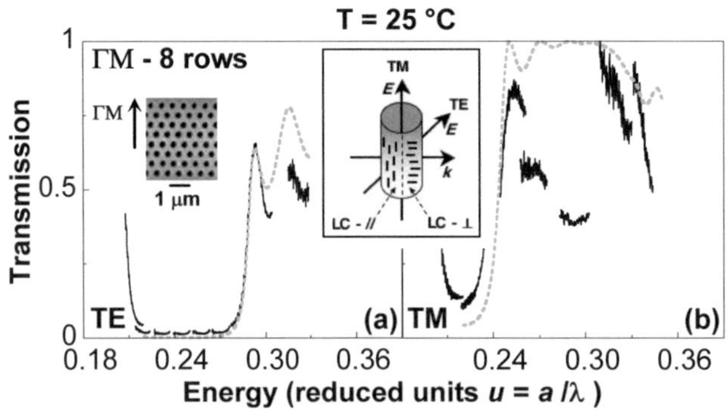

Fig. 14. Measured (black lines) and calculated (grey dotted lines) (a) TE and (b) TM transmission spectra through infiltrated 8-rows thick ΓM-oriented photonic crystal slabs at $T = 25$ °C. The inset shows the *E*-field vector with respect to the light propagation direction (*k*) for both the TE and TM polarizations and the average liquid crystal (LC) orientation parallel (//) and perpendicular (⊥) to the hole axis. (Ferrini et al., 2006).

TE and TM transmission spectra through the infiltrated 8 rows-thick ΓM-oriented PhC slabs (see Sect. 2.1) are shown in Figs. 14a-b, respectively (black lines), for $T = 25$ °C. As in Sects. 3.1-2, the air band transmission edges were fitted by means of a 2D-FDTD model assuming n_{hole} as the free fitting parameter. The calculated spectra are shown in Fig. 14 (grey dotted lines). The fit of the TM spectrum yields $n_{hole} = 1.45$, thus showing that, in contrast to other studies on similar systems (Schuller at al., 2003), most of the LC molecules are aligned perpendicularly to the hole axis (*i.e.* in a planar equilibrium state). This can be due to several factors: the different material system (InP instead of GaAs), the hole shape/surface, etc. We observe that a deeper understanding of the relationship between the n_{hole} values and the molecule orientations may indeed be provided by advanced theoretical calculations taking into account (by means of second-rank tensor indexes) all the possible molecule configurations with respect to the real *E*-field map.

4. Tuning of the optical response of infiltrated planar photonic crystals

4.1 Temperature tuning

As discussed in Sect. 1, the molecular order of LCs infiltrated in planar PhCs can be easily modulated by temperature. When the temperature is increased above the clearing point, the nematic molecular order is destroyed and the LC is in its isotropic phase, where the optical properties are characterized by the isotropic refractive index (n_i).

For instance, transmission spectra through infiltrated FP cavities (see Sect. 2.1) are shown in Fig. 15a for $T = 25$ °C (dotted black line) and $T = 40$ °C (solid black line). The FP resonance energy red-shifts, thus proving the increase of n_{LC} (n_{hole}) with temperature: the fit (see Sect. 3) of the FP resonance at room temperature yields $n_{hole} = 1.47$, *i.e.* for $\eta = 0.87$, $n_{LC} = 1.54$, which is lower than the isotropic n_i value that is found at 40°C (see Fig. 7). The FP resonance wavelength (λ_{peak}) as a function of temperature is shown in Fig. 15b. For $T \approx 25$ °C λ_{peak} is nearly constant as neither the semiconductor nor the LC refractive indices change significantly. However, due to the increase of n_{LC} (n_{hole}) at the nematic-to-isotropic and nematic-to-polycrystalline LC phase transitions, λ_{peak} sharply increases above 30 °C and below 20 °C with $\Delta\lambda_{peak}$ = 7 nm and $\Delta\lambda_{peak}$ = 11 nm, respectively. Above 35°C, λ_{peak} remains constant since the increase of the semiconductor refractive index is compensated by the decrease of $n_{LC} = n_i$ with temperature (see Fig. 7). We observe that a symmetrical tuning of the FP resonance is obtained after LC infiltration by either increasing or decreasing the temperature and, for an equal tuning range ΔT, the measured $\Delta\lambda_{peak}$'s are much larger than those obtained by the temperature tuning of similar empty PhC cavities (Wild et al., 2004). However, it is worth noticing that, while tuning rates as low as 10's μs were achieved by locally tuning the temperature of PhC devices (Tinker & Lee, 2005; Chu et al., 2006), response times in order of ms were measured for infiltrated PhC structures (Busch et al., 2004). Finally, the reversibility of the tuning process was verified by cooling down the heated sample to room temperature and checking that the transmission through the FP cavity agreed with the corresponding spectrum measured before heating (Martz et al., 2005). This shows that the complete heating-cooling cycle affects neither the infiltration efficiency (*i.e.* the LC molecules do not evaporate) nor the LC orientation (*i.e.* the hysteresis effects are negligible).

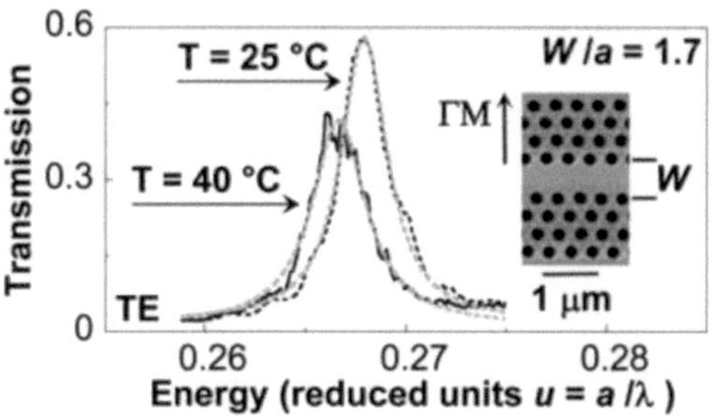

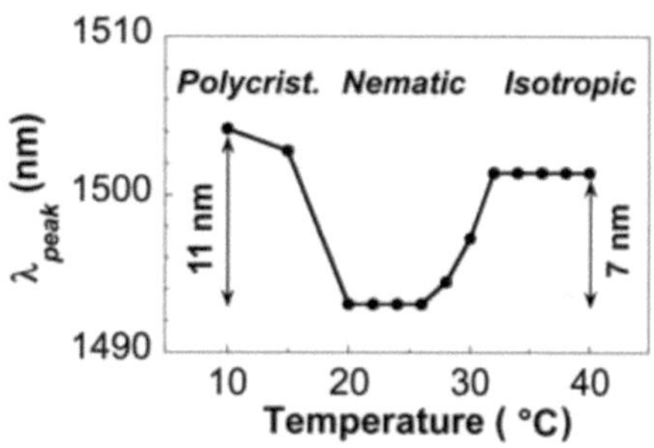

Fig. 15. (a) Measured transmission spectra through an infiltrated Fabry-Pérot (FP) cavity for T = 25 °C (dotted black line) and T = 40 °C (solid black line). The Airy fits of the resonances are shown (grey dotted lines). The inset: scanning electron microscopy top view of the cavity (W = cavity width). (b) Resonance wavelength (λ_{peak}) as a function of temperature for the infiltrated cavity. The liquid crystal phases are indicated. (Ferrini et al., 2006)

4.2 Electric tuning

As we have briefly discussed in Sect. 1, the optical response of a PhC infiltrated with nematic LCs can be adjusted by applying an external electric field that modifies the molecule orientation inside the holes (Maune et al., 2004; Alagappan et al., 2006; Haurylau et al., 2006a-b; Anderson et al., 2007; Reyes et al. 2008). Since semiconductor materials like InP are usually characterized by small electro-optic coefficients (Scrymgeour et al., 2003), this latter mechanism is particularly interesting when relatively moderate electric fields are used for the tuning of the optical properties of semiconductor-based planar PhCs.

We remark that if the semiconductor heterostructure through which the PhC is etched has a good conductivity (*i.e.* much higher than the LC conductivity), the applied electric field is screened out of the infiltrated holes where the field distribution results to be nonuniform, so that only a small part of the LC molecules experience a field large enough to reorient. Several approaches have been thus proposed to compensate for these electric-field screening effects and to maximize the electric tunability of the PhC devices, such as deposing the electric contacts or a conducting glass directly on the PhC surface (Maune et al., 2004; Haurylau et al., 2006a-b). Therefore, before setting-up the electric tuning of our infiltrated PhCs, we characterized the electrical conductivity of the InP-based heterostructure (see Sect. 2.1). First of all, four-probe measurements of the current as a function of the applied voltage were used to measure the substrate conductivity: the obtained resistance value $r = 2 \times 10^{-3}\ \Omega \cdot \mathrm{cm}$ is due to the *N*-doping of the substrate. Then, the current-voltage characteristics of the heterostructure were measured on two non-patterned samples, one of which was treated with HF to remove the residual SiO_2 over-layer that is usually deposited on the InP surface for the fabrication of planar PhCs (see Sect. 3.1; Mulot et al., 2004). The results of these measurements are reported in Fig. 16. On one hand, for the non-treated sample, the current linearly increases up to 140 V, thus showing a resistance-like behaviour due to the presence of the insulating SiO_2 layer: the measured breakdown voltage (*i.e.* 150 V) indicates a layer thickness in the order of 200 nm. On the other hand, a diode-like characteristics is found for the HF-treated sample: the waveguide heterostructure has a resistance-like behaviour up to 10 V with a minimal electric field intensity inside the guiding layers of 7 V/μm, which is comparable to the critical electric field necessary to induce LC reorentation inside the PhC holes (Halevi et al., 2006).

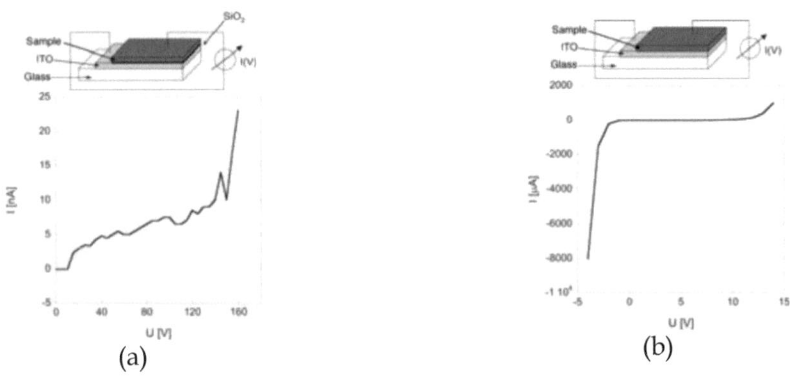

Fig. 16. Measurement of the current-voltage (I-U) characteristics of the InP heterostructures used for the fabrication of planar photonic crystals (see Sect. 2.1) (a) with and (b) without the residual SiO_2 overlayer (El-Kallassi, 2009).

InP-based PhCs similar to those presented in Sect. 2.1 were infiltrated with LCs using the same method as in Sect. 3.1. The infiltration efficiency and the molecule orientation inside the holes were characterized as illustrated in Sect. 3. Differently from what was found in Sect. 3.3, a molecule orientation parallel to the hole axis was found due to a few differences in the PhC fabrication procedure, which may yield different hole morphologies and surface states (*e.g.* rugosity). We remark that, while, when LCs have an equilibrium configuration perpendicular to the hole axis, the application of an electric field parallel to the axis is enough to induce the molecule reorientation (see Fig. 17a), for a purely axial configuration this is not possible. However, if the sample is heated to a temperature close to the clearing point, where the viscosity is reduced and the molecule order is strongly perturbed, the application of an electric field can bring the LCs back to the nematic configuration (see Fig. 17b).

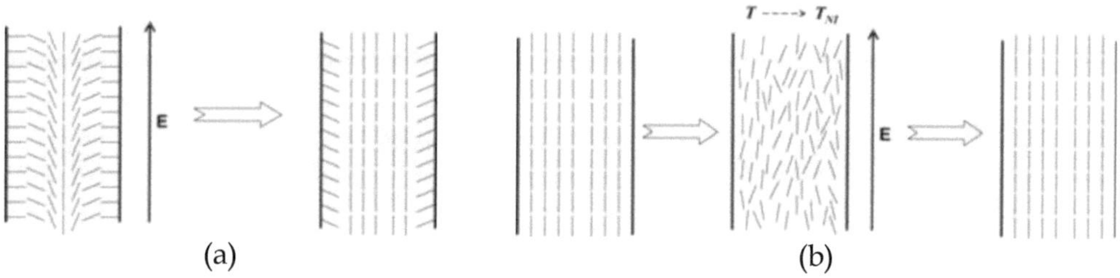

Fig. 17. Sketch of the molecule reorientation induced by an electric field applied along the hole axis when the infiltrated liquid crystals present an equilibrium configuration (a) perpendicular and (b) parallel to the hole axis (El-Kallassi, 2009).

The transmission spectrum through infiltrated FP cavities (see Sect. 2.1) is shown in Fig. 18 (black dotted line) for a temperature T = 32 °C slightly lower than the LC-K15 clearing point (see Sect. 2.2). Due to the application of an electric field (applied tension = +/- 10 V) the FP peak blue-shifts (black dashed line), thus showing a decrease of the hole refractive index n_{hole}. This latter effect is reversible: when switching off the electric field (applied tension = 0 V), the peak comes back to its initial location (grey dashed line). Even without a complete analysis of the LC orientation inside the holes, the decrease of n_{hole} clearly indicates a LC reorentation parallel to the hole axis. Therefore, the possibility of electrically tuning the optical properties of InP-based PhCs infiltrated with nematic LCs is thouroughly demonstrated.

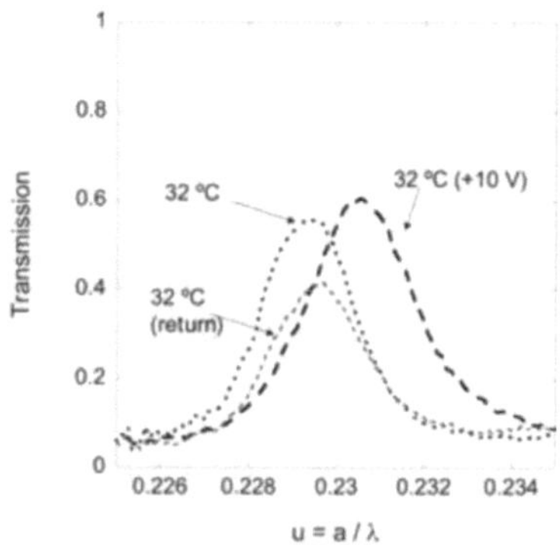

Fig. 18. Measured transmission spectra through an infiltrated Fabry-Pérot (FP) cavity for T = 32 °C (black dotted line) with and without an applied electric field (see Fig. 17): the black and grey dashed lines correspond to an applied voltage V = +/- 10 V and 0 V (return), respectively (El-Kallassi, 2009).

4.3 Optical tuning

In order to optically tune the response of infiltrated planar PhCs, a photo-responsive LC blend was used that consists of nematic LC-K15 as host molecules and 4-butyl-4'-methoxyazobenzene (BMAB) as guest molecules (see Fig. 19a).

BMAB is an azobenzene derivative that possesses an alkoxy substituent and a butyl group at the para positions of the azobenzene. It has a liquid crystalline behavior with a nematic-isotropic phase transition temperature at T_{NI}(BMAB) = 45 °C and a polycrystalline-nematic phase transition temperature at T_{KN}(BMAB) = 35 °C. Azobenzene molecules can undergo a reversible photo-isomerization between their trans (rodlike shape) and cis (bent shape) molecular forms upon irradiation with UV and visible light. In Fig. 19b, the measured absorption spectra of the BMAB trans and cis forms are shown. The trans-isomer has a main absorption band in the UV around 350 nm (π-π* molecular transition), whilst the cis-isomer has an absorption peak in the visible around 450 nm (*n*-π* molecular transition). We remark that the absorption band of the host LC-K15 is located in the UV around 280 nm and does not overlap with those of the BMAB (Legge & Mitchell, 1992). The trans-isomer, which is the thermally stable ground state, can transform into the cis-isomer by absorbing UV light, and the cis-isomer can return to the trans-isomer form either by visible light irradiation or by thermal isomerization. The photo-isomerization reactions have time scales on the order of ps's, while without illumination, the cis form will thermally reconvert to the more stable trans form with a lifetime that, for azobenzenes, is typically on the order of hours (Yager & Barret, 2006). On one hand, upon UV irradiation, a steady state can be reached where 100% of the trans molecules are converted to the cis form. On the other hand, since the *n*-π* band is present in the absorption spectra of both isomers, irradiating a photochromic mixture with visible light will bring the system in a photostationary state, whose composition is based on the competition among the photo-isomerization rates and the thermal decay rate: thus, a 100% cis-to-trans photo-conversion is not possible and the resulting mixture comprises both molecular forms.

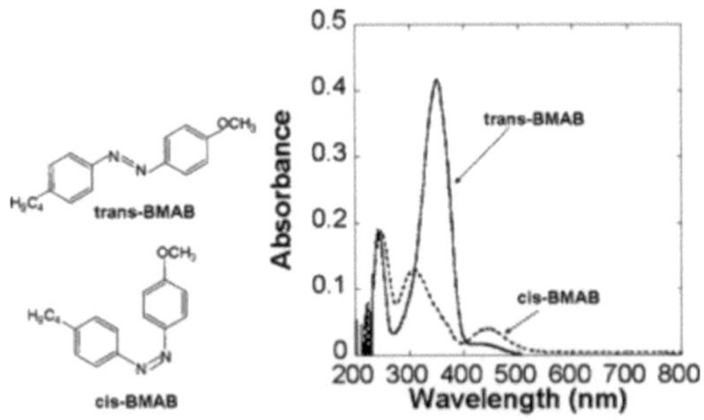

(a) (b)

Fig. 19. (a) Chemical structures of the trans and the cis molecular forms of the azobenzene derivative [4-butyl-4'-methoxyazobenzene (BMAB)]; (b) Measured absorption spectra of the BMAB trans (full curve) and cis (broken curve) forms (El-Kallassi et al., 2007).

The phase transitions in a mixture of LC-K15 doped with BMAB are shown in Figs. 20 and 21 as a function of the BMAB mole fraction ρ (Legge & Mitchell, 1992). In Fig. 20a, the nematic-isotropic phase transition temperature T_{NI} is plotted as a function of ρ for a mixture where all the BMAB molecules are in the trans form, *i.e.* before UV irradiation. Due to its rod-like molecular shape, the trans form stabilizes the LC nematic phase and T_{NI} increases gradually with ρ. Once the mixture is irradiated with UV light and the trans-to-cis photo-isomerization takes place, the bent shape of the cis isomer introduces molecular disorder in the mixture. Therefore, T_{NI} is lowered and its decrease is proportional to the concentration of the cis-isomer. In Fig. 20b, the lowered temperature value T^*_{NI} is plotted as a function of ρ for 100% of the BMAB molecules in the cis form. If the system is held at a temperature between T_{NI} and T^*_{NI}, due to the trans-cis photo-isomerization of the guest BMAB molecule, the nematic-isotropic phase transition of the LC host can be induced isothermally. In the equilibrium phase, the BMAB molecules are in the trans-form and the system is globally in the nematic phase. Upon irradiation with UV light, the photo-isomerization trans-cis brings the system into the isotropic phase. A complete photo-isomerization is not necessary to induce the nematic-to-isotropic transition and response times on the order of ms's can be achieved depending on the BMAB concentration, the temperature, and the irradiation power (Kubo et al., 2005). If the temperature is close to T^*_{NI} the system may show a biphasic behavior (Legge & Mitchell, 1992): while no phase separation is observed for mixtures of LC-K15 doped with BMAB, a phase separation between domains of nematic and isotropic LC-K15+cis-isomers occurs (see Fig. 20b). The photo-induced nematic-isotropic transition is reversible: irradiation with visible light brings part of the cis-isomers back to the trans-form and the system back to its nematic phase. This reverse transition has response times on the order of several 10's seconds up to minutes depending on the temperature and on the percentage of the cis-isomer in the mixture. Finally, if the temperature is close to T_{NI} the system may fall into the photo-stationary state described above and the nematic phase cannot be reached simply by visible light irradiation because the trans-cis and cis-trans photo-isomerization rates are similar. The system is a ternary mixture (LC-K15+cis- and trans-isomers) with a phase transition temperature T_{NI}(PS) lower than T_{NI} (see Fig. 20c).

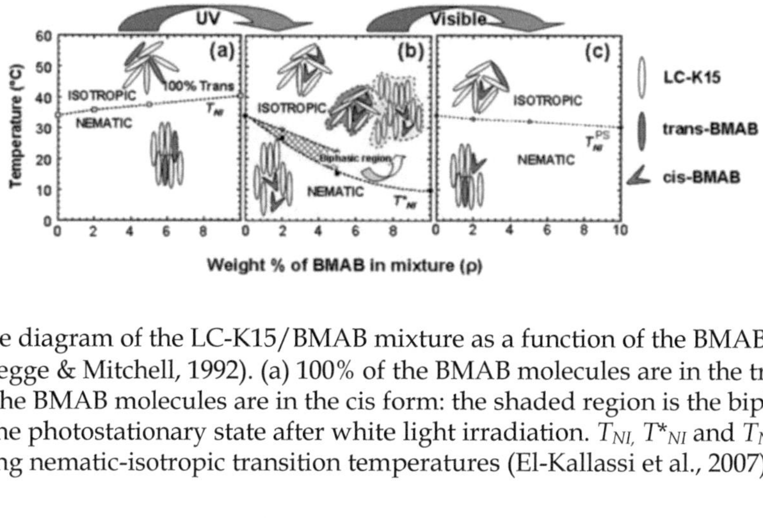

Fig. 20. Phase diagram of the LC-K15/BMAB mixture as a function of the BMAB mole fraction ρ (Legge & Mitchell, 1992). (a) 100% of the BMAB molecules are in the trans form; (b) 100% of the BMAB molecules are in the cis form: the shaded region is the biphasic region; (c) The photostationary state after white light irradiation. T_{NI}, T^*_{NI} and T_{NI}(PS) are the corresponding nematic-isotropic transition temperatures (El-Kallassi et al., 2007).

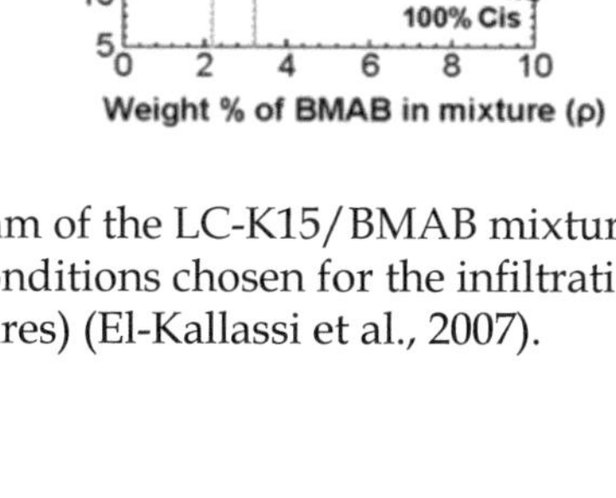

Fig. 21. Complete phase diagram of the LC-K15/BMAB mixture. The photostationary region is hashed. The experimental conditions chosen for the infiltration and the optical tuning of PhCs are indicated (black squares) (El-Kallassi et al., 2007).

The complete phase diagram of the LC-K15/BMAB mixture is summarized in Fig. 21. On one hand, due to the widening of the photo-stationary region with increasing ρ, the BMAB concentration in the mixture must remain well below 10% if one wants to preserve the complete photo-reversibility of the host-guest mixture (Kubo et al., 2005). On the other hand, working conditions close to the photo-stationary limit may shorten the response times of the system. Mixtures with $\rho = 2.2\%$ and 3.2% were prepared. The transition temperature T^*_{NI} is expected to be 5 °C and 7 °C lower than T_{NI} for the 2.2% and 3.2% mixtures, respectively (see Fig. 20). The photo-stationary state is calculated to be located at about 2 °C below T_{NI}.
As in Sect. 3, an infiltration efficiency $\eta = 0.93$ was found by measuring the transmission spectra at a temperature well above the clearing point of the LC-K15/BMAB mixtures (*i.e.* $T = 37$ °C). On the other hand, the fit of the experimental transmission spectra at room temperature yields an average molecule orientation parallel to the hole axis (see Sect. 4.2).

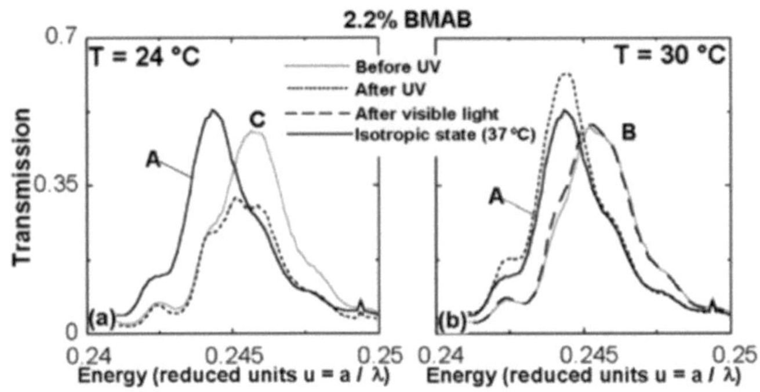

Fig. 22. Measured TE transmission spectra through a Fabry-Perot cavity ($W/a = 1.8$) infiltrated with a LC-K15/BMAB mixture ($\rho = 2.2\%$), before (grey lines) and after irradiation with UV (dotted lines) and visible light (dashed line), (a) at 24 °C and (b) 30 °C. The spectrum for to $T = 37$ °C is represented as reference (black line) (El-Kallassi et al., 2007).

The TE polarized transmission spectra through an infiltrated FP cavity for $\rho = 2.2\%$ at temperatures $T = 24$ °C and 30 °C (points C and B in Fig. 21) are shown in Figs. 22a-b, respectively. The spectra before and after UV irradiation, and after a subsequent irradiation with visible light are reported (grey, dotted and dashed curves, respectively). The spectrum measured at $T = 37$ °C (point A in Fig. 21) is shown as reference (black solid curve). At $T = 37$ °C, since $T > T_{NI}$, the LC-15/BMAB mixture is always in the thermal-isotropic phase even when 100% of trans-isomers are present. Thus, the FP resonance red-shifts with respect to the lower temperatures (nematic phase). Light irradiation has no effect at $T = 24$ °C: since $T < T^*_{NI}$, the system remains in its nematic phase even after a complete photo-isomerization and the shift of the FP resonance after UV irradiation is negligible. At $T = 30$ °C ($T^*_{NI} < T < T_{NI}$) after UV irradiation the resonance peak shifts of the same amount as for $T = 37$ °C: a complete nematic-to-isotropic phase transition is obtained isothermally by light irradiation (photo-isotropic state). Finally, after visible light irradiation, the FP peak blue-shifts to its original position, thus showing the reversibility of the photo-induced phase transition. We remark that response times on the order of ms's to s's might be achieved depending on the BMAB concentration, the temperature, and the irradiation power.

5. Conclusion

In conclusion, we have demonstrated that the combination of semiconductor-based photonic devices with nematic liquid crystals can enable the fabrication of new hybrid structures with increased functionalities. In particular, a reproducible and reliable infiltration procedure for planar PhCs has been developed and systematically validated through optical measurements. With respect to previous studies, this technique allows the accurate control of all the infiltration parameters (*i.e.* temperature and pressure) and, when necessary, the functionalization of the PhC surface. An accurate method based on in-situ optical measurements as a function of temperature and polarization was developed to characterize the infiltration efficiency and the molecule equilibrium organization inside the PhC holes. The temperature, electric and optical tuning of the infiltrated PhC devices were demonstrated. In spite of their slow response time with respect to other tuning approaches (see Sect. 1), PhC devices infiltrated with LCs can be envisaged for reconfiguration applications, switching between different functionalities and adjustment of filter devices. Moreover, recent studies have shown that the selective filling of few holes instead of the whole device further amplifies the potential of PhC infiltration (El-Kallassi et al., 2008). On one hand, the local filling allows one to overcome the decrease of the device performances (*e.g.* the cavity quality factor: see Sect. 3) when the PhC is globally infiltrated due to the reduction of the refractive index contrast between the hole array and the semiconductor matrix. On the other hand, the selective infiltration of few holes can provide an alternative approach for integrated photonic components (Mingaleev et al., 2004; Erickson et al., 2006; Intonti et al., 2006; Tomljenovic-Hanic et al., 2006; Smith et al., 2007; Faraon et al., 2008).

The author thanks Dr. P. El-Kallassi and Prof. L. Zuppiroli from the Ecole Polytechnique Fédérale de Lausanne (Switzerland) for their fundamental contribution to this work. The PhC samples were fabricated within the framework of the EU project Photonic Crystal Integrated Circuits (PCIC) and the EU Network of Excellence on Photonic Integrated Components and Circuits (ePIXnet). The infiltration experiments were performed within the framework of the Swiss National Center of Competence in Research (NCCR) in Quantum Photonics.

6. References

Alagappan, G.; Sun, X.W.; Yu, M.B. & Shum, P. (2006). Controllable polarization splitting in liquid crystal infiltrated phtonic crystals. *Proceedings of SPIE - Photonic Crystal Materials and Devices IV*, Vol. 6128, pp. 61280H-1, San Jose CA, USA, January 2006, SPIE, Bellingham WA

Anderson, S.P.; Haurylau, M.; Zhang, J. & Fauchet, P.M. (2007). Hybrid photonic crystal microcavity switches on SOI. *Proceedings of SPIE - Silicon Photonics II*, Vol. 6477, pp. 647712-1, San Jose CA, USA, January 2007, SPIE, Bellingham WA

Asakawa, K.; Sugimoto, Y.; Watanabe, Y.; Ozaki, N.; Mizutani, A.; Takata, Y.; Kitagawa, Y.; Ishikawa, H.; Ikeda, N.; Awazu, K.; Wang, X.; Watanabe, A.; Nakamura, S.; Ohkouchi, S.; Inoue, K.; Kristensen, M.; Sigmund, O.; Ingo Borel, P. & Baets, R. (2006). Photonic crystals and quantum dots technologies for all-optical switch and logic device. *New Journal of Physics*, Vol. 8, 208

Baba, T.; Shiga, M.; Inoshita, K. & Koyama, F. (2003). Carrier plasma shift in GaInAsP photonic crystal point defect cavity. *Electronics Letters*, Vol. 39, No. 21, 1516 - 1518

Barthelemy, P.; Ghulinyan, M.; Gaburro, Z.; Toninelli, C.; Pavesi, L. & Wiersma, D. (2007). Optical switching by capillary condensation. *Nature Photonics*, Vol. 1, 172-175

Belotelov, V.I. & Zvezdin, A.K. (2005). Magneto-optical properties of photonic crystals. *Journal of the Optical Society of America B*, Vol. 22, No. 1, 286-292

Bristow, A.D.; Kundys, D.O.; García-Déniz, A.Z.; Wells, J.-P. R.; Fox, A.M.; Skolnick, M.S.; Whittaker, D.M.; Tahraoui, A.; Krauss, T.F. & Roberts, J.S. (2006). Enhanced all-optical tuning of leaky eigenmodes in photonic crystal waveguides. *Optics Letters*, Vol. 31, No. 15, 2284-2286

Brugioni, S.; Meucci, R. & Faetti, S. (2006). Refractive indices of liquid crystals E7 and K15 in the mid- and near-IR regions. *Journal of Optical Technology*, Vol. 73, No. 5, 315-317

Burylov, S.V. (1997). Equilibrium configuration of a nematic liquid crystal confined to a cylindrical cavity. *JETP*, Vol. 85, No. 5, 873–886

Busch, K. & John, S. (1999). Liquid-crystal photonic-band-gap materials: the tunable electromagnetic vacuum. *Physical Review Letters*, Vol. 83, No. 5, 967-970

Busch, K.; Lölkes, S.; Wehrspohn, R.B. & Föll, H. (2004). *Photonic Crystals*, Wiley-VCH, ISBN 3-527-40432-5, Weinheim

Chu, T.; Yamada, H.; Gomyo, A.; Ushida, J.; Ishida, S. & Arakawa, Y. (2006). Tunable optical notch filter realized by shifting the photonic bandgap in a silicon photonic crystal line-defect waveguide. *IEEE Photonics Technology Letters*, Vol. 18, No. 24, 2614–2616

Crawford, G.P.; Vilfan, M.; Doane, J.W. & Vilfan, I. (1991a). Escaped-radial nematic configuration in submicrometer-size cylindrical cavities : Deuterium nuclear-magnetic-resonance study. *Physical Review A*, Vol. 43, No. 2, 835–842

Crawford, G.P.; Allender, D.W.; Doane, J.W.; Vilfan, M. & Vilfan, I. (1991b). Finite molecular anchoring in the escaped-radial nematic configuration : A 2 NMR study. *Physical Review A*, Vol. 44, No. 4, 2570–2577

Crawford, G.P.; Stannarius, R. & Doane, J.W. (1991c). Surface-induced orientational order in the isotropic phase of a liquid-crystal material. *Physical Review A*, Vol. 44, No. 4, 2558–2569

Crawford, G.P. & Zumer, S. (1996). *Liquid Crystals in Complex Geometries*, Taylor & Francis, ISBN 0-203-21107-3, London

Du, F. ; Lu, Y.-Q. & Wu, S.-T. (2004). Electrically tunable liquid-crystal photonic crystal fiber. *Applied Physics Letters*, Vol. 85, No. 12, 2181-2183

El-Kallassi, P.; Ferrini, R.; Zuppiroli, L.; Le Thomas, N.; Houdré, R.; Berrier, A.; Anand, S. & Talneau, A. (2007). Optical tuning of planar photonic crystals infiltrated with organic molecules. *Journal of the Optical Society of America B*, Vol. 24, No. 9, 2165-2171

El-Kallassi, P.; Balog, S.; Houdré, R.; Balet, L.; Li, L.; Francardi, M.; Gerardino, A.; Fiore, A.; Ferrini, R. & Zuppiroli, L. (2008). Local infiltration of planar photonic crystals with UV-curable polymers. *Journal of the Optical Society of America B*, Vol. 25, No. 10, 1562-1567

El-Kallassi, P. (2009). *Thesis: Accordabilité de la réponse optique des cristaux photoniques par infiltration de matériaux organiques* (http://library.epfl.ch/en/theses/?nr=4343), EPFL, Lausanne

Erickson, D.; Rockwood, T.; Emery, T.; Scherer, A. & Psaltis, D. (2006). Nanofluidic tuning of photonic crystal circuits. *Optics Letters*, Vol. 31, No. 1, 59-61

Faraon, A.; Englund, D.; Bulla, D.; Luther-Davies, B.; Eggleton; B.J.; Stoltz, N.; Petroff, P. & Vučković, J. (2008). Local tuning of photonic crystal cavities using chalcogenide glasses. *Applied Physics Letters*, Vol. 92, No. 4, 043123

Ferrini, R.; Lombardet, B.; Wild, B.; Houdré, R.; Olivier, S.; Benisty, H.; Djoudi, A.; Legouézigou, L.; Hubert, S.; Sainson, S.; Chandouineau, J.-P.; Fabre, S., Pommereau, F. & Duan, G.-H. (2002a). Optical characterisation of 2D InP-based photonic crystals fabricated by inductively coupled plasma etching. *Electronics Letters*, Vol. 38, No. 17, 962-964

Ferrini, R.; Leuenberger, D.; Mulot, M.; Qiu, M.; Moosbürger, J.; Kamp, M.; Forchel, A.; Anand, S. & Houdré, R. (2002b). Optical study of two-dimensional InP-based photonic crystals by internal light source technique. *IEEE Journal of Quantum Electronics*, Vol. 38, No. 7, 786-799

Ferrini, R.; Lombardet, B.; Wild, B.; Houdré, R. & Duan, G-H. (2003a). Hole depth- and shape-induced radiation losses in two-dimensional photonic crystals. *Applied Physics Letters*, Vol. 82, No. 7, 1009-1011

Ferrini, R.; Houdré, R.; Benisty, H.; Qiu, M. & Moosburger, J. (2003b). Radiation losses in planar photonic crystals : two-dimensional representation of hole depth and shape by an imaginary dielectric constant. *Journal of the Optical Society of America B*, Vol. 20, No. 3, 469-478

Ferrini, R.; Berrier, A.; Dunbar, L.A.; Houdré, R.; Mulot, M.; Anand, S.; de Rossi, S. & Talneau, A. (2004). Minimization of out-of-plane losses in planar photonic crystals by optimising the vertical waveguide. *Applied Physics Letters*, Vol. 85, No. 18, 3998-4000

Ferrini, R.; Martz, J.; Zuppiroli, L.; Wild, B.; Zabelin, V.; Dunbar, L.A.; Houdré, R.; Mulot, M. & Anand, S. (2006). Planar photonic crystals infiltrated with liquid crystals: tuning and optical characterization of molecule orientation. *Optics Letters*, Vol. 31, No. 9, 1238-1240

Gottardo, S.; Wiersma, D.S. & Vos, W.L. (2003). Liquid crystal infiltration of complex dielectrics. *Physica B*, Vol. 338, No. 1, 143-148

Grüning, U.; Lehmann, V.; Ottow, S. & Busch, K. (1995). Macroporous silicon with a complete two-dimensional photonic band gap centered at 5 μm. *Applied Physics Letters*, Vol. 68, No. 6, 747-749

Haché, A. & Bourgeois, M. (2000). Ultrafast all-optical switching in a silicon-based photonic crystal. *Applied Physics Letters*, Vol. 77, No. 25, 4089-4091

Halevi, P.; Reyes-Avendaño, J.A. & Reyes-Cervantes, J. A. (2006). Electrically tuned phase transition and band structure in a liquid-crystal-infilled photonic Crystal. *Physical Review E*, Vol. 73, N. 4, 040701

Haurylau, M.; Anderson, S.P.; Marshall, K.L. & Fauchet, P.M. (2006a). Electrically tunable silicon 2-D photonic bandgap structures. *IEEE Journal of Selected Topics in Quantum Electronics*, Vol. 12, No. 6, 1527-1533

Haurylau, M.; Anderson, S.P.; Marshall, K.L. & Fauchet, P.M. (2006b). Electrical modulation of silicon-based two-dimensional photonic bandgap structures. *Applied Physics Letters*, Vol. 88, No. 6, 061103

Hu, X.; Jiang, P.; Yang, H. & Gong, Q. (2006). All-optical tunable photonic bandgap microcavities with a femtosecond time response. *Optics Letters*, Vol. 31, No. 18, 2777-2779

Hu, X.; Jiang, P.; Ding, C. & Gong, Q. (2007). All-optical tunable narrow-band organic photonic crystal filters. *Applied Physics B*, Vol. 87, No. 2, 255-258

Ikeda, T. (2003). Photomodulation of liquid crystal orientations for photonic applications. *Journal of Materials Chemistry*, Vol. 13, 2037-2057

Intonti, F.; Vignolini, S.; Türck, V.; Colocci, M.; Bettotti, P.; Pavesi, L.; Schweizer, S.L.; Wehrspohn, R. & Wiersma, D. (2006). Rewritable photonic circuits. *Applied Physics Letters*, Vol. 89, No. 21, 211117

Joannopolous, J.D.; Johnson, S.G.; Winn, J.N. & Meade, R.D. (2008). *Photonic Crystals: Molding the Flow of Light – Second Edition*, Princeton University Press, ISBN 978-0-691-12456-8, Princeton

Kanamori, Y.; Kitani, T. & Hane, K. (2007). Control of guided resonance in a photonic crystal slab using microelectromechanical actuators. *Applied Physics Letters*, Vol. 90, No. 3, 031911

Kang, D.; Maclennan, J.E.; Clark, N.A.; Zakhidov, A.A. & Baughman, R.H. (2001). Electro-optic behaviour of liquid-crystal-filled silica opal photonic crystals: effect of liquid-crystal alignment. *Physical Review Letters*, Vol. 86, No. 18, 4052-4055

Kee, C.-S.; Kim, J.-E.; Park, H.Y.; I. Park & Lim, H. (2000). Two-dimensional tunable magnetic photonic crystals. *Physical Review B*, Vol. 61, No. 23, 15523-15525

Kosmidou, E.P.; Kriezis, E.E. & Tsiboukis, T.D. (2005). Analysis of tunable photonic crystal devices comprising liquid crystal materials as defects. *IEEE Journal of Quantum Electronics*, Vol. 41, No. 5, 657-665

Kubo, S.; Gu, Z.-Z.; Takahashi, K.; Ohko, Y.; Sato, O. & Fujishima, A. (2002). Control of the optical band structure of liquid crystal infiltrated inverse opal by a photoinduced nematic−isotropic phase transition. *Journal of the American Chemical Society*, Vol. 124, No. 37, 10950-10951

Kubo, S.; Gu, Z.-Z.; Takahashi, K.; Fujishima, A.; Segawa, H. & Sato, O. (2004). Tunable photonic band gap crystals based on a liquid crystal-infiltrated inverse opal structure. *Journal of the American Chemical Society*, Vol. 126, No. 26, 8314-8319

Kubo, S.; Gu, Z.-Z.; Takahashi, K.; Fujishima, A.; Segawa, H. & Sato, O. (2005). Control of the optical properties of liquid crystal-infiltrated inverse opal structures using photo irradiation and/or an electric field. *Chemistry of Materials*, Vol. 17, No. 9, 2298-2309

Legge, C.H. & Mitchell, G.R. (1992). Photo-induced phase transitions in azobenzene-doped liquid crystals. *Journal of Physics D: Applied Physics*, Vol. 25, No. 3, 492-499

Leonard, S.W.; Mondia, S.W.; van Driel, H.M.; Toader, O.; John, S.; Busch, K.; Birner, A.; Gösele, U. & Lehmann, V. (2000). Tunable two-dimensional photonic crystals using liquid crystal infiltration. *Physical Review B*, Vol. 61, No. 4, R2389-R2392

Leonard, S.W.; van Driel, H.M.; Schilling, J. & Wehrspohn, R.B. (2002). Ultrafast band-edge tuning of a two-dimensional silicon photonic crystal via free-carrier injection. *Physical Review B*, Vol. 66, No. 16, 161102-161105

Li, J. & Wu, S.-T. (2004). Extended Cauchy equations for the refractive indexes of liquid crystals. *Journal of Applied Physics*, Vol. 95, No. 3, 896-901

Lourtioz, J.M.; Benisty, H.; Berger, V.; Gérard, J.M.; Maystre, D. & Tchelnokov, A. (2005). *Photonic Crystals*, Springer-Verlag, ISBN 3-540-24431-X, Berlin Heidelberg

Lyubchanskii, I.L.; Dadoenkova, N.N.; Lyubchanskii, M.I.; Shapovalov, E.A. & Rasing, Th. (2003). Magnetic photonic crystals. *Journal of Physics D: Applied Physics*, Vol. 36, No. 18, R277-R287

Mansare, J.T.; Decressain, R.; Gors, C. & Dolganov, V.K. (2002). Phase transformations and dynamics of 4-cyano-4'-pentylbiphenyl (5CB) by nuclear magnetic resonance, analysis differential scanning calorimetry, and wide-angle X-ray diffraction analysis. *Molecular Crystals and Liquid Crystals Science and Technology A: Molecular Crystals and Liquid Crystals*, Vol. 382, 97-111

Märki, I.; Salt, M. & Herzig, H.P. (2006). Tuning the resonance of a photonic crystal microcavity with an AFM probe. *Optics Express*, Vol. 14, No. 7, 2969-2978

Marroum, R.-M.; Iannacchione, G.S.; Finotello, D. & Lee, M.A. (1995). Numerical study of cylindrically confined nematic liquid crystals. *Physical Review E*, Vol. 51, No. 4, R2743-R2746

Martz, J.; Zuppiroli, L. & Nüesch, F. (2004). Benzoic and aliphatic carboxylic acid monomolecular layers on oxidized GaAs surface as a tool for two-dimensional photonic crystal infiltration. *Langmuir* Vol. 20, No. 26, 11428-11432

Martz, J.; Wild, B.; Ferrini, R.; Dunbar, L.A.; Mulot, M.; Anand, S.; Houdré, R. & Zuppiroli, L. (2005). Tuning the optical properties of planar photonic crystals by liquid crystal infiltration. *Proceedings of SPIE - Tuning the Optical Response of Photonic Bandgap Structures II*, Vol. 5926, pp. 592601-1, San Diego CA, USA, August 2005, SPIE, Bellingham WA

Martz, J.; Ferrini, R.; Nüesch, F.; Zuppiroli, L.; Wild, B.; Dunbar, L.A.; Houdré, R.; Mulot, M. & Anand, S. (2006). Liquid crystal infiltration of InP-based planar photonic crystals. *Journal of Applied Physics*, Vol. 99, No. 10, 103105

Maune, B.; Loncar, M.; Witzens, J.; Hochberg, M.; Baehr-Jones, T.; Psaltis, D.; Scherer, A. & Qiu, Y. (2004). Liquid-crystal electric tuning of a photonic crystal laser. *Applied Physics Letters*, Vol. 85, No. 3, 360-362

Maune, B.; Witzens, J.; Baehr-Jones, T.; Kolodrubetz, M.; Atwater, H.; Scherer, A.; Hagen, R. & Qiu, Y. (2005). Optically triggered Q-switched photonic crystal laser. *Optics Express*, Vol. 13, No. 12, 4699-4707

Mertens, G.; Röder, T.; Schweins, R.; Huber, K. & Kitzerow, H.-S. (2002). Shift of the photonic band gap in two photonic crystal/liquid crystal composites. *Applied Physics Letters*, Vol. 80, No. 11, 1885-1887

Mertens, G.; Röder, T.; Matthias, H.; Marsmann, H.; Kitzerow, H.-S.; Schweizer, S.L.; Jamois, C.; Wehrspohn, R.B. & Neubert, M. (2003). Two- and three-dimensional photonic crystals made of macroporous silicon and liquid crystals. *Applied Physics Letters* Vol. 83, No. 15, 3036-3038

Mingaleev, S.; Schillinger, M.; Hermann, D. & Busch, K. (2004). Tunable photonic crystal circuits: concepts and designs based on single-pore infiltration. *Optics Letters*, Vol. 29, No. 24, 2858-2860

Mulot, M.; Ferrini, R.; Wild, B.; Moosburger, J.; Forchel, A.; Houdré, R. & Anand, S. (2004). Fabrication of two-dimensional InP-based photonic crystals by chlorine based chemically assisted ion beam etching. *Journal of Vacuum Science & Technology B*, Vol. 22, No. 2, 707-709

Myers, D. (1991). *Surfaces, Interfaces and Colloids*, Wiley-VCH, ISBN 0471330604, Weinheim

Ndi, F.C.; Toulouse, J.; Hodson, T. & Prather, D.W. (2005). All-optical switching in silicon photonic crystal waveguides by use of the plasma dispersion effect. *Optics Letters*, Vol. 30, No. 17, 2254-2256

Ndi, F.C.; Toulouse, J.; Hodson, T. & Prather, D.W. (2006). Optically tunable silicon photonic crystal microcavities. *Optics Express*, Vol. 14, No. 11, 4835-4841

Nielsen, K.; Noordegraaf, D.; Sørensen, T.; Bjarklev, A. & Hansen, T.P. (2005). Selective filling of photonic crystal fibres. *Journal of Optics A: Pure Applied Optics*, Vol. 7, No. 8, L13-L20

Qiu, M.; Jaskorzynska, B.; Swillo, M. & Benisty, H. (2002). Time-domain 2D modeling of slab-waveguide based photonic-crystal devices in the presence of radiation losses. *Microwave and Optical Technology Letters*, Vol. 34, No. 5, 387-393

Raineri, F.; Cojocaru, C.; Raj, R.; Monnier, P.; Levenson, A.; Seassal, C.; Letartre, X. & Viktorovitch, P. (2005). Tuning a two-dimensional photonic crystal resonance via optical carrier injection. *Optics Letters*, Vol. 30, No. 1, 64-66

Reyes, J.A.; Reyes-Avendaño, J.A. & Halevi, P. (2008). Electrical tuning of photonic crystals infilled with liquid crystals. *Optics Communications* Vol. 281, No. 9, 2535–2547

Shimoda, Y.; Ozaki, M. & Yoshino, K. (2001). Electric field tuning of a stop band in a reflection spectrum of synthetic opal infiltrated with nematic liquid crystal. *Applied Physics Letters*, Vol. 79, No. 22, 3627-3629

Schuller, Ch.; Klopf, F.; Reithmaier, J.P.; Kamp, M. & Forchel, A. (2003). Tunable photonic crystals fabricated in III-V semiconductor slab waveguides using infiltrated liquid crystals. *Applied Physics Letters*, Vol. 82, No. 17, 2767-2769

Scrymgeour, D.; Malkova, N.; Kim, S. & Gopalan, V. (2003). Electro-optic control of the superprism effect in photonic crystals. *Applied Physics Letters*, Vol. 82, No. 19, 3176-3178

Smith, C.L.C.; Wu, D.K.C.; Lee, M.W.; Monat, C.; Tomljenovic-Hanic, S.; Grillet, C.; Eggleton, J.; Freeman, D.; Ruan, Y.; Madden, S.; Luther-Davies, B.; Giessen, H. & Lee, Y-H. (2007). Microfluidic photonic crystal double heterostructure. *Applied Physics Letters*, Vol. 91, No. 12, 121103

Song, B-S.; Noda, S.; Asano, T. & Akahane, Y. (2005). Ultra-high-Q photonic double-. heterostructure nanocavity. *Nature Materials*, Vol. 4, No. 3, 207-210

Sugimoto, Y.; Tanaka, Y.; Ikeda, N.; Nakamura, Y.; Asakawa, K. & Inoue, K. (2004). Low propagation loss of 0.76 dB/mm in GaAs-based single-line-defect two-dimensional photonic crystal slab waveguides up to 1 cm in length. *Optics Express*, Vol. 12, No. 6, 1090-1096

Sung, J.-H.; Hirano, S.; Tsutsumi, O.; Kanazawa, A.; Shiono, T. & Ikeda, T. (2002). Dynamics of photochemical phase transition of guest/host liquid crystals with an azobenzene derivative as a photoresponsive chromophore. *Chemistry of Materials*, Vol. 14, No. 1, 385-391

Takeda, H. & Yoshino, K. (2003). Tunable light propagation in Y-shaped waveguides in two-dimensional photonic crystals composed of semiconductors depending on temperature. *Optics Communications*, Vol. 219, No. 1-6, 177-182

Tanabe, T.; Nishiguchi, K.; Shinya, A.; Kuramochi, E.; Inokawa, H.; Notomi, M.; Yamada, K.; Tsuchizawa, T.; Watanabe, T.; Fukada, H.; Shinoijma, H. & Itabashi, S. (2007). Fast all-optical switching using ion-implanted silicon photonic crystal nanocavities. *Applied Physics Letters*, Vol. 90, No. 3, 031115

Tay, S.; Thomas, J.; Momeni, B.; Askari, M.; Adibi, A.; Hotchkiss, P.J.; Jones, S.C.; Marder, S.R.; Norwood, R.A. & Peyghambarian, N. (2007). Planar photonic crystals infiltrated with nanoparticle/polymer composites. *Applied Physics Letters*, Vol. 91, No. 22, 221109

Teo, S.H.G.; Liu, A.Q.; Zhang, J.B. & Hong, M.H. (2006). Induced free carrier modulation of photonic crystal optical intersection via localized optical absorption effect. *Applied Physics Letters*, Vol. 89, No. 9, 091910

Tinker, M.T. & Lee, J.-B. (2005). Thermo-optic photonic crystal light modulator. *Applied Physics Letters*, Vol. 86, No. 22, 221111

Tomljenovic-Hanic, S.; Martijn de Sterke, C. & Steel, M.J. (2006). Design of high-Q cavities in photonic crystal slab heterostructures by air-holes infiltration. *Optics Express*, Vol. 14, No. 25, 12451-12456

van der Heijden, R.; Carlström, C-F.; Snijders, J.; van der Heijden, R.W.; Karouta, F.; Nötzel, R.; Salemink, H.; Kjellander, C.; Bastiaansen, C.; Broer, D. & van der Drift, E. (2006a). InP-based two-dimensional photonic crystals filled with polymers. *Applied Physics Letters*, Vol. 88, No. 16, 161112

van der Heijden, R.; Kjellanderb, C.; Carlströma, C.F.; Snijders, J.; Kickena, H.; van der Heijden, R.W.; Cees Bastiaansen, C.; Broerc, D.; Karoutaa, F.; Nötzel, R.; van der Driftd, E. & Salemink, H. (2006b). InP-based planar photonic crystals infiltrated with solid polymers and liquid crystals. *Proceedings of* SPIE - *Tuning the Optic Response of Photonic Bandgap Structures III*, Vol. 6322, pp. 632205-1, San Diego CA, USA, August 2006, SPIE, Bellingham WA

Vlasov, Y.A.; Bo, X-Z.; Sturm, J.C. & Norris, D.J. (2001). On-chip natural assembly of silicon photonic bandgap crystals. *Nature*, Vol. 414, No. 6861, 289-293

Yoshino, K.; Shimoda, Y.; Kawagishi, Y.; Nakayama, K. & Ozaki, M. (1999b). Temperature tuning of the stop band in transmission spectra of liquid-crystal infiltrated synthetic opal as tunable photonic crystal. *Applied Physics Letters*, Vol. 75, No. 7, 932-934

Weiss, S.M. & Fauchet, P.M. (2003). Electrically tunable porous silicon active mirrors. *Physica Status Solidi (a)*, Vol. 197, No. 2, 556-560

Weiss, S.M.; Ouyang, H.; Zhang, J. & Fauchet, P.M. (2005a). Electrical and thermal modulation of silicon photonic bandgap microcavities containing liquid crystals. *Optics Express*, Vol. 13, No. 4, 1090-1097

Weiss, S.M.; Haurylau, M. & Fauchet, P.M. (2005b). Tunable photonic bandgap structures for optical interconnects. *Optical Materials*, Vol. 27, No. 5, 740-744

Wild, B.; Ferrini, R.; Houdré, R.; Mulot, M.; Anand, S. & Smith, C. J. M. (2004). Temperature tuning of the optical properties of planar photonic crystal microcavities. *Applied Physics Letters*, Vol. 84, No. 6, 846-848

Wong, C.W.; Rakich, P.T.; Johnson, S.G.; Qi, M.; Smith, H.I.; Ippen, E.P.; Kimerling, L.C.; Jeon, Y.; Barbastathis, G. & Kim, S-G. (2004). Strain-tunable silicon photonic band gap microcavities in optical waveguides. *Applied Physics Letters,* Vol. 84, No. 8, 1242-1244

Yager, K.G. & Barrett, C.J. (2006). Novel photo-switching using azobenzene functional materials. *Journal of Photochemistry and Photobiology A: Chemistry*, Vol. 182, 250-26

3

Manipulating Nematic Liquid Crystals-based Magnetophotonic Crystals

Hai-Xia Da and Z.Y. Li
Department of Physics and Jiangsu Key Laboratory of Thin films, Soochow University, Suzhou, 215006, China

1. Introduction

Magnetic materials have attracted considerable attention for their applications in high density magneto-optical storage devices, which is not only appealing scientifically, but also makes the magnetic materials promising for a wide range of applications [1-11]. Due to their unique optical properties, magnetic materials have been introduced into the photonic crystals, forming magnetophotonic crystals (MPCs), i.e., the photonic crystals with at least one magnetic material component [2,3]. Now the optical properties can be mediated and controlled by the external electric or magnetic field due to the existence of magnetic materials. The special importance of MPCs can be ascribed to the existence of magneto-optical effects, for example, Kerr effect and Faraday rotation [12-28]. They were discovered by Kerr and Faraday, respectively, and are now widely used in integrated optics and magneto-optical devices for magnetic domain imaging, mapping of hysteresis loops and high density recording [6].

Magnetic materials with large magneto-optical responses are always the attractive ones used in magnetophotonic crystals. In contrast to the corresponding three-dimensional magnetic structures, magnetic materials exhibit a larger magneto-optical effects due to the light's confinement in the MPCs, offering a genuine chance to put their optical responses into applications. The optimized MPCs have been shown to behave like mixed systems, with a coexistence of high transmittance and large magneto-optical effects [12-28]. All possible configurations are proposed to achieve strong magneto-opitcal effects, including the ordinary cavity-based, multilayered periodic and aperiodic structures. Furthermore, the diffracted magneto-optical enhancement is also demonstrated in the grating structures theoretically and experimentally, which greatly reduces the thickness of the device contrary to multilayered structures and miniaturizes magneto-optical devices in integrated optics [29,30].

The ability to tune the otpical properties by an external stimulus is a key issue of modern optoelectronics. Although most attention has so far been focused on the magneto-optical properties of the given structures, tunable magneto-optical devices have important applications and will be respected in optical switches and displays. There are few reports concerning with the tunable magnetophotonic crystals [31-33]. For example, it is possible to manipulate the magnetic order of magnetic conducting spheres using the magnetic field, thus forming the tunable magnetophotonic crystals [31]. Semiconductor quantum well has

Source: New Developments in Liquid Crystals, Book edited by: Georgiy V. Tkachenko,
ISBN 978-953-307-015-5, pp. 234, November 2009, I-Tech, Vienna, Austria

been discussed as possible candidates for achieving artificial tunable MPCs [32]. From the application-oriented perspective it would be very desirable if tunable magneto-optical effects could be achieved by controlling applied electric fields, thus making such effect a potential proposition. Therefore, to search the alternate scheme is of great importance in tunable magneto-optical electronics. It was known that nematics liquid crystals (NLCs) are good choices for tunable photonic crystals due to their unique sensitivity to temperature, the electric, magnetic field, or lights itself [34]. However, seldom reports concerning tunable magneto-optical effects based on the NLCs have been presented in the literatures. Therefore, it would be interesting to investigate a possible way of creating electrically controlled magneto-optical effects in the MPCs with the NLCs. The application of liquid crystals in the MPCs offers new opportunities for the tunable optoelectronic devices.

Liquid crystals are materials that display a phase of matter whose properties lie between those of a conventional liquid and a solid crystal. They are a class of materials particularly attractive for liquid crystal displays and optical electronic applications due to their high sensitivity to the external stimulus and have been studied experimentally extensively [35-37]. There are three basis kinds of liquid crystals: NLCs, cholesteric liquid crystals and smectic liquid crystals [38-47]. Here we only focus attention on the NLCs, which is characterized by molecules that have no positional order but tend to align along the same direction. Due to thermal random motion, friction and collision between molecules, not all molecules align along a certain direction and their directions vary around the average direction randomly. This average direction is referred to as the orientation of the liquid crystal, which stands for the average direction of most molecules. This parameter, i.e., the director, is an important factor to denote the liquid crystal's properties. Normally, the light waves with electric fields perpendicular or parallel to the director of a NLC have ordinary n_o or extraordinary n_e refractive indices, respectively. Therefore, the refractive index of a NLC may be changed between n_o and n_e by controlling the orientation of the directors using the applied electric field or adjusting the temperature [34]. Specifically, the refractive index of NLC is especially easy to control using the external electric field in one-dimensional cases [38-40]. Previous studies have demonstrated, experimentally as well as theoretically, how tunability is brought to electro-optical systems by employing NLCs, such as the realization of tunable band gaps in one-, two-, and three-dimensional photonic crystals based on NLCs [34, 38-40]. Tunable negative refraction is also achieved at the interface between air and the NLC, and that the refraction can be adjusted by an applied electric field or by the temperature [48-50]. Multistable all-optical switching have been demonstrated by utilizing an unique nonlinear coupling between light and the NLC in a periodic dielectric structure [51]. The absorption peak of nanoparticles doped in the NLC can be controlled by the external electric field [52]. The light-induced reorientational effects have been observed in a one-dimensional photonic crystal with the NLCs [53]. Especially, the orientation of the magntic anisotropy from ferromagnetic nanorods have been manipulated by means of electric fields in the composite of liquid crystals and ferromagnetic nanorods [54]. While the investigations of the magnetic properties in a electrically controlled way is still in infancy, the current experimental work suggests a possible chance of realizing the tunable MPCs with liquid crystals for fundamental research and practical applications.

Moreover, the particular nonreciprocity of magnetic materials, i.e., the polarized state (right/left handed polarized light) will switch into the opposite one (left/right handed polarized light) when the direction of the incident light is reversed, makes the MPCs remarkable workbenches for the unidirectional electromagnetic transmission [55-61]. Such a character shown in the cavity-based, periodic and aperiodic multilayered structures has made them good candidates to be isolators, unidirectional transmitted MPCs and one way waveguides with MPCs [57-59]. It has been unraveled that both ternary and microcavity configurations combined with waveguides can generate the isolators and circulators [55]. The asymmetric isolators are shown to be much efficient compared with symmetric ones [62]. The performance of small antennas embedded within MPCs constructed from periodic arrangements of homogenous and anisotropic material layers was demonstrated that the extraordinary high gain and enhanced power reception can be achieved [63]. Furthermore, it has been demonstrated that the optical Tamm states (OTS) can be observed in one dimensional MPCs in recent theoretical and experimental works [64-70]. There will be considerable interest to exploit and identify the OTS in the MPCs. The ability of creating and manipulating the surface states is central to the development of subwavelength microscopy. However, few reports concerning the tunable OTS in MPCs have been presented [66]. Therefore, it is interesting to find whether the NLCs can be used to realize the OTS in MPCs, or more importantly, whether such a configuration will support the controllable OTS. Controlling the OTS by electrical means is particularly interesting, as that would allow the integration of magnetic materials and the NLC with conventional optical electronics. Due to these unique optical properties, the MPCs with the NLC will hold substantial promise for possible optoelectronic devices.

In this chapter, we present a theoretical investigation on the NLC-based MPCs by performing a 4×4 transfer matrix method [71-73]. The optical properties of the MPCs with the NLCs will be reminiscent of both components. The coupling between the incident waves and the magnetic materials creates the magneto-optical effects, while the component of the NLCs is now responsible for its tunability occurring in the present configurations. The chapter is organized as follows: In Sec. II, we give a general description of the 4×4 transfer matrix method, which is a consequence of the combination of Maxwell equations and constitutive relations. The relations between the reflected, transmitted field vectors and the incident field vectors are listed to elucidate the magnitude of magneto-optical effects. In Section III, the case of periodic structure is considered, i.e., one dimensional MPC whose unit cell is composed of alternating NLCs and magnetic materials. In this part, the NLC is treated as a simple isotropic dielectric material approximately [74]. In Sec. IV, one-dimensional MPC infiltrated with the NLC is investigated, where the intrinsic anisotropic properties of the NLC are taken into account. Combined with transfer matrix method and a piecewise homogeneity approximation method for the NLC [75], the magneto-optical effects of MPCs with the NLC is definitely achieved. In Sec. V, the tunable OTS is reported theoretically in the MPCs with the NLC. Finally, we give a summary about this work.

Our investigations reveal a variety of interesting results which are the main characteristics of one-dimensional MPCs with the NLC. The magneto-optical effects appear the peaks due to the abnormal dispersion relations at the edge of the band gaps. It is shown the shift of the peaks and the enhanced values in the magneto-optical spectrum as the permittivity of NLCs increases, which demonstrates that the magneto-optical effects can be altered by changing the permittivity in the NLCs. When the NLC-based cavity structure is subject to the applied electric field, the magneto-optical effects are observed to exhibit the similar trends as those of the periodic one treating the NLC as an approximate isotropic dielectric material. A significant shift of the peaks in the magneto-optical spectrum is observed with the external voltage due to the directors' reorientation. In addition, the existence of intra-Brillouin-zone band gaps is demonstrated and they are predicted to be dependent on the applied voltages in the MPC with the NLC. If the present MPC is attached to the other photonic crystals with two dielectric materials, the OTS is also observed in the transmission spectrum and can be controlled by the applied voltages. Our results provide an approach to modify a interface state in a controllable manner, which is significant to its potential applications. Moreover, it is possible to design a larger tunable magneto-optical effects by simply selecting the appropriate liquid crystals in the MPCs with the NLC. The ability of creating the electrically-controlled magneto-optical effects can provide an extra space to be explored in the future. Such tunability of the magneto-optical effect may be useful for future application in electro-optical devices.

2. Theoretical treatment

A 4×4 transfer matrix method is an effective way to explore the optical properties of one dimensional multilayered configurations, which has been proved to be accurate enough to tackle the band structures of the uniaxial, bi-axially anisotropic dielectric materials and liquid crystals [71-72]. For the sake of convenience, we describe the theoretical method and detailed treatment in this section.

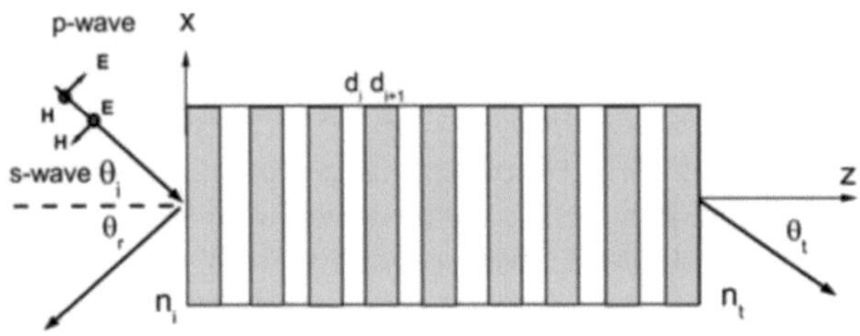

Fig. 1. Schematic view of the multilayered structure.

Fig. 1 shows a typical multilayered structure, which is sandwiched to the incident medium with the refractive index n_i and transmitted medium with the refractive index n_t, respectively. The incident wave vector is expressed as $\vec{k_i} = k_0(v_x, v_y, v_z)$, where $k_0 = \omega/c$, ω is the angular frequency of the incident wave, and c stands for the velocity of light in the vacuum. Here an electromagnetic wave impinges on the multilayered structure with an angle θ_i, which is confined in the x –z plane indicating $k_y = 0$, $v_x = n_i \sin\theta$ and $v_z = n_i \cos\theta$. The dielectric permittivity and magnetic permeability of any component are characterized by the tensors:

$$\vec{\varepsilon} = \begin{pmatrix} \varepsilon_{11} & \varepsilon_{12} & \varepsilon_{13} \\ \varepsilon_{21} & \varepsilon_{22} & \varepsilon_{23} \\ \varepsilon_{31} & \varepsilon_{32} & \varepsilon_{33} \end{pmatrix}, \quad \vec{\mu} = \begin{pmatrix} \mu_{11} & \mu_{12} & \mu_{13} \\ \mu_{21} & \mu_{22} & \mu_{23} \\ \mu_{31} & \mu_{32} & \mu_{33} \end{pmatrix}. \tag{1}$$

For a monochrome light propagating along the z - axis with a given frequency ω, based on Maxwell's equations, we have [71]

$$\frac{d\Psi(z)}{dz} = \frac{i\omega}{c}\Delta(z)\Psi(z), \tag{2}$$

where $\Psi(z)$ is the field vector provided as $\Psi(z) = (E_x, H_y, E_y, -H_x)^t$, and $\Delta(z)$ is the Berrmeman matrix which is the function of the permittivity, permeability of the material and the incident wave vector. The subscript t denotes the transpose operator. The components of the Berreman matrix are provided with [71]

$$\begin{aligned}
t_{11} &= -v_x\tfrac{\varepsilon_{31}}{\varepsilon_{33}}\\
t_{12} &= -\tfrac{v_x^2}{\varepsilon_{33}}+\mu_{22}-\tfrac{\mu_{23}\mu_{32}}{\mu_{33}}\\
t_{13} &= -v_x\tfrac{\varepsilon_{32}}{\varepsilon_{33}}+v_x\tfrac{\mu_{23}}{\mu_{33}}\\
t_{14} &= -\mu_{21}+\tfrac{\mu_{23}\mu_{31}}{\mu_{33}}\\
t_{21} &= \varepsilon_{11}-\tfrac{\varepsilon_{13}\varepsilon_{31}}{\varepsilon_{33}}\\
t_{22} &= -v_x\tfrac{\varepsilon_{13}}{\varepsilon_{33}}\\
t_{23} &= \varepsilon_{12}-\tfrac{\varepsilon_{13}\varepsilon_{32}}{\varepsilon_{33}}\\
t_{24} &= 0\\
t_{31} &= 0\\
t_{32} &= -\mu_{12}+\tfrac{\mu_{13}\mu_{32}}{\mu_{33}}\\
t_{33} &= -v_x\tfrac{\mu_{13}}{\mu_{33}}\\
t_{34} &= \mu_{11}-\tfrac{\mu_{13}\mu_{31}}{\mu_{33}}\\
t_{41} &= \varepsilon_{21}-\tfrac{\varepsilon_{23}\varepsilon_{31}}{\varepsilon_{33}}\\
t_{42} &= -v_x\tfrac{\varepsilon_{23}}{\varepsilon_{33}}+v_x\tfrac{\mu_{32}}{\mu_{33}}\\
t_{43} &= \varepsilon_{22}-\tfrac{\varepsilon_{23}\varepsilon_{32}}{\varepsilon_{33}}-\tfrac{v_x^2}{\mu_{33}}\\
t_{44} &= -v_x\tfrac{\mu_{31}}{\mu_{33}}
\end{aligned} \tag{3}$$

Eq. (2) has a solution of the form:

$$\Psi(z) = \Psi(z_0)e^{i\frac{\omega}{c}\Delta(z-z_0)}, \tag{4}$$

Within the frame of transfer matrix method, the field vector can be expressed as,

$$\Psi(z) = T(z, z_0)\Psi(z_0). \tag{5}$$

where $T(z, z_0)$ is the transfer matrix provided with the Cauchy and Hamilton theorem [71]:

$$T(z, z_0) = \beta_0 I + \beta_1\Delta + \beta_2\Delta^2 + \beta_3\Delta^3, \tag{6}$$

These coefficients β_i are:

$$
\begin{aligned}
\beta_0 &= -\sum_{i=1}^{4} \lambda_j \lambda_k \lambda_l \frac{f_i}{\lambda_{ij}\lambda_{ik}\lambda_{il}}, \\
\beta_1 &= \sum_{i=1}^{4} (\lambda_j \lambda_k + \lambda_j \lambda_l + \lambda_k \lambda_l) \frac{f_i}{\lambda_{ij}\lambda_{ik}\lambda_{il}}, \\
\beta_2 &= -\sum_{i=1}^{4} (\lambda_j + \lambda_k + \lambda_l) \frac{f_i}{\lambda_{ij}\lambda_{ik}\lambda_{il}}, \\
\beta_3 &= \sum_{i=1}^{4} \frac{f_i}{\lambda_{ij}\lambda_{ik}\lambda_{il}}.
\end{aligned}
\tag{7}
$$

where

$$
\begin{aligned}
\lambda_{ij} &= \lambda_i - \lambda_j, \\
f_i &= e^{i\frac{\omega}{c}\lambda_i(z-z_0)}, \\
i, j, k, l &= 1, 2, 3, 4.
\end{aligned}
\tag{8}
$$

λ_i are the eigenvalues of the Berreman matrix. In contrast with the isotropic materials, it is more complicate due to the existence of the non-diagonal elements in the dielectric permittivity or magnetic permeability tensors. Even the incident plane wave is a linear polarization, it will form the non-linear polarized waves. Therefore, it typically yields the four solutions for the anisotropic materials, two of which represent left- and right- handed polarized forward propagating waves and another two are the corresponding backward propagating waves. If the individual layer occupies the spatial region of $z_0 < z < z_0+d$, the field vectors $\Psi(z_0 + d)$ and $\Psi(z_0)$ can be connected with

$$\Psi(z_0 + d) = T_i \Psi(z_0). \tag{9}$$

where T_i is the single-layer transfer matrix and d is the thickness of this layer. For the finite photonic crystals, the total transfer matrix T_{total} of the system can be achieved by multiplying the individual transfer matrix together. The NLC is of spatially inhomogeneity, whose treatment needs to adapt the piecewise homogeneity approximation method [75].
(E_i^p, E_i^s), (E_r^p, E_r^s) and $(E_t^p, E_t^s)$are used to describe the electric field components of the incident, reflected and transmitted waves, respectively. Subscript $p(s)$ denotes electromagnetic wave parallel (perpendicular) to the incident plane x –z, respectively. The incident, reflected and transmitted field vectors are expressed as

$$
\begin{aligned}
\Psi_i &= \left(E_i^p \cos\theta_i, n_i E_i^p, E_i^s, n_i E_i^s \cos\theta_i\right)^t; \\
\Psi_r &= \left(-E_r^p \cos\theta_i, n_i E_r^p, E_r^s, -n_i E_r^s \cos\theta_i\right)^t; \\
\Psi_t &= \left(E_t^p \cos\theta_t, n_t E_t^p, E_t^s, n_t E_t^s \cos\theta_t\right)^t.
\end{aligned}
\tag{10}
$$

Due to the continuous boundary conditions of electronic and magnetic fields, the equation $\Psi_t = T(\Psi_i + \Psi_r)$ should be satisfied. Accordingly, the relations between the reflected, transmitted and incident waves are connected with [79]

$$\begin{pmatrix} E_r^p \\ E_r^s \end{pmatrix} = R \begin{pmatrix} E_i^p \\ E_i^s \end{pmatrix} = \begin{pmatrix} r_{pp} & r_{ps} \\ r_{sp} & r_{ss} \end{pmatrix} \begin{pmatrix} E_i^p \\ E_i^s \end{pmatrix}, \tag{11}$$

$$\begin{pmatrix} E_t^p \\ E_t^s \end{pmatrix} = T \begin{pmatrix} E_i^p \\ E_i^s \end{pmatrix} = \begin{pmatrix} t_{pp} & t_{ps} \\ t_{sp} & t_{ss} \end{pmatrix} \begin{pmatrix} E_i^p \\ E_i^s \end{pmatrix}, \tag{12}$$

where R (T) is the Jones reflectance (transmitted) matrices and r_{ij} (t_{ij}) is the ratio of the incident j polarized electric field and the reflected (transmitted) i polarized electric field. Then, the magneto-optical effects, i.e., Kerr effect and Faraday effect can be determined by [79]

$$\Theta_K^p = \theta_p + i\eta_p = \frac{r_{ps}}{r_{pp}}, \tag{13}$$

$$\Theta_K^s = \theta_s + i\eta_s = \frac{r_{sp}}{r_{ss}}. \tag{14}$$

$$\Theta_F^p = \theta_p + i\eta_p = \frac{t_{ps}}{t_{pp}}, \tag{15}$$

$$\Theta_F^s = \theta_s + i\eta_s = \frac{t_{sp}}{t_{ss}}. \tag{16}$$

where $\theta_{K,F}^p$ ($\theta_{K,F}^s$) and $\eta_{K,F}^p$ ($\eta_{K,F}^s$) are the Kerr/Faraday rotation angle and the ellipticity for the $p(s)$ polarized wave, respectively. Based on these equations, the magneto-optical effects and the corresponding transmission/reflectance coefficients can be obtained directly.

3. Electrically-controlled magneto-optical effects in magnetophotonic crystals consisting of magnetic materials and NLC

In this part, we investigate the magneto-optical effects of a one dimensional MPC whose unit cell is composed of a magnetic material and NLC and give a detailed explanation to the obtained results.

Generally, the permittivity of NLC is provided by $\varepsilon_{LC} = \frac{\varepsilon_o \varepsilon_e}{\varepsilon_o \cos^2\theta + \varepsilon_e \sin^2\theta}$ for linear polarized light incident as an extraordinary wave onto it, where ε_o and ε_e are the respective dielectric permittivities for light polarized parallel and perpendicular to the director axis $\hat{n}$ [74]. It is noted that the only factor which affects the value of the permittivity of NLC is the director axis orientation angle θ with respect to the optical wave vector. Normally, losses are negligible for typical micron thick NLC in the optical frequency region. In the following numerical simulation, we treat the NLC as a homogeneous isotropic dielectric layer with the permittivity of ε_{LC} approximately, which simplify the calculations greatly [74]. In our model, we apply an electric field to control the director axis orientation θ of the aligned NLC with respect to the wave vector, which results in the variation of the extraordinary refractive index from $n_{\perp}$ ($\theta = 0°$) to $n_{\|}$ ($\theta = \frac{\pi}{2}$). The corresponding range of dielectric permittivity of NLC, ε_{LC}, is taken as $2 \leq \varepsilon_{LC} \leq 3$ from nematic liquid crystal $5CB$ ($\Delta n_{LC} \simeq 0.18$) [34].

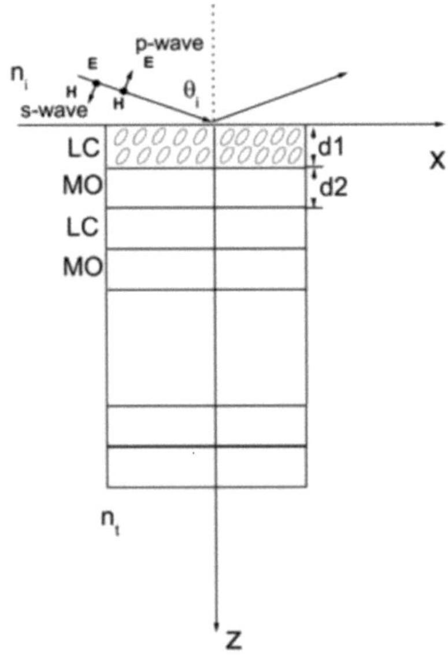

Fig. 2. One dimensional MPC consisting of the NLC and magnetic materials.

Fig. 2 shows a schematic illustration of a one-dimensional finite MPC, i.e., (NLC/magnetic materials). In the calculations, the total periodic number N is taken as 8 and the thicknesses of NLC and magnetic layers are d_1 = 500nm and d_2 = 20nm, respectively. The input and output media are supposed to be vacuum and glass, i.e., n_i = 1 and n_t = 1.5, respectively. The dielectric tensor of magnetic layers has an antisymmetric form and is denoted as [80]

$$\tilde{\varepsilon} = \begin{pmatrix} \varepsilon_{xx} & \cos\Theta\varepsilon_{xy} & -\sin\Theta\sin\Phi\varepsilon_{xy} \\ -\cos\Theta\varepsilon_{xy} & \varepsilon_{yy} & \sin\Theta\cos\Phi\varepsilon_{xy} \\ \sin\Theta\sin\Phi\varepsilon_{xy} & -\sin\Theta\cos\Phi\varepsilon_{xy} & \varepsilon_{zz} \end{pmatrix}. \quad (17)$$

where the magnetization vector makes an angle Θ with the z axis and its projection on the $x-y$ plane makes the azimuth Φ with the x axis. As an illustrative example, the permittivity parameters of magnetic material are taken as $\varepsilon_{xx} = \varepsilon_{yy} = \varepsilon_{zz} = -4.8984 + 19.415\,i$, $\varepsilon_{xy} = -\varepsilon_{yx} = 0.4322 + 0.0058i$ [81] while the permeability of magnetic material is nearly 1 at the optical frequency. Note that there are three cases according to the different values of Θ and Φ: the polar Kerr effect (Θ = Φ = 0), the longitudinal Kerr effect ($\Theta=\frac{\pi}{2}$, Φ = 0) and the transverse Kerr effect (Θ, $\Phi = \frac{\pi}{2}$). Here we only focus attention on the polar Kerr effect since the other two cases can be treated in the same way. The Kerr rotation angles as a function of the incident wavelength for different ε_{LC} are calculated when the light is incident normally to MPC. Since normal incident light leads to decoupling between p- and s-polarized waves and the two polarized lights are degenerate, only results for the p-wave are presented in the following figures.

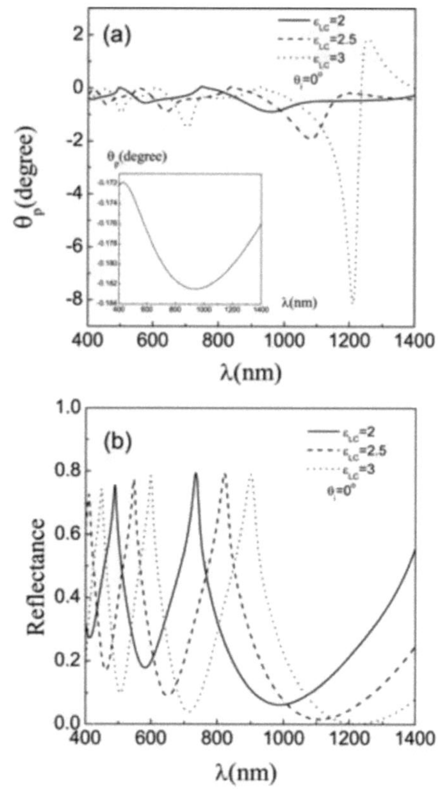

Fig. 3. (a) θ_p versus the incident wavelength under the dielectric permittivities of LCs $\varepsilon_{LC} = 2$, 2.5, 3 at the normal incident angle, are shown by solid, dash and dotted lines, respectively. (b) The dependence of the reflectance on the wavelength with the same conditions as Fig. 3(a). The inset in (a) shows θ_p in a single magnetic material with the thickness $d = 20nm$.

The solid, dash and dotted lines in Fig. 3 (a) describe the behaviors of θ_p for $\varepsilon_{LC} = 2$, 2.5, 3, respectively. Though the choice of the value of ε_{LC} seems arbitrary, the general tendency can still be seen clearly in the figures. Compared with the wavelength dependence of θ_p of a single magnetic layer as shown in the inset of Fig. 3(a), the Kerr rotation angles exhibited in the NLC-based MPC structures are enhanced dramatically, especially at certain wavelengths. The maximum value of θ_p increases from $\theta_{pmax} = -0.86^\circ$ for $\varepsilon_{LC} = 2$ up to -8° for $\varepsilon_{LC} = 3$. The latter is nearly 40 times larger than that of the single magnetic layer($\theta_{pmax} = -0.19^\circ$). The enhancement of θ_p compared with the single magnetic layer originates from the weak localization of light caused by the multiple interference due to the nonreciprocal properties of the Kerr effect [2]. Furthermore, the emergence of the θ_p-peaks is also an important feature of the multilayered structure in our MPC model. It is well-known that the Kerr effect is strengthened dramatically due to the abnormal dispersion relation at the edges of the photonic band gaps [28]. To make it more clear, Fig. 3 (b) shows the dependence of the reflectance of MPC on the wavelength. We clearly observe that the positions of the remarkable enhancement of θ_p are coincident with the positions of photonic band gap exactly.

Note that all the peaks in Fig. 3(a) shift towards longer wavelengths when the NLC's permittivity increases, which is particularly interesting since it reflects the good tunability of the Kerr effect in our model. A similar shift is also observed for the photonic band gaps as shown in Fig. 3(b). This result is reasonable since the positions of photonic band gaps depend on the wave impedance ratio of two components [82, 83], while the latter changes with the NLC's permittivity in the present case. For a rough estimation, the central wavelength of the NLC-based MPC structure can be evaluated by $\lambda_c = 2(n_{LC}d_1+n_m d_2)$ approximately [84], where n_m is the complex refractive index of magnetic layers and invariable with temperature or applied field. So the increase of the refractive index of NLC will directly lead to the enhancement of the central wavelength of the MPC structure. Since the refractive index of liquid crystal may be tuned between the ordinary and extraordinary refractive indices by controlling the orientation of the molecules under the influence of the applied electric field, the upper (lower) limit of the red shift is determined by the extraordinary (ordinary) permittivity, i.e., $\varepsilon_{LC} = 3$ ($\varepsilon_{LC} = 2$). Therefore, the NLC with larger optical anisotropy may behave as a better candidate for realizing larger tunability.

Fig. 3(a) also shows the significant increase of peak height with the permittivity of NLC in the spectrums of the Kerr rotation angle, which can be explained according to the circular birefringence. Different refractive indices due to different localization conditions, i.e., the localization wavelengths for left- and right- circular polarized lights, leads to the rotation of the reflected light [2]. In photonic crystals, the effective parameters of the unit cell can be expressed by the effective homogeneous anisotropic medium method [85]. The effective permittivity of MPC consisting of magnetic materials and NLC is given by

$$\varepsilon_{eff} = \begin{pmatrix} f_a\varepsilon_{LC} + f_b\varepsilon_{xx} & f_b\varepsilon_{xy} & 0 \\ f_b\varepsilon_{yx} & f_a\varepsilon_{LC} + f_b\varepsilon_{yy} & 0 \\ 0 & 0 & (f_a\varepsilon_{LC}^{-1} + f_b\varepsilon_{zz}^{-1})^{-1} \end{pmatrix}. \tag{18}$$

where $f_a = \frac{d_1}{d_1+d_2}$ and $f_b = \frac{d_2}{d_1+d_2}$. Note that the diagonal components of the effective permittivity depend on the permittivity of NLC but the non-diagonal components do not. Therefore, with the increase of ε_{LC}, the difference between the dielectric permittivity of the left- and right- circular polarized lights is enlarged, which leads to a relatively larger Kerr rotation angle at a longer wavelength.

Generally, refractive indices of liquid crystal are fundamentally interesting and practically useful parameters in the calculation of optical properties. By changing the permittivity of the liquid crystals, the magneto-optical effects varied in a controlled fashion, indicating that the engineering of magneto-optical devices relies on the control of the applied electric field. Here we focus attention on the change of the orientation of the molecules under the influence of the applied electric field at a fixed temperature. Actually, besides the external electric field, temperature is also an important factor affecting the liquid crystal refractive indices, which offers an alternative approach to control the magneto-optical effects, i.e., a temperature tunable MPC.

4. Voltage-controlled magneto-optical effects in cavity- based magnetophotonic crystal with the NLC

We have predicted that the opportunity of creating electrically controlled Kerr effect in magnetic multilayered structures [76]. It was suggested that a one-dimensional MPC composed of alternating NLC and magnetic materials can create tunable Kerr effect by considering the properties of liquid crystals and thus provide for tunable MPCs. However, we just employed an approximate isotropic treatment of NLC to analyze magneto-optical effects, which is a rough theoretical evaluation. It is generally known that the directors of NLCs exhibit inhomogeneous distribution under the influence of the applied electrical field [40]. Therefore, a rigorous anisotropic treatment of NLCs is employed to consider the NLC director's spatially inhomogeneous property upon an applied external voltage, which is based on the Newton method and continuous elastic theories. Although we expect such the tunable magneto-optical effect to be seen with a variety of patterns containing NLCs, we investigated a multilayered structure infiltrated with NLC, as shown schematically in Fig. 4. The defect of the NLC ensures that the present structure is sensitive to the external electric field and therefore shows the tunability of magneto-optical effects for observations at normal incidence.

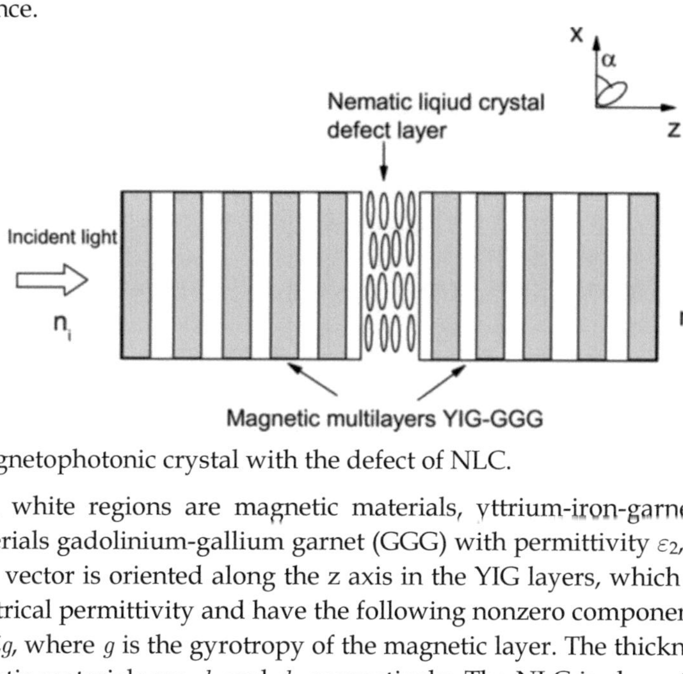

Fig. 4. The magnetophotonic crystal with the defect of NLC.

The gray and white regions are magnetic materials, yttrium-iron-garnet(YIG) and non-magnetic materials gadolinium-gallium garnet (GGG) with permittivity ε_2, respectively. The magnetization vector is oriented along the z axis in the YIG layers, which are characterized by antisymmetrical permittivity and have the following nonzero components, $\varepsilon_{xx} = \varepsilon_{yy} = \varepsilon_{zz} = \varepsilon_1$, $\varepsilon_{xy} = -\varepsilon_{yx} = ig$, where g is the gyrotropy of the magnetic layer. The thicknesses of magnetic and nonmagnetic materials are d_1 and d_2, respectively. The NLC is characterized by a long-range uniaxial orientational order of rod-like anisotropic molecules along a common director, whose dielectric tensor is given by [86]

$$\varepsilon_{LC} = \begin{pmatrix} n_0^2 + \Delta\varepsilon\cos^2\alpha & 0 & \Delta\varepsilon\cos\alpha\sin\alpha \\ 0 & n_0^2 & 0 \\ \Delta\varepsilon\cos\alpha\sin\alpha & 0 & n_0^2 + \Delta\varepsilon\sin^2\alpha \end{pmatrix}. \tag{19}$$

where α is the tilt angle between the principal axis of NLC molecule and the x axis. $\Delta\varepsilon$ is the optical dielectric anisotropy given as $n_e^2 - n_0^2$, n_0 and n_e are ordinary and extraordinary refractive indices, respectively. Upon an applied external voltage, the distribution of NLC director exhibits spatially inhomogeneous property. The distribution of α is given in the NLC slab by the Ossen Frank elastic theory when an electric field is applied along the z axis. The relations between α and the voltage V applied to NLC slab can be achieved in the light of the following equations [87]

$$(k_1 \cos^2 \alpha + k_3 \sin^2 \alpha)(\frac{dz}{d\alpha})^{-2} = \Delta\varepsilon_0 E^2 (\sin^2 \alpha_m - \sin^2 \alpha), \tag{20}$$

$$\frac{V}{V_c} = \frac{2}{\pi} \int_0^{\pi/2} \sqrt{\frac{1 + k\eta^2 \sin^2 \psi}{1 - \eta^2 \sin^2 \psi}} d\psi, \tag{21}$$

with $\eta = \sin \alpha_m$ and $k = (k_3 - k_1)/k_1$. The parameters k_1 and k_3 are the Frank elastic constants for splay and bend modes of the nematic director distortion, respectively. α_m is the maximum tilt angle in the NLC slab, $\Delta\varepsilon_0$ is the dielectric anisotropy at zero frequency. Note that the direction of NLC molecules is affected by the external voltage only when the applied voltage V exceeds a threshold value V_c $(= \pi\sqrt{k_1/\Delta\varepsilon_0})$.

For the one-dimensional multilayered structure $(GGG|YIG)^n(YIG|GGG)^m NLC(YIG|GGG)^n$, at the near infrared wavelength $\lambda = 1.55\mu m$, YIG and GGG are characterized by $\varepsilon_1 = 5.5 + i0.0025$, $g = (1 - i0.15) \times 10^{-2}$, and $\varepsilon_2 = 3.709$, respectively [80]. The thicknesses are $d_1 = 0.466D$ and $d_2 = 0.534D$, respectively. $k_1 = 7.4$pN, $k_3 = 10.2$pN, $\Delta\varepsilon_0 = 1.43 \times 10^{-10}$F/m, $n_e = 1.75$, $n_0 = 1.54$ for typical NLC (5CB) [72], $d_{LC} = \frac{20}{6}D$. The repetition numbers are $n = 16$ and $m = 29$, respectively.

Fig. 5(a) corresponds to the distribution of α with different voltage $V = 0,\ 2V_c,\ 5V_c$ for *p*-polarized wave. The anisotropic character of NLC can be considered exactly by utilizing the obtained distribution of α. The calculated results for θ_p as functions of normalized frequency $\omega D/(2\pi c)$ with different voltages are plotted in Fig. 5(b). The solid, dashed, and dot-dashed lines correspond to θ_p with $V = 0,\ 2V_c,\ 5V_c$, respectively. It can be seen clearly that the spectrum of θ_p moves at different voltages, which is a direct manifestation of voltage-controlled Kerr effect. The frequency dependence of θ_p in the present case exhibits the following features. First, the positions of θ_p-peaks are dependent on the applied voltage due to the reorientation of the director. It is seen that the positions of peaks move toward high frequencies with the applied voltages. This is because of NLC's high sensitivity to the applied voltage. When the applied voltage is very small, the distribution of the director aligns x-direction due to the surface anchoring. With the increase of applied voltage, the energy of electric field adds gradually. When the applied voltage is bigger than the threshold value, the function of electric field plays a leading role. In such a case, the alignment of the director along the electric field can be observed. The reorientation of the NLC's director will lead to the change of its permittivity, which directly affects the Kerr rotation angles. Second, the amplitudes of θ_p depend strongly on the applied voltage, i.e., the heights of θ_p-peaks increase considerably with the increase of V. For example, the maximum of θ_p at $V = 5V_c$ ($\theta_{p\ max} = -30°$) is approximately 3 times larger than that of θ_p ($\theta_{p\ max} = -10°$) at $V = 0$, which describes the large difference in the indices of refraction between the right and left circularly polarized light. This can be understood according to the circular birefringence, in which different refractive indices due to different localization wavelengths for left- and right- circularly polarized lights, leads to the large rotation of the reflected light [2]. In addition, the positions of θ_p-peaks agree with those of the dips in the reflection spectrum exactly. Take $V = 5V_c$ as an example, we calculated θ_p and the corresponding reflection spectrum, which are exhibited in Fig. 5(c) and (d). Such a phenomenon is directly related to the edges of photonic band gaps, where the abnormal dispersion relation leads to strong photonic localization. As a result, θ_p can be enhanced at these specific frequencies in the present structure [28].

Thus, the controllable magneto-optical effects have been predicted in the MPCs with the NLC, which will be a critical problem in the optoelectronic applications.

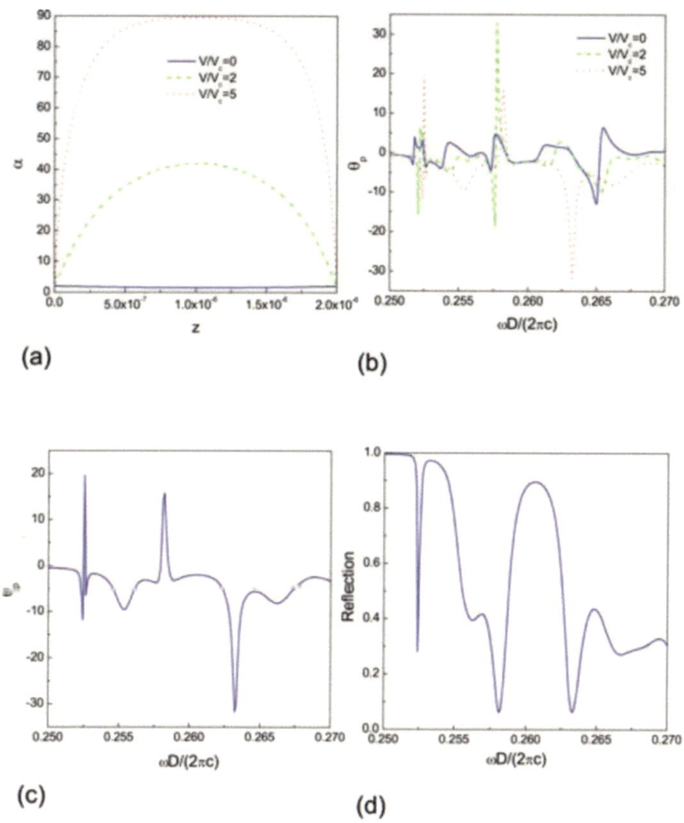

Fig. 5. (Color online) (a) α versus the distance at V =0,2Vc,5Vc, respectively; (b) θ_p versus the normalized frequency under V=0,2Vc,5Vc at the normal incident angle, are shown by solid, dashed, and dotted curves, respectively. (c) The dependence of θ_p and (d) the reflectance on the normalized frequency at V=5Vc.

5. Voltage-controlled tamm state in periodic magnetophotonic crystal with the NLC

The Tamm state proposed by Tamm is one of the most fundamental physical properties of the interface, which means electron states can occur in the energy band gap at a crystal surface [68]. In analogy with the electronic case, the OTS can be realized at the interface between two different photonic crystals with all isotropic dielectric materials. Conventional surface waves with a wave vector exceeding that of light in an incident medium decay exponentially away from the surface. In comparison with the conventional surface waves, the OTS can be formed for both the s– and p–polarized waves and occurs even at normal incidence [70]. In this part, we consider a one-dimensional MPC composed of the NLC and magnetic material, as illustrated in Fig. 6 (a). A linearly p-polarized monochromatic light impinges normally onto the structure. The thicknesses of the NLC and magnetic materials are d_1 and d_2, respectively. Due to the periodicity of the structure the eigenwaves satisfy Bloch's theorem, which leads to the equation $\det[P - e^{ikd}] = 0$. Here P is the total propagation matrix of unit cell. Based on these equations, we may illustrate the characteristics of the normalized angular frequency and kd in the present configuration.

The parameters used in our calculations are taken as $\varepsilon_1 = 3$ and $g = 0.5$, where a larger value of g is used to make the band gaps easily observable. $k_1 = 7.4$pN, $k_3 = 10.2$pN, $\Delta\varepsilon_0 = 1.43 \times 10^{-10}$ F/m, $n_e = 1.75$, $n_0 = 1.54$ for the NLC (5CB)[72], with the thicknesses $d_1 = 0.5D$, $d_2 = 0.5D$, respectively. The spatially inhomogeneous property of the directors is also taken into account in this calculation. In Fig. 7 (a), we present the $\omega d\ /\ c$ - kd characteristics under the applied voltages $V = 0,\ 2V_c,\ 5V_c$, respectively. From the figure it is clear that the intra-Brillouin-zone band gaps do occur and obviously depend on the applied voltages. This is a remarkable result in itself, as it indicates the existence and tunability of the OTS with the NLC. To understand qualitatively the origin of the appearance of intra-Brillouin-zone band gaps, we start with the simple case, i.e., we set the applied voltage as zero, where the distribution of the directors aligns along x-direction due to the surface anchoring. Now the NLC is a general anisotropic dielectric layer, where the eigenmodes are the ordinary and extraordinary waves when the incident wave enters into the NLC [88], then they become left- and right-circularly polarized waves in the magnetic materials. It appears an approximate standing wave due to the strong interference between these forward and backward waves, which leads to the formation of the intra-Brillouin-zone band gaps [65]. More interesting, the kd dependence on $\omega d/c$ in the present case exhibits the following feature, i.e., the intra-Brillouin-zone band gap moves with the applied voltages. The driving mechanism for manipulating the OTS arise from the intrinsic properties of the NLC. It is mainly caused by the sensitivity of the NLC on the applied voltage, which has significantly influence on the distribution of the NLC's directors. As widely known, the directors of the NLC molecules are affected by the external electric fleld only when the voltage V exceeds a threshold value V_c, where V_c is given by $V_c = \pi\sqrt{k_1/\Delta\varepsilon_0}$. The directors rotate parallel to the applied electric fleld to minimize the total free energy for $V/\ V_c > 1$. The reorientation of the NLC's directors leads to the variation of the effective refractive index, which change the optical path directly. Subsequently, the positions of standing waves will be finally changed with the applied voltages.

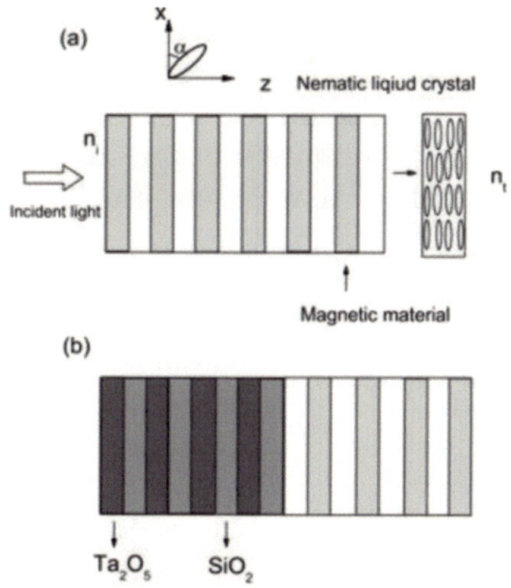

Fig. 6. (a) magnetophotonic crystal; (b) two adjoining photonic crystals, one of which consists of two dielectric materials and the other is composed of the NLC and magnetic material, i.e., $(Ta2O5\,|\,SiO2)^8(NLC|\,Bi:DyIG)^8$.

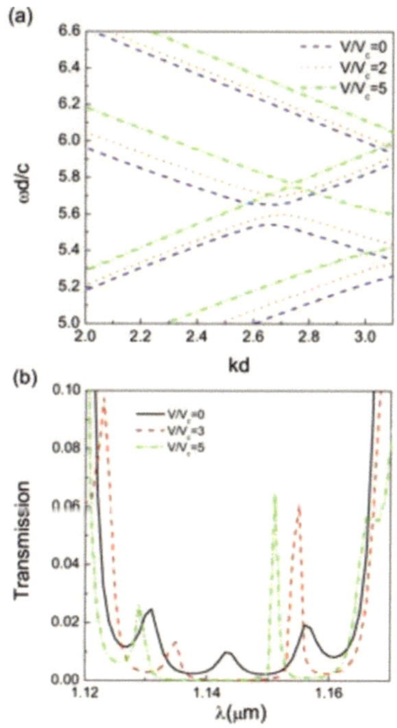

Fig. 7. (a) Normalized angular frequency $\omega * d/c$ versus kd at the normal incident angle under $V = 0$ (dash line), $V = 2Vc$ (dot line) and $V = 5Vc$ (dash dot line), respectively. $\varepsilon_1 = 3$ and $g = 0.5$ are used for the magnetic material. (b) The transmission of Fig. 6(b) versus λ at the normal incident angle under $V = 0$; $3Vc$; $5Vc$ (solid, dash, and dash dot line), respectively. Here $\varepsilon_{Ta2O5} = 4.41$, $\varepsilon_{SiO2} = 2.07$, the realistic values $\varepsilon_1 = 5.58$ and $g = -0.00198$ are taken for Bi:DyIG.

Furthermore, two stacked photonic crystals are generally used to illuminate the existence of the OTS in experiments [64], which can be observed by studying the transmission/reflection spectra with a sharp narrow peak/dip in the finite photonic crystal. Thus, it is necessary for us to provide the transmission spectrum. We design the two stacked photonic crystals, shown in Fig. 6(b), one of which consists of two dielectric materials and the other is composed of the NLC and magnetic material. Now we use the realistic material parameters at the wavelength λ = 1.55μm, $\varepsilon_{Ta_2O_5}$ = 4.41, ε_{SiO_2} = 2.07, ε_1 = 5.58 and g = −0.00198 for Bi:DyIG. The thicknesses are $d_{Ta_2O_5}$ = 0.696μm, d_{SiO_2} = 1.37μm, d_{NLC} = 1μm and $d_{Bi:DyIG}$ = 1.77μm, respectively [67]. Fig. 7 (b) depicts the transmission of the two photonic crystals as a function of wavelength at normal incidence. It is observed that a sharp narrow peak indeed appears in the transmission spectrum. In particular, we notice two photonic crystals generate photonic gaps nearly from 1.12 - 1.17μm and 1.13 - 1.16μm, respectively. However, the peak appears in the band gaps of two photonic crystals, which significantly demonstrates the sharp peak comes from the resonant tunneling of the electromagnetic wave through the OTS at the interface. In addition, the peak undergoes slightly blueshift with the applied voltages, which manifest itself a good candidate to be a controlled OTS structure. Finally, it is noted that the existence of OTS survives for s-polarized wave but the tunability disappears because ε_{yy} is not dependent on the external voltage.

6. Conclusion

The tunable magneto-optical effects in magnetophotonic crystals with the NLC have been reviewed in this chapter. Two types of MPCs are studied, i.e., MPCs consisting of alternate magnetic materials and the NLC; the cavity-based MPCs with the NLC. Both of them exhibit the tunable magneto-optical properties qualitatively different from those of the unvariable MPCs without the NLC. It is predicted theoretically that the magneto-optical effects can be manipulated by the applied voltages in the theoretical treatments, i.e., the 4×4 transfer matrix method in combination with a piecewise homogeneity approximation for liquid crystals. A significant shift of the peaks' positions in magneto-optical spectrum is observed toward high frequencies with the applied voltages due to the high sensitivity of the directors on the external electric fields. By applying a tunable electric field, the voltage-induced reorientation of the directors of the NLCs alters the values of the dielectric permittivity in the NLC, thus leading to alternating magneto-optical effects. In addition, the tunable intra-Brillouin-zone band gaps can be realized in the MPCs with the NLC, which implies the existence of the OTS at the interface between two photonic crystals. The methods described here may possible to be implemented on many other periodic multilayered structures, cavity-based periodic structure, quasi-periodic and aperiodic structures, thus, opening the possibility to systematically investigate the magneto-optical effects in a controlled way. The structures we propose here is not only interesting in itself but also allows easy access to fabricate it in the experiments due to the mature techniques for one-dimensional systems. In fact, apart from the external applied electric field, the variation of temperature can also change the permittivity in the NLC component, and then alter the magneto-optical effects. It is obvious that there are a great space for the potential application of the MPCs with NLC. These results highlight an intriguing avenue for future investigations in the development of tunable liquid crystal-based magnetophotonic crystals optoelectronic devices.

7. Acknowledgment

This project was supported by the National Natural Science Foundation of China under Grant No.10774107 and Jiangsu provincial innovation project under Grant No. ZY320607.

8. References

[1] M. Mansuripur, *The Principles of Magneto-Optical Recording* (Cambridge University Press, Cambridge, 1995).

[2] M Inoue, R Fujikawa, A Baryshev, A Khanikaev, P B Lim, H Uchida, O Aktsipetrov, A. Fedyanin, T Murzina and A. Granovsky, "Magnetophotonic crystals", J. Phys. D: Appl. Phys. 39, R151 (2006).

[3] Mitsuteru Inoue, Alexander V. Baryshev, Alexander B. Khanikaev Maxim E. Dokukin Kwanghyun Chung, Jinheo Hiroyuki Takagi, Hironaga Uchida, Pang Boey Lim, and Jooyoung Kim, "Magnetophotonic Materials and Their Applications", IEICE TRANS. ELECTRON, 91, 1630 (2008).

[4] R. Fujikawa, K. Tanizaki, A.V. Baryshev, P.B. Lim, K.H. Shin, H. Uchida, and M. Inoue, "Magnetic field sensors using magnetophotonic crystals", Proc. SPIE, 6369, 63690, (2006).

[5] J.H. Park, H. Takagi, K. Nishimura, et al., "Magneto-optic spatial light modulators driven by an electric field", J. Appl. Phys., 93, 8525, (2003).

[6] H.J. Park, J.K. Cho, K. Nishimura, and M. Inoue, "Magneto-optic spatial light modulator for volumetric digital recording system", Jpn. J. Appl. Phys., 41, 1813, (2002).

[7] Z.Wang and S. Fan, "Optical circulators in two-dimensional magneto-optical photonic crystals", Opt. Lett., 30, 1989, (2005).

[8] A.M. Merzlikin, A.P. Vinogradov, M. Inoue, et al., "The Faraday effect in two-dimensional magneto-photonic crystals", Journal of Magnetism and Magnetic Materials 300, 108, (2006).

[9] A.M. Merzlikin and A.P. Vinogradov, "Superprism effect in 1D photonic crystal", Opt. Commun. 259, 700, (2006).

[10] Belotelov VI, Kotov VA and Zvezdin AK, "New magnetooptical materials on a nanoscale", Phase Transitions, 79, 1135 (2006).

[11] Belotelov VI and Zvezdin AK, "Magneto-optical properties of photonic crystals", J. Opt. Soc. Am. B 22, 286 (2005).

[12] M. Inoue, K. Arai, T. Fujii and M. Abe, "Magneto-optical properties of one-dimensional photonic crystals composed of magnetic and dielectric layers", J. Appl. Phys. 83, 6768 (1998).

[13] M. Inoue and T. Fujii, "A theoretical analysis of magneto-optical Faraday effect of YIG films with random multilayer structures", J. Appl. Phys. 81, 5659, (1997).

[14] H. Kato and M. Inoue, "Reflection-mode operation of oned imensional magnetophotonic crystals for use in film-based magneto-optical isolator devices", J. Appl. Phys. 91, 7017, (2002).

[15] H. Kato, T. Matsushita, A. Takayama, et al., "Properties of one dimensional magnetophotonic crystals for use in optical isolator devices", IEEE Trans. Magn. 38, 3246, (2002).
[16] H. Kato, T. Matsushita, A. Takayama, M. Egawa, K. Nishimura, M. Inoue, "Theoretical analysis of optical and magneto-optical properties of one-dimensional magnetophotonic crystals", J. Appl. Phys. 93, 3906 (2003).
[17] Miguel Levy, "Normal modes and birefringent magnetophotonic crystals", J. Appl. Phys. 99, 073104 (2006).
[18] Miguel Levy, Rong Li, "Polarization rotation enhancement and scattering mechanisms in waveguide magnetophotonic crystals", Appl. Phys. Lett. 89, 121113 (2006).
[19] M. Levy and A.A. Jalali, "Band structure and Bloch states in birefringent one-dimensional magnetophotonic crystals: An analytical approach, J. Opt. Soc. Am. B 24, 1603, (2007).
[20] Khartsev SI and Grishin AM, "[Bi3Fe5O12/Gd3Ga5O12](m) magneto-optical photonic crystals", Appl. Phys. Lett. 87, 122504 (2005).
[21] S.I. Khartsev and A.M. Grishin, "High performance [Bi3Fe5O12/ Sm3Ga5O12]m magneto-optical photonic crystals", J. Appl. Phys. 101, 053906, (2007).
[22] A.B. Khanikaev, A.V. Baryshev, M. Inoue, et al., "Two-dimensional magnetophotonic crystal: Exactly solvable model", Phys. Rev. B 72, 035123, (2005).
[23] S. Sakaguchi and N. Sugimoto, "Multilayer films composed of periodic magneto-optical and dielectric layers for use as Faraday rotators", Opt. Commun. 162, 64 (1999).
[24] A.M. Merzlikin, A.P. Vinogradov, A.V. Dorofeenko, et al., "Controllable Tamm states in magnetophotonic crystal", Physica B 394, 277, (2007).
[25] A.P. Vinogradov, A.V. Dorofeenko, S.G. Erokhin, et al., "Surface state peculiarities in one-dimensional photonic crystal interfaces", Phys. Rev. B 74, 045128, (2006).
[26] I.L. Lyubchanskii, N.N. Dadoenkova, M.I. Lyubchanskii, et al., "Response of two-defect magnetic photonic crystals to oblique incidence of light: Effect of defect layer variation", J. Appl. Phys. 100, 096110, (2006).
[27] Kahl S, Grishin AM, "Enhanced Faraday rotation in all-garnet magneto-optical photonic crystal", Appl. Phys. Lett. 84, 1438 (2004).
[28] A. G. Zhdanov, A. A. Fedyanin, O. A. Aktsipetrov, D. Kobayashi, H. Uchida and M. Inoue, "Enhancement of Faraday rotation at photonic-band-gap edge in garnet-based magnetophotonic crystals", Journal of Magnetism and Magnetic Materials 300, 253 (2006).
[29] Yuehui Lu, Min Hyung Cho, JinBae Kim, YoungPak Lee, Jooyull Rhee, and Jae-Hwang Lee, "Control of Diffracted Magneto-Optical Enhancement in Ni Gratings", IEEE TRANSACTIONS ON MAGNETICS 44, 3300, (2008).
[30] Y. H. Lu, M. H. Cho, J. B. Kim,1 G. J. Lee, Y. P. Lee, and J. Y. Rhee, "Magneto-optical enhancement through gyrotropic gratings", Optics Express 16, 5378 (2008).
[31] M. Golosovsky, Y. Neve-Oz, and D. Davidov, "Magnetic-field-tunable photonic stop band in a three-dimensional array of conducting spheres", Phys. Rev. B 71, 195105 (2005).
[32] Jiang-Tao Liu and Kai Chang, "Tunable giant Faraday rotation of exciton in semiconductor quantum wells embedded in a microcavity", Appl. Phys. Lett. 90, 061114 (2007).

[33] S. V. Chernovtsev, D. P. Belozorov and S. I. Tarapov, "Magnetically controllable 1D magnetophotonic crystal in millimetre wavelength band", J. Phys. D: Appl. Phys. 40, 295C299 (2007).

[34] E. Graugnard, J. S. King, S. Jain, C. J. Summers, Y. Zhang-Williams, and I. C. Khoo, "Electricfield tuning of the Bragg peak in large-pore TiO2 inverse shell opals", Phys. Rev. B 72, 233105, (2005).

[35] P. Mach, P. Wiltzius, M. Megens, D. A. Weitz, K.-H. Lin, T. C. Lubensky, and A. G. Yodh, "Electro-optic response and switchable Bragg diffraction for liquid crystals in colloid-templated materials", Phys. Rev. E 65, 65, (2002).

[36] T. T. Larsen, A. Bjarklev, D. S. Hermann, and J. Broeng, "Optical devices based on liquid crystal photonic bandgap fibers", Optics Express 11, 2589, (2003).

[37] W. Hu, R. Dickie, R. Cahill, H. Gamble, Y. Ismail, V. Fusco, D. Linton, N. Grant, and S. Rea, "Liquid crystal tunable mm wave frequency selective surface", IEEE Microw. Wireless Compon. Lett. 17, 667, (2007).

[38] J. Cos, J. Ferre-Borrull, J. Pallares, L.F. Marsal, "Tunable FabryCProt filter based on one-dimensional photonic crystals with liquid crystal components", Optics Communications 282, 1220, (2009).

[39] V.G. Arkhipkin, V.A. Gunyakov, S.A. Myslivets, V.Ya. Zyryanov, and V.F. Shabanov, "Angular tuning of defect modes spectrum in the one-dimensional photonic crystal with liquid-crystal layer", Eur. Phys. J. E 24, 297 (2007).

[40] Ryotaro Ozakia and Hiroshi Moritake, Katsumi Yoshino, Masanori Ozaki, "Analysis of defect mode switching response in one-dimensional photonic crystal with a nematic liquid crystal defect layer", J. Appl. Phys. 101, 033503 (2007).

[41] Jiun-Yeu Chen and Lien-Wen Chen, "Polarization-dependent filters based on chiral photonic structures with defects", J. Opt. A: Pure Appl. Opt. 7 558, (2005).

[42] Hiroyuki Yoshida, Chee Heng Lee, Akihiko Fujii, and Masanori Ozaki, "Tunable Chiral Photonic Defect Modes in Locally Polymerized Cholesteric Liquid Crystals", Mol. Cryst. Liq. Cryst. 477, 255, (2007).

[43] Yuhua Huang, Ying Zhou, Charlie Doyle, and Shin-Tson Wu, "Tuning the photonic band gap in cholesteric liquid crystals by temperature-dependent dopant solubility", Optics Express, 14, 1236 (2006).

[44] N. Scaramuzza, C. Ferrero, B. V. Carbone, and C. Versace, "Dynamics of selective reflections of cholesteric liquid crystals subject to electric fields", J. Appl. Phys. 77, 572 (1995).

[45] Kuniaki Konishi, Benfeng Bai, Xiangfeng Meng, Petri Karvinen, Jari Turunen, Yuri. P. Svirko, and Makoto Kuwata-Gonokami, "Observation of extraordinary optical activity in planar chiral photonic crystals", Optics Express, 16, 7189 (2008).

[46] M V Gorkunov and M A Osipov, "Molecular theory of layer contraction in smectic liquid crystals, J. Phys. Condens. Mater 20, 465101 (2008).

[47] Barbero G and Komitov L, "Temperature-induced tilt transition in the nematic phase of liquid crystal possessing smectic C-nematic phase sequence", J. Appl. Phys. 105, 064516 (2009).

[48] Qian Zhao, Lei Kang, Bo Li, Ji Zhou, Hong Tang and Baizhe Zhang, "Tunable negative refraction in nematic liquid crystals", Appl. Phys. Lett. 89, 221918 (2006).

[49] Lei Kang, Qian Zhao, Bo Li, and Ji Zhoua, Hao Zhu, "Experimental verification of a tunable optical negative refraction in nematic liquid crystals", Appl. Phys. Lett. 90, 181931 (2007).
[50] Jeremy A. Bossard, Xiaotao Liang, Ling Li, Seokho Yun, Douglas H. Werner, E, Brian Weiner, Theresa S. Mayer, Paul F. Cristman, Andres Diaz, and I. C. Khoo, "Tunable Frequency Selective Surfaces and Negative-Zero-Positive Index Metamaterials Based on Liquid Crystals", IEEE Transactions on Antennas and Propagation, 56, 1308, (2008).
[51] Miroshnichenko AE, Brasselet E, Kivshar YS, "All-optical switching and multistability in photonic structures with liquid crystal defects", Appl. Phys. Lett. 92, 253306 (2008).
[52] Li-Hsuan Hsu, Kuang-Yao Lo, Shih-An Huang, Chi-Yen Huang, and Chung-Sung Yang, "Irreversible redshift of transmission spectrum of gold nanoparticles doped in liquid crystals", Appl. Phys. Lett. 92, 181112 (2008).
[53] Laudyn U A, Miroshnichenko A E, Krolikowski W, et al., "Observation of light-induced reorientational effects in periodic structures with planar nematic-liquid-crystal defects", Appl. Phys. Lett. 92, 203304 (2008).
[54] Lin TJ, Chen CC, Lee W, et al., "Electrical manipulation of magnetic anisotropy in the composite of liquid crystals and ferromagnetic nanorods", Appl. Phys. Lett. 93, 013108 (2008).
[55] Alexander B. Khanikaev and M. J. Steel, "Low-symmetry magnetic photonic crystals for non-reciprocal and unidirectional devices", Optics Express 17, 5265 (2009).
[56] Z. Wang, Y. D. Chong, J. D. Joannopoulos, and M. Soljacic, "Reflection-Free One-Way Edge Modes in a Gyromagnetic Photonic Crystal", Phys. Rev. Lett. 100, 01390501 (2008).
[57] Z. Yu, G. Veronis, Z. Wang, and S. Fan, "One-Way Electromagnetic Waveguide Formed at the Interface between a Plasmonic Metal under a Static Magnetic Field and a Photonic Crystal", Phys. Rev. Lett. 100, 02390201 (2008).
[58] Z. Yu, Z. Wang, and S. Fan, "One-way total reflection with one-dimensional magneto-optical photonic crystals", Appl. Phys. Lett. 90, 121133 (2007).
[59] A. Figotin and I. Vitebskiy, "Electromagnetic unidirectionality in magnetic photonic crystals", Phys. Rev. B 67, 165210 (2003).
[60] A. Figotin and I. Vitebskiy, "Nonreciprocal magnetic photonic crystals", Phys. Rev. E 63, 066609 (2001).
[61] K.-Y. Jung, B. Donderici, and F. L. Teixeira, "Transient analysis of spectrally asymmetric magnetic photonic crystals with ferromagnetic losses", Phys. Rev. B 74, 165207 (2006).
[62] Ruiyi Chen, Dongjie Tao, Haifeng Zhou, Yinlei Hao, Jianyi Yang, Minghua Wang and Xiaoqing Jiang, "Asymmetric multimode interference isolator based on nonreciprocal phase shift", Optics Communications 282 (2009) 862.
[63] Mumcu G., Sertel K., Volakis J.L., "Miniature Antennas and Arrays Embedded Within Magnetic Photonic Crystals", IEEE 5, 168 (2006).
[64] T. Goto, A.V. Dorofeenko, A. M. Merzlikin, A.V. Baryshev, A. P. Vinogradov, M. Inoue, A. A. Lisyansky, and A. B. Granovsky, "Optical Tamm States in One-Dimensional Magnetophotonic Structures", Phys. Rev. Lett. 101, 113902 (2008).

[65] Fei Wang, Akhlesh Lakhtakiab, "Intra-Brillouin-zone bandgaps due to periodic misalignment in one-dimensional magnetophotonic crystals", Appl. Phys. Lett. 92, 011115 (2008).

[66] A.M. Merzlikina, A.P. Vinogradova, A.V. Dorofeenkoa, M. Inoueb, M. Levyc, A.B. Granovskyd, "Controllable Tamm states in magnetophotonic crystal", Phys. B 394, 277 (2007).

[67] A. P. Vinogradov, A. V. Dorofeenko, S. G. Erokhin, M. Inoue, A. A. Lisyansky, A. M. Merzlikin, and A. B. Granovsky, "Surface state peculiarities in one-dimensional photonic crystal interfaces", Phys. Rev. B 74, 045128 (2006).

[68] I. E. Tamm, Phys. Z. Sowjetunion, "On the possible bound states of electrons on a crystal surface ", 1, 733 (1932).

[69] M. Kaliteevski, I. Iorsh, S. Brand, R. A. Abram, J. M. Chamberlain, A. V. Kavokin, and I. A. Shelykh, "Tamm plasmon-polaritons: Possible electromagnetic states at the interface of a metal and a dielectric Bragg mirror", Phys. Rev. B 76, 165415 (2007).

[70] A. V. Kavokin, I. A. Shelykh, and G. Malpuech1, "Lossless interface modes at the boundary between two periodic dielectric structures", Phys. Rev. B 72, 233102 (2005).

[71] H. Wohler, G. Haas, M. Fritsch, and D. A. Mlynski, "Faster 4 x 4 matrix method for uniaxial inhomogeneous media", J. Opt. Soc. Am. A 5, 1554 (1988).

[72] D. W. Berreman, "Optics in smoothly varying anisotropic planar structures: Application to liquid-crystal twist cells", J. Opt. Soc. Am. 63, 1374 (1973).

[73] H.X. Da, C. Xu, and Z.Y. Li, "Magneto-optical effect of left-handed material", Eur. Phys. J. B 45, 347 (2005).

[74] Yao-Yu Wang, and Lien-Wen Chen, "Tunable negative refraction photonic crystals achieved by liquid crystals ", Optics Express 14, 10580 (2006).

[75] A. Lakhtakia and R. Messier, "Sculptured Thin Films: Nanoengineered Morphology and Optics", SPIE Press, Bellingham, WA, USA, 2005.

[76] H. X. Da, P. Xu, J. C. Wu, and Z. Y. Li, "Electrically controlled Kerr effect in magnetophotonic crystals based on nematic liquid crystals", J. Appl. Phys. 104, 033911 (2008).

[77] Hai-xia Da, Zi-qiang Huang, and Z. Y. Li, "Voltage-controlled Kerr effect in magnetophotonic crystal", Opt. Lett. 34, 356 (2009).

[78] Hai-xia Da, Zi-qiang Huang, and Z. Y. Li, "Electrically-controlled Optical Tamm States in magnetophotonic crystal based on nematic liquid crystals", IN PRESS, OPTICS LETTERS.

[79] Chun-Yeol You and Sung-Chul Shin, "Novel method to determine the off-diagonal element of the dielectric tensor in a magnetic medium", Appl. Phys. Lett. 70, 2595 (1997).

[80] A. Zvezdin and V. Kotov, Modern Magneto–optics and Magneto–optical Materials (IOP, Bristol, 1997).

[81] K. Balasubramanian, A. Marathay, and H. A. Macleod, "Modeling Magneto-optical thin-film media for optical-data storage", Thin Solid Films 164, 391 (1988).

[82] C.-S. Kee, J. Y. Park, S. J. Kim, H. C. Song, Y. S. Kwon, N. H. Myung, S. Y. Shin, and H. Lim, "Essential parameter in the formation of photonic band gaps", Phys. Rev. E 59, 4695 (1999);

[83] C.-S. Kee, J.-E. Kim, H. Y. Park, and H. Lim, "Roles of wave impedance and refractive index in photonic crystals with magnetic and dielectric properties", IEEE Trans. Microwave Theory Tech. 47, 2148 (1999).
[84] A. Saib, D. Vanhoenacker-Janvier, I. Huynen Laboratoire, A. Encinas, L. Piraux, E. Ferain, and R. Legras, "Magnetic photonic band-gap material at microwave frequencies based on ferromagnetic nanowires", Appl. Phys. Lett. 83, 2378 (2003).
[85] R. H. Tarkanyan and D. G. Niarchos, "Effective negative refractive index in ferromagnet-semiconductor superlattices", Optics Express 14, 5433 (2006).
[86] I.-C. Khoo, *Liquid Crystals*, 1st ed (Wiley, New York, 1995), pp. 21-24.
[87] L. Liebert, *Liquid Crystals*, 1st ed (Academic, New York, 1978), pp. 81.
[88] P. Yeh, "Electromagnetic propagation in birefringent layered media", J. Opt. Soc. Am. 69, 742 (1979).

4

A New Method of Generating Atmospheric Turbulence with a Liquid Crystal Spatial Light Modulator

Christopher C Wilcox and Dr. Sergio R Restaino
Naval Research Laboratory
United States of America

1. Introduction

Light traveling from a star, or any point source, will propagate spherically outward. After a long distance, the wavefront, or surface of equal phase, will be flat; as is illustrated in Fig. 1.

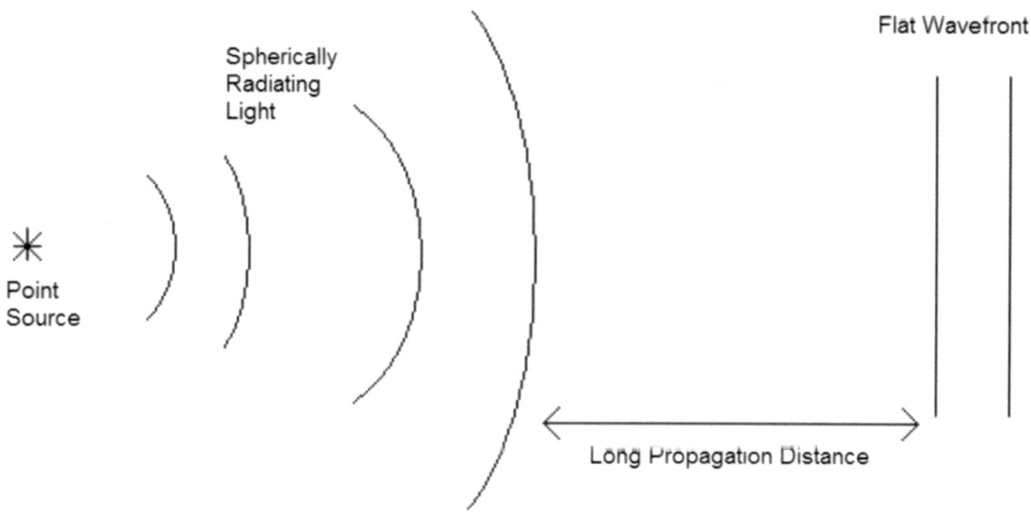

Fig. 1. Flat wavefront after a long propagation distance from a point source

When the light begins to propagate through Earth's atmosphere, the varying index of refraction will alter the optical path, as shown in Fig. 2. The Earth's atmosphere can be described as a locally homogeneous medium in which its properties vary with respect to temperature, pressure, wind velocities, humidity and many other factors. Also, the Earth's atmosphere temporally changes in a quasi-random fashion. All of these processes are usually simply refered to as ``atmospheric turbulence''. The Kolmogorov model of energy distribution in a turbulent medium is a useful statistical model to describe the fluctuation in refractive index due to mostly the humidity and pressure changes. This model was first

Source: New Developments in Liquid Crystals, Book edited by: Georgiy V. Tkachenko,
ISBN 978-953-307-015-5, pp. 234, November 2009, I-Tech, Vienna, Austria

proposed by a Russian mathematician named Andreï Kolmogorov in 1941 and describes how in a fully turbulent media the kinetic energy of large scale motions is transfered to smaller and smaller scale motions (Kolmogorov, 1941). It is supported by a variety of experimental measurements and is quite widely used in simulations for the propagation of electromagnetic waves through a random medium. The first author to fully describe such phenomena was Tatarski in his textbook ``Wave propagation in a turbulent medium'' (Tatarski, 1961). The complex and random nature of the Earth's atmospheric turbulence effect on wave propagation is currently a subject of active research and experimental measurements. Many of the parameters of Earth's atmospheric turbulence can be, at best, described statistically.

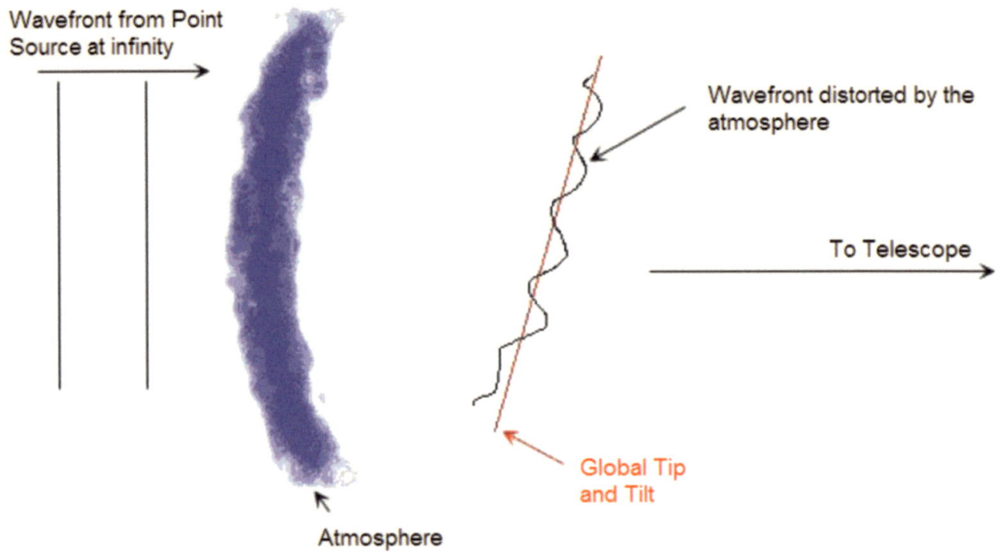

Fig. 2. Propagation of light from a distant source that then passes through the atmosphere

These statistical parameters represent the strength and changeability of the atmospheric turbulence, these conditions are customarily refered to as the ``astronomical seeing'', as they are widely used for astronomical applications. It is with this statistical information about a certain astronomical site and the specifications of the telescope that an Adaptive Optics (AO) system can be designed to correct the wavefront distortions caused by the atmosphere at that site. As telescopes continue to be manufactured larger and larger, the need for AO is increasing because of the limiting factors caused by atmospheric turbulence. In order to adequately characterize the performance of a particular AO system, an accurate spatial and temporal model of the Earth's atmosphere is required.

AO is the term used for a class of techniques dealing with the correction of wavefront distortions in an optical system in real time. Some wavefront distortions may include those caused by the atmosphere. Astronomical applications of AO particularly include the correction of atmospheric turbulence for a telescope system. Other possible applications include Free Space Laser Communications, High Energy Laser Applications, and Phase-Correction for Deployable Space-Based Telescopes and Imaging systems. However, prior to deployment, an AO system requires calibration and full characterization in a laboratory environment.

Many techniques are currently being used with AO systems for simulating atmospheric turbulence. Some static components use glass phase screens with holograms etched into them. In addition, it is also important to simulate the temporal transitions of atmospheric turbulence. Some of these methods include the use of a static aberrator, such as a clear piece of plastic or glass etched phase screen, and rotating it. Rotating filter wheels with etched holographic phase screens can simulate temporal transitions, as well. Also, simply using a hot-plate directly under the beam path in an optical system can simulate temporally the atmospheric turbulence.

However, etching holographic phase screens into glass can be quite costly and not very flexible to simulate different atmospheric characteristics. Thus, one would need more than one phase screen. A testbed that simulates atmospheric aberrations far more inexpensively and with greater fidelity and flexibility can be achieved using a Liquid Crystal (LC) Spatial Light Modulator (SLM). This system allows the simulation of atmospheric seeing conditions ranging from very poor to very good and different algorithms may be easily employed on the device for comparison. These simulations can be dynamically generated and modified very quickly and easily.

2. Background

2.1 Brief history of the study of atmospheric turbulence

Ever since Galileo took a first look at the moons of Jupiter through one of the first telescopes, astronomers have strived to understand our universe. Within the last century, telescopes have enabled us to learn about the far reaches of our universe, even the acceleration of the expansion of the universe, itself. The field of building telescopes has been advancing much in recent years. The twin Keck Telescopes on the summit of Hawaii's dormant Mauna Kea volcano measure 10 meters and are currently the largest optical telescopes in the world. Plans and designs for building 30 and 100 meter optical telescopes are underway. As these telescope apertures continue to grow in diameter, the Earth's atmosphere degrades the images we try to capture more and more. As Issac Newton said is his book, *Optiks* in 1717, "... the air through which we look upon the stars is in perpetual tremor; as may be seen by the tremulous motion of shadows cast from high towers, and by the twinkling of the fixed stars.... The only remedy is a most serene and quiet air, such as may perhaps be found on the tops of high mountains above grosser clouds." It was at this time when we first realized that the Earth's atmosphere was the major contributor to image quality for ground-based telescopes. The light arriving from a distant object, such as a star, is corrupted by turbulence-induced spatial and temporal fluctuations in the index of refraction of the air.

In 1941, Kolmogorov published his treatise on the statistics of the energy transfer in a turbulent flow of a fluid medium. Tatarskii used this model to develop the theory of electro-magnetic wave propagation through such a turbulent medium. Then, Fried used Tatarskii's model to introduce measurable parameters that can be used to characterize the strength of the atmospheric turbulence.

The theory of linear systems allows us to understand how a system transforms an input just by defining the characteristic functions of the system itself. Such a characteristic function is represented by a linear operator operating on an impulse function. The characteristic system function is generally called the ``impulse response function''. Very often, such an operator is the so-called Fourier transform. An imaging system can be approximated by a linear, shift-invariant system over a wide range of applications. The next few sections will explain the use of a Fourier transform in such an optical imaging system and its applications with optical aberrations.

2.2 Brief overview of fourier optics and mathematical definitions

A fantastic tool for the mathematical analysis of many types of phenomena is the Fourier transform. The 2-dimensional Fourier transform of the function $g(x,y)$ is defined as,

$$G(f_x, f_y) = \mathbb{F}\{g(x,y)\} = \int_{-\infty}^{\infty}\int_{-\infty}^{\infty} g(x,y)e^{-j2\pi(f_x x + f_y y)}dxdy \tag{1}$$

where, for an imaging system, the x-y plane is the entrance pupil and the f_x-f_y plane is the imaging plane. A common representation of the Fourier transform of a function is by the use of lower case for the space domain and upper case for the Fourier transform, or frequency domain. Similarly, the inverse Fourier transform of the function $G(f_x,f_y)$ is defined as,

$$g(x,y) = \mathbb{F}^{-1}\{G(f_x, f_y)\} = \int_{-\infty}^{\infty}\int_{-\infty}^{\infty} G(f_x, f_y)e^{j2\pi(f_x x + f_y y)}df_x df_y \tag{2}$$

There exist various properties of the Fourier transform. The linearity property states that the Fourier transform of the sum of two or more functions is the sum of their individual Fourier transforms and is shown by,

$$\mathbb{F}\{ag(x,y) + bf(x,y)\} = a\mathbb{F}\{g(x,y)\} + b\mathbb{F}\{f(x,y)\} \tag{3}$$

where a and b are constants. The scaling property states that stretching or skewing of a function in the x-y domain results in skewing or stretching of the Fourier transform, respectively, and is shown by,

$$\mathbb{F}\{g(ax,by)\} = \frac{1}{|ab|}G\left(\frac{f_x}{a}, \frac{f_y}{b}\right) \tag{4}$$

where a and b are constants. The shifting property states that the translation of a function in the space domain introduces a linear phase shift in the frequency domain and is shown by,

$$\mathbb{F}\{g(x-a, y-b)\} = G(f_x, f_y)e^{-j2\pi(af_x + bf_y)} \tag{5}$$

where a and b are constants. This property is of particular interest in the mathematical analysis of tip and tilt in an optical system, as it describes horizontal or vertical position in the imaging plane. Parsaval's Theorem is generally known as a statement for the conservation of energy and is shown as,

$$\int_{-\infty}^{\infty}\int_{-\infty}^{\infty}|g(x,y)|^2 dxdy = \int_{-\infty}^{\infty}\int_{-\infty}^{\infty}|G(f_x, f_y)|^2 df_x df_y \tag{6}$$

The convolution property states that the convolution of two functions in the space domain is exactly equivalent to the multiplication of the two functions' Fourier transforms, which is usually a much simpler operation. The convolution of two functions is defined as,

$$g(x,y)*f(x,y)=\int_{-\infty}^{\infty}\int_{-\infty}^{\infty}g(\xi,\eta)f(\xi-x,\eta-y)d\xi d\eta \tag{7}$$

The convolution property is shown as,

$$\mathbb{F}\{g(x,y)*f(x,y)\}=G(f_x,f_y)F(f_x,f_y) \tag{8}$$

A special case of the convolution property is known as the autocorrelation property and is shown as,

$$\mathbb{F}\{g(x,y)*g^*(x,y)\}=\left|G(f_x,f_y)\right|^2 \tag{9}$$

where the superscript * denotes the complex conjugate of the function $g(x,y)$. The autocorrelation property gives the Power Spectral Density (PSD) of a function and is a useful way to interpret a spatial function's frequency content. The square of the magnitude of the $G(f_x,f_y)$ function is also referred to as the Point Spread Function (PSF). The PSF is the imaging equivalent of the impulse response function. It is easy to see that the PSF represents the spreading of energy on the output plane of a point source at infinity.
The spatial variation as a function of spatial frequency is described by the Optical Transfer Function (OTF). The OTF is defined as the Fourier transform of the PSF written as,

$$\text{OTF}=\mathbb{F}\left\{\left|G(f_x,f_y)\right|^2\right\}=\mathbb{F}\{\text{PSF}\} \tag{10}$$

The Modulation Transfer Function (MTF) is the magnitude of the OTF and is written as,

$$\text{MTF}=\left|\mathbb{F}\left\{\left|G(f_x,f_y)\right|^2\right\}\right|=\left|\mathbb{F}\{\text{PSF}\}\right| \tag{11}$$

Two common aperture geometries, or pupil functions, that will be discussed are the rectangular and circular apertures. The rectangular aperture is defined as,

$$rect\left(\frac{x}{k},\frac{y}{l}\right)=\begin{cases}1 & |x|\le k/2 \text{ and } |y|\le l/2\\ 0 & \text{otherwise}\end{cases} \tag{12}$$

where k and l are positive constants that refer to the length and width of the aperture, respectively. The circular aperture is defined as,

$$circ\left(\frac{\rho}{l}\right)=\begin{cases}1 & \rho\le l \text{ and } \rho=\sqrt{x^2+y^2}\\ 0 & \text{otherwise}\end{cases} \tag{13}$$

where l is a positive constant referring to the radius of the aperture. These pupil functions become of great use when analyzing an imaging system with these apertures. For the purposes of this discussion, a circular aperture will be considered as it is of particular use with Zernike polynomials and Karhunen-Loeve polynomials which will be discussed later.
In order to include the effects of aberrations, it is useful to introduce the concept of a "generalized pupil function". Such a function is complex in nature and the argument of the imaginary exponential is a function that represents the optical phase aberrations by,

$$\mathbb{P}(x,y) = P(x,y)e^{j\frac{2\pi}{\lambda}W(x,y)} \tag{14}$$

where $P(x,y) = circ(\rho)$, λ is the wavelength, and $W(x,y)$ is the effective path length error, or error in the wavefront. It is in this wavefront error that atmospheric turbulence induces and degrades image quality of an optical system and induces aberrations. This wavefront error can be induced in an optical system through the use of a LC SLM. The next several sections will describe methods of simulating atmospheric turbulence in an optical system and introduce the new method of simulating atmospheric turbulence developed at the Naval Research Laboratory.

2.3 Optical aberrations as Zernike polynomials

The primary goal of AO is to correct an aberrated, or distorted, wavefront. A wavefront with aberrations can be described by the sum of an orthonormal set of polynomials, of which there are many. One specific set is the so called Zernike polynomials, $Z_i(\rho,\theta)$, and they are given by,

$$Z_i(\rho,\theta) = \begin{cases} \sqrt{n+1}R_n^m(\rho)\cos(\theta) & \text{for } m \neq 0 \text{ and } i \text{ is even} \\ \sqrt{n+1}R_n^m(\rho)\sin(\theta) & \text{for } m \neq 0 \text{ and } i \text{ is even} \\ R_n^0(\rho) & \text{for } m = 0 \end{cases} \tag{15}$$

where

$$R_n^m(\rho) = \sum_{s=0}^{\frac{n-m}{2}} \frac{(-1)^s(n-s)!}{s!\left(\frac{n+m}{2}-s\right)!\left(\frac{n-m}{2}-s\right)!}\rho^{n-2s} \tag{16}$$

The azimuthal and radial orders of the Zernike polynimials, m and n, respectively, satisfy the conditions that $m \leq n$ and n-m = even, and i is the Zernike order number (Roggemann & Welsh, 1996). The Zernike polynomials are used because, among other reasons, the first few terms resemble the classical aberrations well known to lens makers. The Zernike order number is related to the azimuthal and radial orders via the numerical pattern in Table 1.

i	*n*	*m*	*i*	*n*	*m*	*i*	*n*	*m*	*i*	*n*	*m*
1	0	0	**8**	3	-1	**15**	4	-4	**22**	6	0
2	1	1	**9**	3	3	**16**	5	1	**23**	6	2
3	1	-1	**10**	3	-3	**17**	5	-1	**24**	6	-2
4	2	0	**11**	4	0	**18**	5	3	**25**	6	4
5	2	2	**12**	4	2	**19**	5	-3	**26**	6	-4
6	2	-2	**13**	4	-2	**20**	5	5	**27**	6	6
7	3	1	**14**	4	4	**21**	5	-5	**28**	6	-6

Table 1. Relationship between Zernike order and azimuthal and radial orders

Zernike polynomials represent aberrations from low to high order with the order number. A wavefront can generally be represented by,

$$\text{Wavefront}(\rho,\theta)=\sum_{i=1}^{M}a_i Z_i(\rho,\theta) \tag{17}$$

where the a_i's are the amplitudes of the aberrations and M is the total number of Zernike orders the wavefront is represented by. This wavefront can be substituted into Equation (14) as it represents the phase in an imaging system.

2.4 Kolmogorov's statistical model of atmospheric turbulence

The Sun's heating of land and water masses heat the surrounding air. The buoyancy of air is a function of temperature. So, as the air is heated it expands and begins to rise. As this air rises, the flow becomes turbulent. The index of refraction of air is very sensitive to temperature. Kolmogorov's model provides a great mathematical foundation for the spatial fluctuations of the index of refraction of the atmosphere. The index of refraction of air is given by,

$$n(\vec{r},t)=n_0+n_1(\vec{r},t) \tag{18}$$

where $\vec{r}$ is the 3-dimensional space vector, t is time, n_0 is the average index of refraction, and $n_1(\vec{r},t)$ is the spatial variation of the index of refraction. For air, we may say $n_0 = 1$. At optical wavelengths the dependence of the index of refraction of air upon pressure and temperature is $n_1 = n - 1 = 77.6\text{x}10^{-6}P/T$, where P is in millibars and T is in Kelvin. The index of refraction for air can now be given as,

$$n(P,T)=1+\frac{77.6\times10^{-6}P}{T} \tag{19}$$

Differentiating the index of refraction with respect to temperature gives,

$$\frac{\partial}{\partial T}n(P,T)=-\frac{77.6\times10^{-6}P}{T^2} \tag{20}$$

From Equation (20), we can see that the change in index of refraction with respect to temperature cannot be ignored (Roggemann & Welsh, 1996). These slight variances of temperature, of which the atmosphere constantly has many, will affect the index of refraction enough to affect the resolution of an imaging system.

As light begins to propagate through Earth's atmosphere, the varying index of refraction will alter the optical path slightly. To a fairly good approximation, the temperture and pressure can be treated as random variables. Unfortunately, because of the apparent random nature of Earth's atmosphere, it can at best be described statistically. It is with this statistical information about a certain astronomical site and the specifications of the telescope that an adaptive optics system can be designed to correct the wavefront distortions caused by the atmosphere at that site.

The quantity C_n^2 is called the structure constant of the index of refraction fluctuations with units of $m^{-2/3}$ (Roggemann & Welsh, 1996), it is a measurable quantity that indicates the strength of turbulence with altitude in the atmosphere. The value C_n^2 can vary from ~10^{-17} $m^{-2/3}$ or less and ~10^{-13} $m^{-2/3}$ or more in weak and strong conditions, respectively

(Andrews, 2004). C_n^2 can have peak values during midday, have near constant values at night and minimum values near sunrise and sunset. These minimum values' occurrence at sunrise and sunset is known as the diurnal cycle.

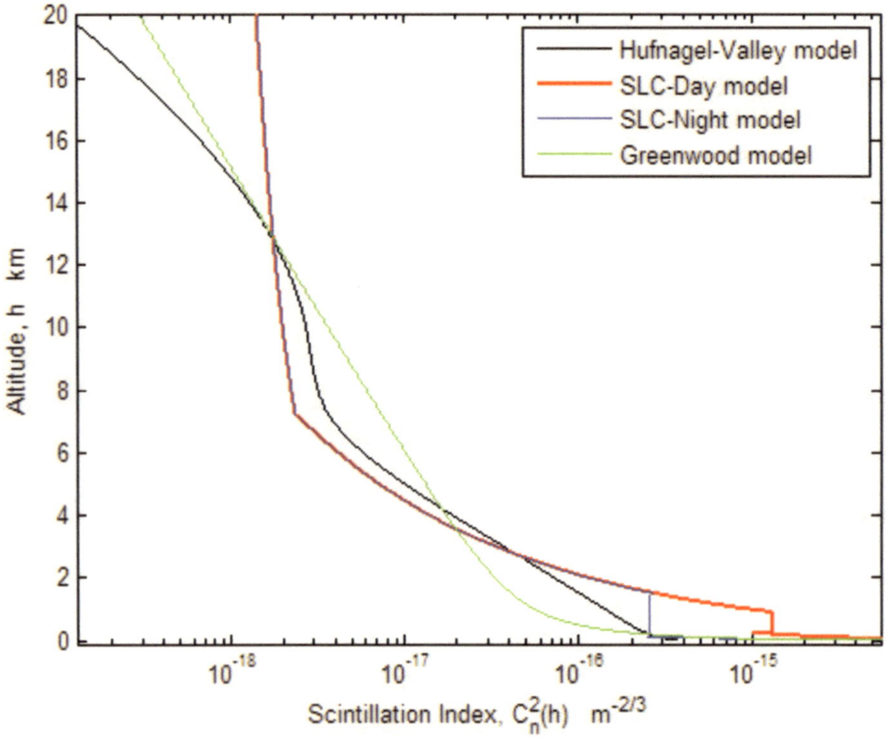

Fig. 3. Plots of the Hufnagel-Valley, SLC-Day, SLC-Night, and Greenwood models for C_n^2 with respect to altitude.

Some commonly accepted models of $C_n^2(h)$ as functions height are the Hufnagel-Valley, SLC-Day, SLC-Night and Greenwood models. The Hufnagel-Valley model is written as,

$$C_n^2(h) = 0.00594\left(v_w/27\right)^2\left(10^{-5}h\right)^{10} e^{-h/1000} + 2.7\times10^{-16} e^{-h/1500} + C_n^2(0) e^{-h/100} \tag{21}$$

where v_w is the rms wind speed and $C_n^2(0)$ is the ground-level value of the structure constant of the index of refraction. The SLC-Day model is written as,

$$C_n^2(h) = \begin{cases} 1.7\times10^{-14} & 0 < h < 18.5 \\ 3.13\times10^{-13} h^{-1.05} & 18.5 < h < 240 \\ 1.3\times10^{-15} & 240 < h < 880 \\ 8.87\times10^{-7} h^{-3} & 880 < h < 7200 \\ 2.0\times10^{-16} h^{-1/2} & 7200 < h < 20000 \end{cases} \tag{22}$$

The SLC-Night model is written as,

$$C_n^2(h)=\begin{cases}8.4\times10^{-15} & 0<h<18.5\\ 2.87\times10^{-12}h^{-2} & 18.5<h<110\\ 2.5\times10^{-16} & 110<h<1500\\ 8.87\times10^{-7}h^{-3} & 1500<h<7200\\ 2.0\times10^{-16}h^{-1/2} & 7200<h<20000\end{cases} \tag{23}$$

The Greenwood model is written as,

$$C_n^2(h)=\left[2.2\times10^{-13}(h+10)^{-1.3}+4.3\times10^{-17}\right]e^{-h/4000} \tag{24}$$

In each of these models, h may be replaced by $\frac{h}{\cos(\theta_z)}$ if the optical path is not vertical, or at zenith, and θ_z is the angle away from zenith.

2.5 Fried and Noll's model of turbulence

The fact that a wavefront can be expressed as a sum of Zernike polynomials is the basis for Noll's analysis on how to express the phase distortions due to the atmosphere in terms of Zernike polynomials.

Fried's parameter, also known as the coherence length of the atmosphere and represented by r_0, is a statistical description of the level of atmospheric turbulence at a particular site. Fried's parameter is given by,

$$r_0=\left[0.423k^2\sec\zeta\int_{\text{Path}}C_n^2(z)dz\right]^{-3/5} \tag{25}$$

where $k=\frac{2\pi}{\lambda}$ and λ is the wavelength, ζ is the zenith angle, the Path is from the light source to the telescope's aperture along the z axis and it is expressed in centimeters. The value of r_0 ranges from under 5 cm with poor seeing conditions to more than 25 cm with excellent seeing conditions in the visible light spectrum. The coherence length limits a telescope's resolution such that a large aperture telescope without AO does not provide any better resolution than a telescope with a diameter of r_0 (Andrews, 2004). In conjunction with r_0, another parameter that is important is the isoplanatic angle, θ_0, given and approximated by,

$$\theta_0=\left[2.91k^2\sec^{8/3}\zeta\int_{\text{Path}}C_n^2(z)z^{5/3}dz\right]^{-3/5}\approx0.4125r_0 \tag{26}$$

and is expressed in milli-arcseconds. The isoplanatic angle describes the maximum angular difference between the paths of two objects in which they should traverse via the same atmosphere. This is illustrated in Fig. 4.

It is also important to remember that the atmosphere is a statistically described random medium that has temporal dependence as well as spatial dependence. One common simplification is to assume that the wind causes the majority of the distortions, temporally. The length of time in which the atmosphere will remain roughly static is represented by τ_0 and is approximated by,

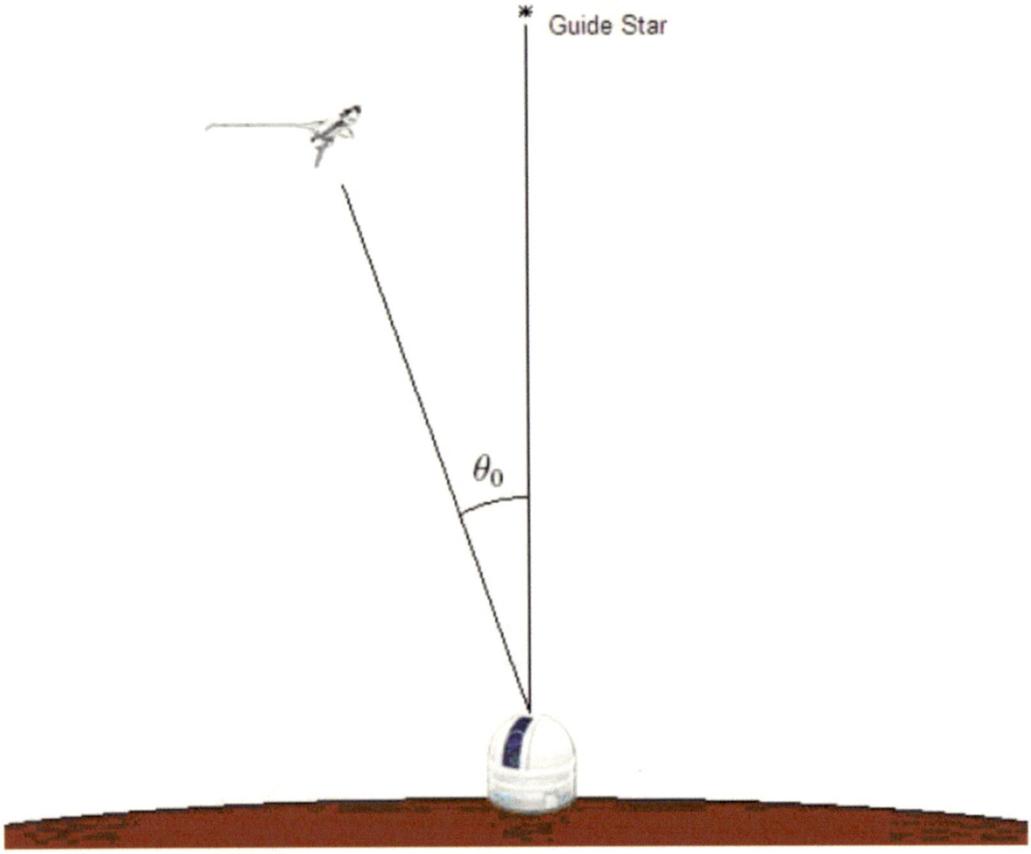

Fig. 4. Illustration of isoplanatic angle

Zernike Mode	**Zernike-Kolmogorov residual error**
Tip	$\Delta_1 = 1.0299\ (D/r_0)^{5/3}$
Tilt	$\Delta_2 = 0.5820\ (D/r_0)^{5/3}$
Focus	$\Delta_3 = 0.1340\ (D/r_0)^{5/3}$
Astigmatism X	$\Delta_4 = 0.0111\ (D/r_0)^{5/3}$
Astigmatism Y	$\Delta_5 = 0.0880\ (D/r_0)^{5/3}$
Coma X	$\Delta_6 = 0.0648\ (D/r_0)^{5/3}$
Coma Y	$\Delta_7 = 0.0587\ (D/r_0)^{5/3}$
Trefoil X	$\Delta_8 = 0.0525\ (D/r_0)^{5/3}$
Trefoil Y	$\Delta_9 = 0.0463\ (D/r_0)^{5/3}$
Spherical	$\Delta_{10} = 0.0401\ (D/r_0)^{5/3}$
Secondary Astigmatism X	$\Delta_{11} = 0.0377\ (D/r_0)^{5/3}$
Secondary Astigmatism Y	$\Delta_{12} = 0.0352\ (D/r_0)^{5/3}$
Higher orders ($J > 12$)	$\Delta_J = 0.2944\ J^{\sqrt{3}/2}\ (D/r_0)^{5/3}$

Table 2. Zernike-Kolmogorov residual errors, Δ_J, and their relation to D/r_0

$$\tau_0 = \left[2.91k^2 \sec\zeta \int_{\text{Path}} C_n^2(z) v_w^{5/3} dz \right]^{-3/5} \approx \frac{0.314 r_0}{v_w} \left(\frac{D}{r_0} \right)^{1/6} \tag{27}$$

where v_w is the average wind speed at ground level, and D is the telescope aperture. The three parameters r_0, θ_0, and τ_0 are required to know the limitations and capabilities of a particular site in terms of being able to image objects through the atmosphere.

To make a realization of a wavefront after being distorted by the Earth's atmosphere, Fried derived Zernike-Kolmogorov residual errors (Fried, 1965, Noll, 1976, Hardy, 1998). The a_i's in Equation (17) are calculated from the Zernike-Kolmogorov residual errors, Δ_J, measured through many experimental procedures and calcutated by Fried (Fried, 1965) and by Noll (Noll, 1976) and are given in Table 2. Thus, a realization of atmospheric turbulence can be simulated for different severities of turbulence and for different apertures.

2.6 Frozen Seeing model of atmospheric turbulence

Time dependence of atmospheric turbulence is very complex to simulate and even harder to generate in a laboratory environment. One common and widely-accepted method of simulating temporal effects of atmospheric turbulence is by the use of Frozen Seeing, also known as the Taylor approximation (Roggemann & Welsh, 1996). This approximation assumes that given a realization of a large portion of atmosphere, it drifts across the aperture of interest with a constant velocity determined by local wind conditions, but without any other change, whatsoever (Roddier, 1999). This technique has proved to be a good approximation given the limited capabilities of simulating accurate turbulence conditions in a laboratory environment. For example, a large holographic phase screen can be generated and may be simply moved across an aperture and measurements can then be made. A sample realization of atmospheric turbulence with a ratio of $D/r_0 = 2.25$ can be seen in Fig. 5.

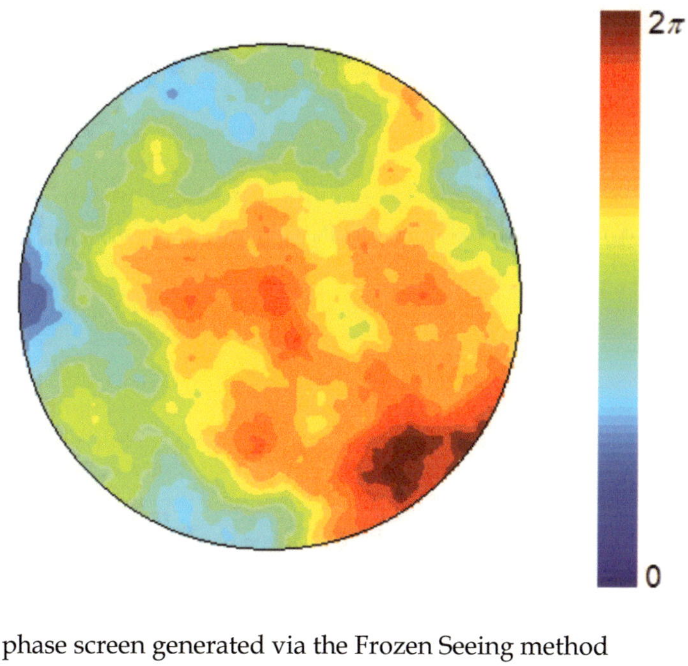

Fig. 5. A sample phase screen generated via the Frozen Seeing method

3. New method of generating atmospheric turbulence with temporal dependence

In this next section, a new method of generating atmospheric turbulence is introduced. This method takes into account the temporal and spatial effects of simulating atmospheric turbulence with the thought in mind of being able to use this method in a laboratory with a LC SLM. Some advantages of this method include far less computational constraints than using the Frozen Seeing model in software. In addition, the use of Karhunen-Loeve polynomials is introduced rather than using Zernike polynomials, as they are a statistically independent set of orthonormal polynomials.

3.1 Karhunen-Loeve polynomials

Karhunen-Loeve polynomials are each a sum of Zernike polynomials, however, they have statistically independent coefficients (Roddier, 1999). This is important due to the nature of atmospheric turbulence as described by the Kolmogorov model following Kolmogorov statistics. The Karhunen-Loeve polynomials are given by,

$$K_p(\rho,\theta)=\sum_{j=1}^{N} b_{p,j} Z_j(\rho,\theta) \tag{28}$$

where the $b_{p,j}$ matrix is calculated and given by Wang and Markey (Wang & Markey, 1978), and N is the number of Zernike orders the Karhunen-Loeve order j is represented by. Thus, to represent a wavefront, Equation (17) can be rewritten as,

$$\text{Wavefront}(\rho,\theta)=\sum_{i=1}^{M} a_i K_i(\rho,\theta) \tag{29}$$

and now the wavefront is now represented as a sum of Karhunen-Loeve polynomials with the Zernike-Kolmogorov resitual error weights in the a_i's.

3.2 Spline technique

Tatarski's model describes the phase variances to have a Gaussian random distribution (Tatarski, 1961). So, by taking Equation (29) and modifying it such that there is Gaussian random noise factored in gives,

$$\text{Wavefront}(\rho,\theta)=\sum_{i=1}^{M} X_i a_i K_i(\rho,\theta) \tag{30}$$

where X_i is the amount of noise for the ith mode based on a zero-mean unitary Gaussian random distribution and the a_i's are the amplitudes of the aberrations calculated from Zernike-Kolmogorov residual errors in Table 2.

The X_i's in Equation (30) can be generated by just using randomly generated numbers. But generating a continuous transition for the atmospheric turbulence realization temporally will require another method. The X_i's can be modified from being just random numbers to a continuous function of time for each mode. Thus, Equation (30) can be rewritten as,

$$\text{Wavefront}(\rho,\theta,t) = \sum_{i=1}^{M} X_i(t) a_i K_i(\rho,\theta) \tag{31}$$

where the $X_i(t)$ function here is generated by, first, creating a vector with a few random numbers with the zero mean unitary Gaussian distribution, as in Fig. 6.

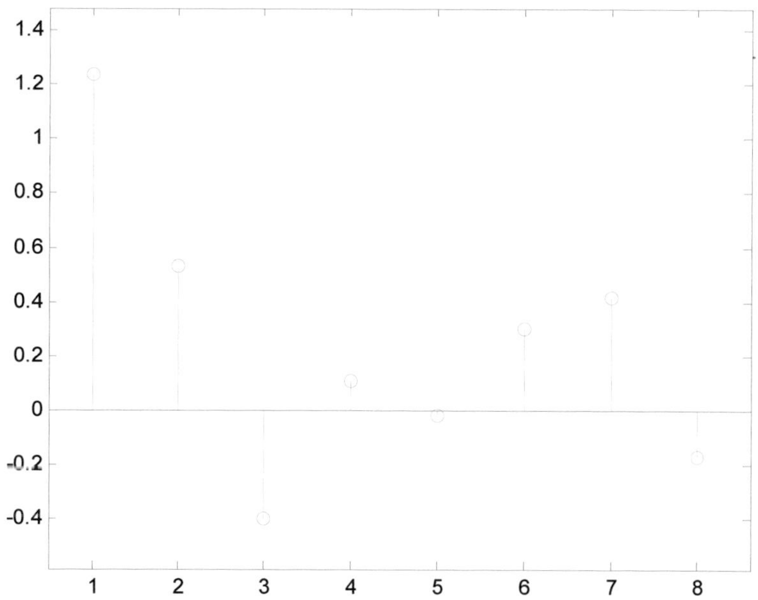

Fig. 6. Sample vector with a few random numbers with the zero mean unitary Gaussian distribution.

Next, a spline curve is fit to this vector of a few random numbers, shown in Fig. 7, and this spline curve is now the $X_i(t)$ temporal function for generating the wavefronts in the atmospheric turbulence simulation. Without this splining technique, the change between phase screens would be discontinuous and would not provide an accurate representation of the atmosphere for testing an adaptive optics system. In reality, the Earth's atmosphere is a continuous medium. With this technique, the temporal transition of the wavefronts in the atmospheric turbulence simulation is continuous and smooth. Also, in conjunction with the use of Karhunen-Loeve polynomials, a statistically independent realization of the atmosphere is preserved.

It has been shown in various experiments that the first order aberrations, ie tip and tilt, are larger in magnitude and vary less with respect to time (Born & Wolf, 1997, Wilcox, 2005). To futher validate the Spline technique, one can take this into account by using a vector of fewer numbers than for the higher order aberrations for tip and tilt and the larger magnitude is taken care of by the Zernike-Kolmogorov residual errors used with the a_i's in Equation (31). Fig. 8 illustrates the temporal difference between the transitions of tip and tilt and those of some higher order aberrations. In the next section, a comparison between this technique and the Frozen Seeing model will be analyzed and discussed.

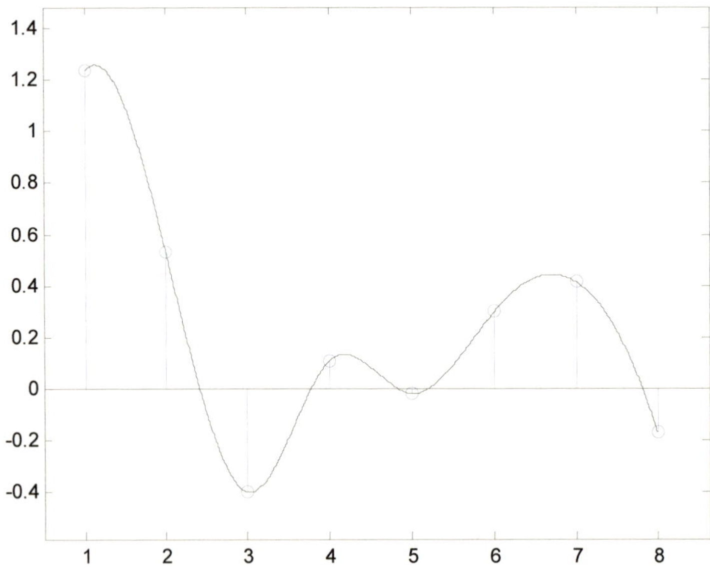

Fig. 7. Sample $X_i(t)$ temporal function generated from a vector of a few random elements.

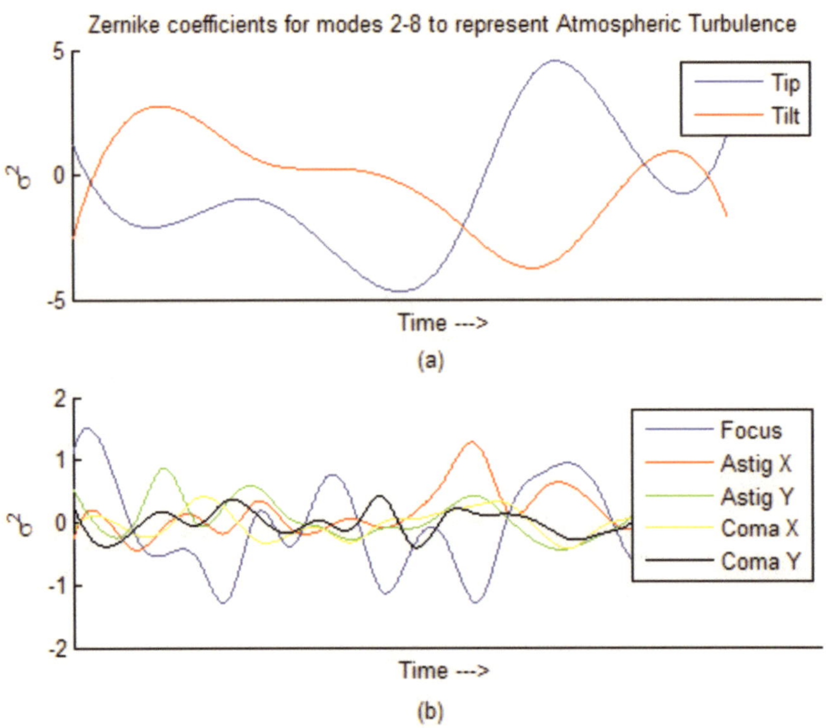

Fig. 8. Temporal and magnitude difference between (a) tip and tilt and (b) higher order aberrations.

4. Comparison of spline technique to the Frozen Seeing model

The method of generating atmospheric turbulence with temporal evolution as described in the previous section proposes various advantages compared to the Frozen Seeing model. The computational time to generate a phase screen of atmosphere of size *NxN* increases exponentially. Fig. 9 illustrates the number of seconds required to generate a phase screen of atmosphere using the Frozen Seeing model.

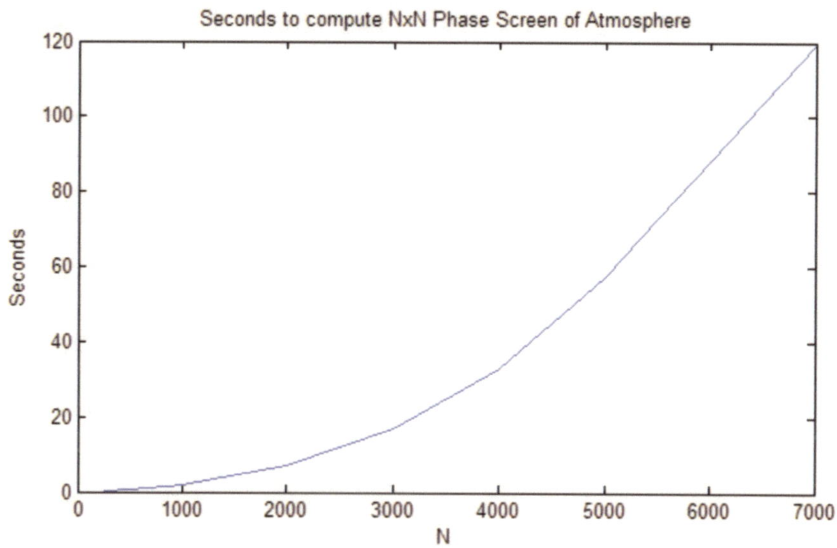

Fig. 9. Seconds to compute an *NxN* phase screen of atmosphere.

Fig. 10. Simulation of the Frozen Seeing model

Once the phase screen is generated, to be able to simulate the atmospheric turbulence on a SLM with the Frozen Seeing model, a subsection of that image of appropriate size is taken and used at the phase screen to represent the atmosphere at a moment in time. Then, that subsection is drifted across the large phase screen and that represents the next moment in time, to simulate the behavior of wind. This process is repeated until the edge of the NxN phase screen is reached, as shown in the illustration in Fig. 10.

One can clearly see that by generating atmospheric turbulence in this fasion will last for only a few seconds. Increasing the size of the large phase screen, N, would allow for a longer simulation, but the computational requirement to generate that phase screen would be a computational burden. In addition, with an NxN array, the number of bytes in that array will be N^2. This will quickly lead to an image size of dozens of megapixels which will eventually lead to a software overflow. What can also be done is rather than drifting the subsection of the large phase screen across in a stright line is drifting in a circular motion about the large phase screen, but this will lead to a simulation of atmosphere that is very repetitious. Using the Spline technique outlined in the previous section, one can realistically simulate atmospheric turbulence for a longer period of time with far less computational requirements.

To compare the Frozen Seeing model to the Spline technique outlined in the previous section, each subsection of the larger phase screen can be analyzed with a single value decomposition (SVD) of the numerical values and calculate the Zernike coefficients, a_i's, of Equation (17) with M = 24. Fig. 11 illustrates the SVD (b) of a sample wavefront from a realization of atmosphere with a D/r_0 = 2.25 (a), and the values of the SVD are listed in Table 3. Next, this process is repeated as the subsection is drifted across the large phase screen, to show the temporal transition of the a_i's.

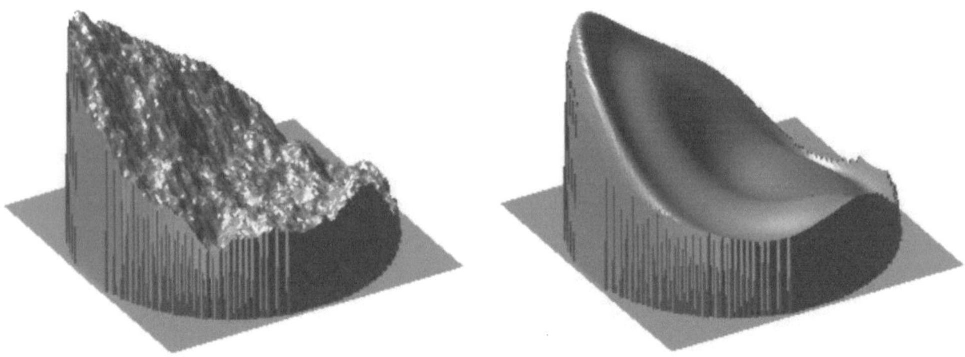

(a) (b)

Fig. 11. (a) Sample wavefront of atmosphere with D/r_0 = 2.25 and its (b) SVD of Zernike polynomials

The SVD representation in Fig. 11 (b) of the wavefront in Fig. 11 (a) has a fitted percent error of less than 2%. The a_i's progression over time are expected to change in a quasi-random fasion. It can be seen in Fig. 12 (a) and (b) that the temporal transitions of the a_i's resemble the temporal transitions as in the Spline technique, as shown in Fig. 13. Furthermore, the tilt components, a_2 and a_3, are larger in magnitude than higher orders, which is consistant with the turbulence model outlined by the Zernike-Kolmogorov residual errors and the Spline technique.

i	a_i	i	a_i
1	-3.74173	13	0.344473
2	1.3423	14	-0.09171
3	0.384392	15	0.045092
4	-0.47419	16	0.179839
5	-1.51239	17	0.003466
6	0.398136	18	-0.01032
7	-0.2772	19	-0.14574
8	-0.4476	20	0.052195
9	-0.21973	21	0.121335
10	0.249914	22	0.354472
11	-0.06698	23	-0.18282
12	0.267201	24	-0.24076

Table 3. Zernike polynomial coefficients that make up a sample representation of atmosphere with a $D/r_0 = 2.25$

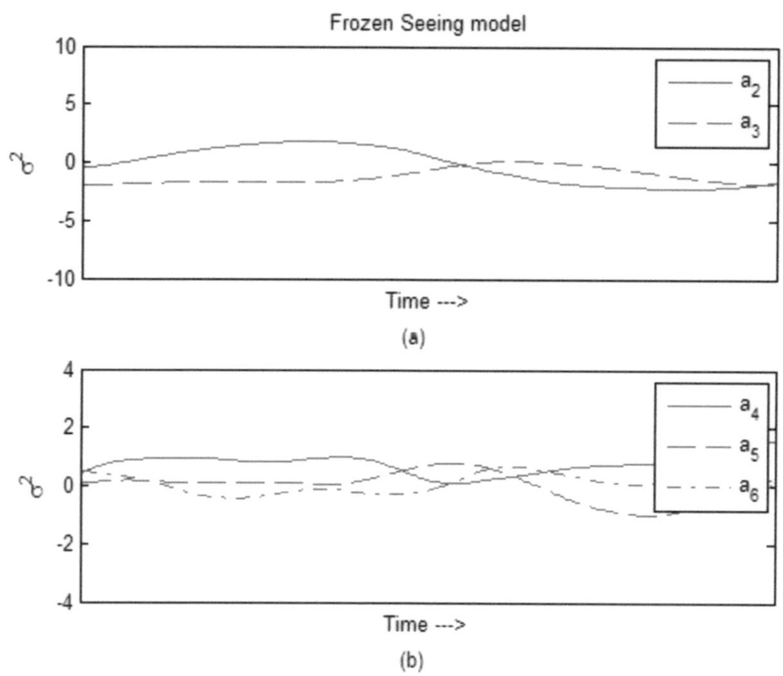

Fig. 12. The a_i's progression over time for (a) tip and tilt and (b) higher order Zernike terms in a simulation of atmospheric turbulence generated via the Frozen Seeing method.

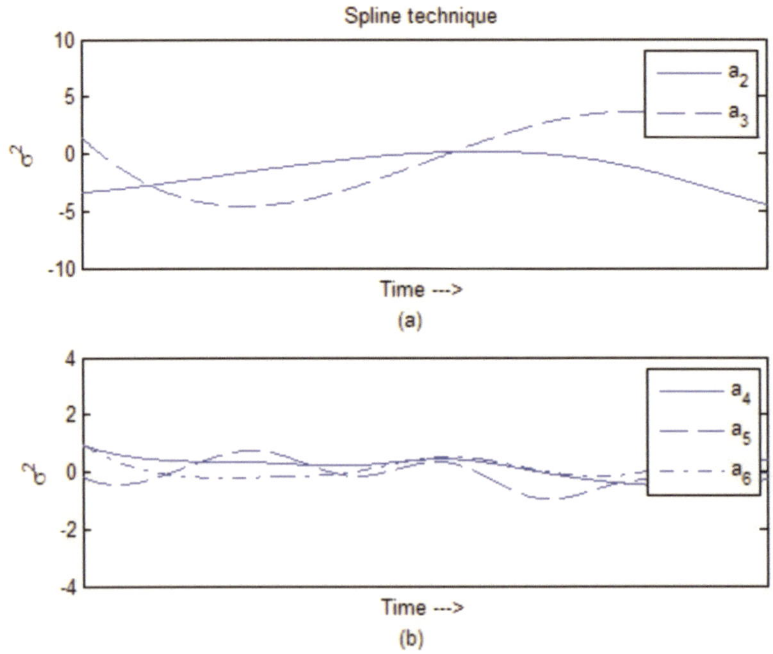

Fig. 13. The a_i's progression over time for (a) tip and tilt and (b) higher order Zernike terms in a simulation of atmospheric turbulence generated via the Spline Technique.

By visual inspection, the two methods simulate atmospheric turbulence in a similar way. A statistical measure of the similarity of these two methods can be described by the cross-correlation of the respective a_i's.

Zernike order	Aberration	Average Cross-Correlation
a_2	Tip	0.7004
a_3	Tilt	0.7471
a_4	Focus	0.6686
a_5	Astigmatism X	0.7433
a_6	Astigmatism Y	0.5937
a_7	Coma X	0.5981
a_8	Coma Y	0.6703
a_9	Trefoil X	0.6909
a_{10}	Trefoil Y	0.4878
a_{11}	Spherical	0.5910
a_{12}	Sec. Astigmatism X	0.5277

Table 4. Average cross-correlation values for each a_i

After generating and analyzing ten realizations of atmosphere from the Frozen Seeing method, the average cross-correlation values are summarized in Table 4. An overall average of these cross-correlation values is 0.6381. This shows a consistancy between the generally accepted Frozen Seeing model and the new Spline technique outlined here.

5. System performance and results

The Holoeye LC2002 SLM device used in this example is a diffractive device that can directly modulate the phase of an incoming wavefront by π radians. In order to utilize the full 2π radian phase modulation on the impinging wavefront, one can set up a Fourier Filter and use either the +1 or -1 diffractive order through the rest of the system. The graphical user interface (GUI) of the software developed for controlling this system, written in Matlab, can be seen in Fig. 14.

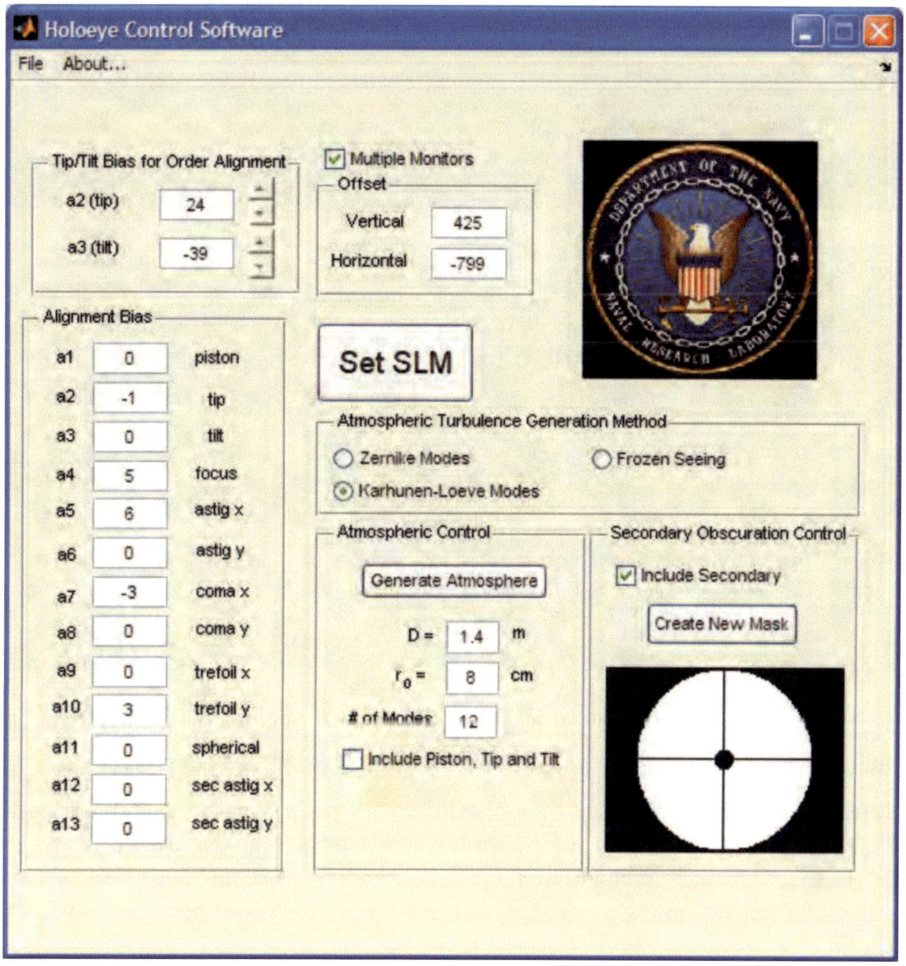

Fig. 14. Graphical user interface for Holoeye Atmospheric Turbulence System

This software is capable of controlling any LC SLM. The Holoeye LC2002 is a device with 800x600 pixels and can accept a beam of 0.82" in diameter. This software developed can set up the alignment of the diffraction orders and set alignment biases that may be entered to compensate misalignments in the optical components of the overall system for maximum performance. Different algorithms of generating turbulence can be used if desired and the parameters for the simulated telescope diameter and Fried parameter can control the severity of turbulence, as well. If desired, a secondary annular obscuration can be included in the simulation to simulate a telescope's secondary mirror.

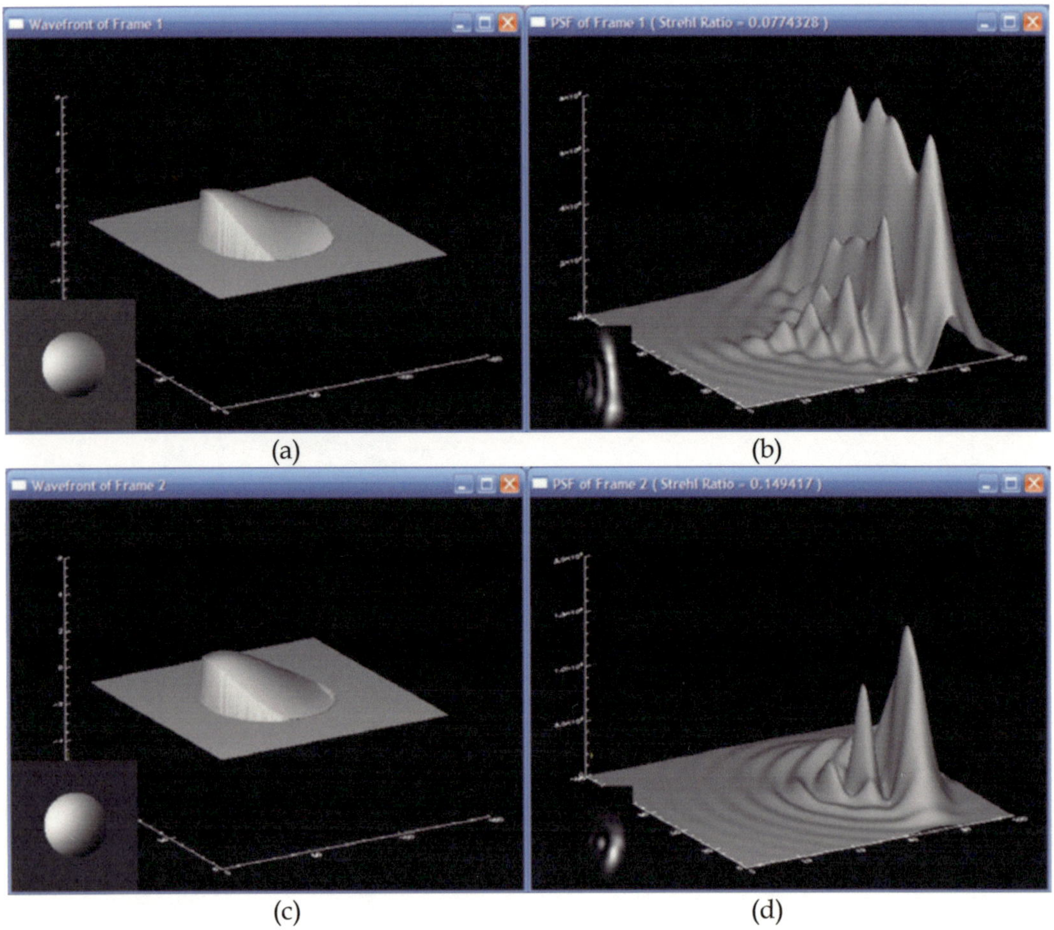

Fig. 15. Sample (a) and (c) wavefronts and (b) and (d) their corresponding PSFs due to atmospheric turbulence with a 0.4 meter telescope and an r_0 of 1 cm.

Using the GUI developed, sample atmospheric conditions have been calculated and put on the SLM and then theire PSFs are measured with an imaging camera. The simulated atmospheric turbulence was calculated for a 0.4 meter telescope with seeing conditions having an r_0 of 1 cm. Sample wavefronts from the simulation and their theoretical PSFs can be seen in Fig. 15.

The measured PSFs from the wavefronts in Fig. 15 (a) and (c) can be seen in Fig. 16 (a) and (b) and they are similar to that of the calculated PSFs in Fig. 15 (b) and (d), respectively. The 2-dimenstional cross-correlation factors between the two frames and their theoretical components are 0.9589 and 0.8638, respectively, showing that the system performs quite well and the measured and theoretical values are consistent with each other.

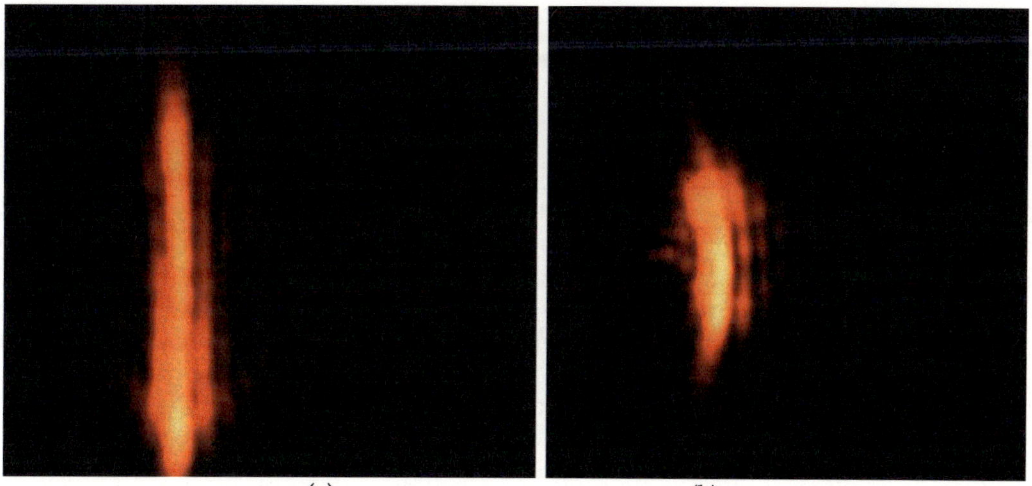

(a) (b)

Fig. 16. PSF measurements of the two sample wavefronts in the optical system

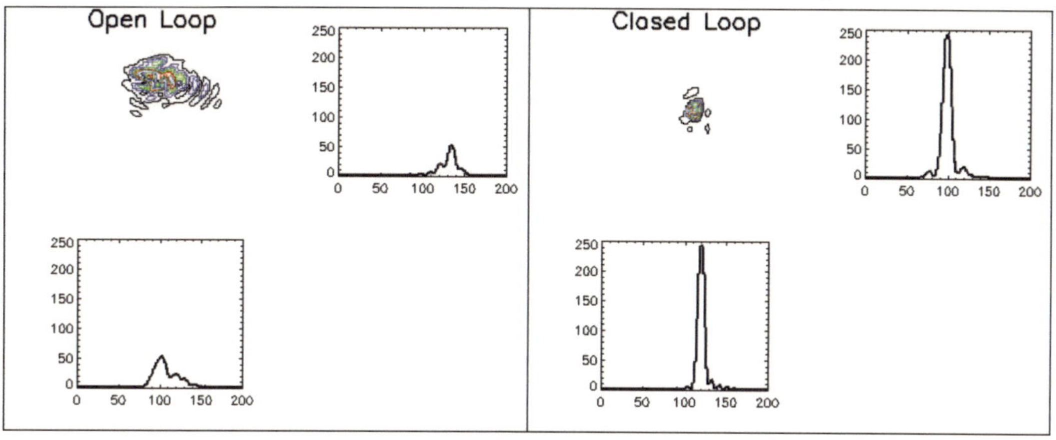

Fig. 17. PSFs of (a) open-loop and (b) closed loop frame with x and y cross section plots.

At the Naval Research Laboratory, we have developed an AO system for use in astronomical applications (Restaino, S.R., et. al., 2008). We have simulated atmospheric turbulence with the system outlined in the previous sections and caused distortions on a laser beam for our AO system to correct. Simulating fairly reasonable seeing contitions with $D/r_0 = 1.5$ lead to roughly a time-averaged Strehl ratio of 0.32. The Strehl ratio is a common way of measuring the effect that aberrations have on the imaging system (Born and Wolf, 1997). The typical definition of the Strehl ratio is the ratio of the peak intensity between the unaberrated system PSF and the system PSF with aberrations. Thus, a diffraction limited system, or a system limited only by the diffraction at the edge of the entrance pupil, will have a Strehl ratio of 1, and any aberration present in the system will cause the Strehl ratio to be less than 1.

Fig. 17 (a) and (b) show the PSFs of a frame taken during open-loop and closed-loop operation with their respective x and y cross sections. There is noticeable increase in peak intensity of the PSF and the other feature that is very important is the formation of the first ring of the Airy function in the corrected PSF. After closing the loop and allowing the AO system to begin correction, the time-averaged Strehl ratio for the simulation was increased to 0.84.

6. Summary

The method of generating atmospheric turbulence via the Spline technique is virtually the same as the Frozen Seeing method with the added feature of being far less computationally intensive on a computer system. This advantage can be exploited in the development of a software package that can drive any SLM to simulate atmospheric turbulence in almost any wavelength for any telescope diameter and adaptive optical and laser communication systems can be tested for performance evaluations. At the Naval Research Laboratory, a current system is being used with software written in the programming language Matlab and various tests are ongoing. Currently, two SLMs from Holoeye and Boulder Non-Linear Systems are being investigated and various wavelengths are being ustilized for different applications. Future work will include the investigation of other new liquid crystal devices as the field of liquid crystal technology is a very rapidly moving and growing field.

7. References

Kolmogorov, A. (1941). The local structure of turbulence in incompressible viscous fluid for very large Reynold's Numbers *Rendus de l'Acad. de Sci de l'URSS*, 30, 301-305

Tatartski, V.I. (1961). *Wave Propagation in a Turbulent Medium*, McGraw-Hill Books

Roggemann M.C. & Welsh, B. (1996). *Imaging Through Turbulence*, CRC Press LLC

Andrews, L.C. (2004). *Field Guide to Atmospheric Optics*, SPIE Press

Fried, D.L. (1965). Statistics of a Geometric Representation of Wavefront Distortion, *J. Opt. S. Am.*, 55, 1427

Noll, R.J. (1976). Zernike polynomials and atmospheric turbulence, *J. Opt. S. Am.*

Hardy, J.W. (1998). *Adaptive Optics for Astronomical Telescopes*, Ox. Ser. in Opt. & Imag. Sci.

Roddier, F. (1999). *Adaptive Optics in Astronomy*, Cambridge University Press

Wang, J.Y. & Markey, J.K. (1978). "Modal compensation of atmospheric turbulence phase distortion," *J. Opt. Soc. Am.* 68, 78-87

Wilcox, C.C. (2005). "The Design of an Adaptive Tip/Tilt Mirror for Adaptive Optics", Master's Thesis, *The University of New Mexico*

Goodman, J.W. (1968). *Introduction to Fourier Optics* McGraw Hill

Goodman, J.W. (1985).*Statistical Optics* John Wiley & Sons, Inc.

Born, M. & Wolf, E. (1997). *Principles of Optics, Sixth Ed.*, Cambridge University Press

Restaino, S.R., et. al. (2008) Adaptive Optics with MEMS and Liquid Crystals *J. Opt. A: Pure and Applied Optics* Vol 10, Num 6

Three Dimensional Temperature Distribution Analysis of Ultrasound Therapy Equipments Using Thermochromic Liquid Crystal Films

Gerardo A. López Muñoz and Gerardo. A. Valentino Orozco
Department of Bionics,
Interdisciplinary Professional Unit of Engineering and Advance Technology,
National Polytechnic Institute
Mexico

1. Introduction

As a coherent ultrasonic wave propagates through biological tissue, it is attenuated due to absorption and scattering. Absorption results from the irreversible conversion of acoustic energy to local heat, and it is the primary mode of attenuation in tissue. Ultrasound ability to interact with tissue to produce local heating has been known for a long time and nowadays ultrasound physiotherapy is widely used in health care to treat tissue injuries (Haar, 1999).

However, a large number of therapeutic equipment does not meet international standards. Rigorous quality control to verify whether physiotherapy ultrasound equipment performance is within acceptable range of acoustic intensity output plays a very important role in this context (Artho et al., 2002). Thermographic method of color analysis with thermochromic liquid crystal films is a fast and simple way to evaluate the bidimensional thermal distribution in an acoustic field produced by the ultrasound physiotherapy equipment (Jones & Carnochan, 1986).

Thermochromic liquid crystal films temperature visualization is based on the properties of some cholesteric liquid crystal materials that reflect definite colors at specific temperatures and viewing angles. These properties depend on their molecular organization; they are composed of molecular layers, where each one has a light rotation respect to the closest adjacent plane around an axis. This propriety generates a helicoidal structure that reflexes the incident white light (Ireland & Jones,2000). Correspondence between color and temperature is possible since liquid crystals emit narrow centered bands around one wave length and these bands change regularly for others colors with temperature rise (Cristoforetti et al., 1993).

Color changes in thermochromic liquid crystal films are repeatable and reversible as long as the films are not physically or chemically damaged. The time response of thermochromic liquid crystal films is approximately 10 ms. Due to the reversibility and the repeatability of color changes, thermochromic liquid crystal films can be calibrated accurately and used in this way as a temperature indicators (Stasiek & Kowalewski, 2002).

Source: New Developments in Liquid Crystals, Book edited by: Georgiy V. Tkachenko,
ISBN 978-953-307-015-5, pp. 234, November 2009, I-Tech, Vienna, Austria

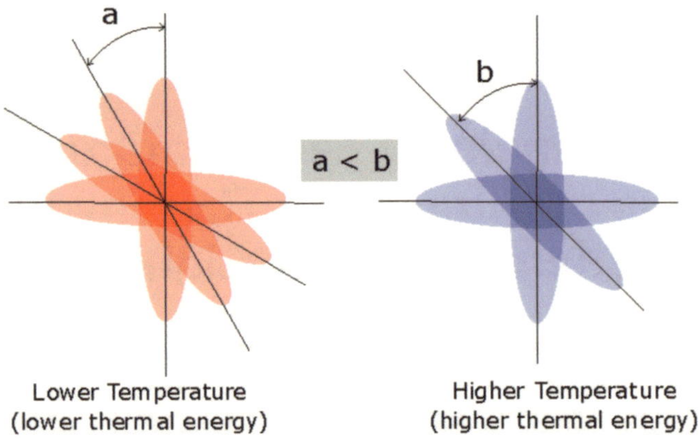

Fig. 1. Relative angles between the molecules with the change of temperature

The use of thermochromatic materials for thermal distribution mapping is based on the fact that the acoustic energy absorption by one medium and the temperature rise, according to the temporal average intensity of the ultrasound, is given by (Martin & Fernandez, 1997):

$$A = 2\alpha_a I, \tag{1}$$

where A is the energy absorption rate in the material, a_a is the absorption coefficient and I is the temporal average intensity of ultrasound.

As the intensity distribution is non-uniform in a cross-section of the ultrasonic beam (Fish, 1994), it is possible to observe changes in the intensity distribution or temperature in a plane where the beam is intercepted by the thermochromatic film (Macedo et al., 2003).

Digital image processing allows visualizing and extracting information from an image and it has many advantages over analog image processing. It allows a much wider range of algorithms to be applied to the input data, and can avoid problems such as the noise and signal distortion during processing (Castelman, 1996). Three-dimensional reconstruction is the process in which a sequence of two-dimensional images taken from a common scene is processed to create the planar projections of a volume (Gómez et al., 2006). Changes produced by the acoustic wave passing through a medium might be interpreted for obtaining the geometrical and mechanical characteristics of the medium.

Temperature qualitative measurements using thermochromic liquid crystal films have been described. They use digital image processing algorithms based on the conversion of color images to gray-scale images (Gómez et al., 2006). By using color models, an abstract mathematical model describing the way colors can be represented as tuples of numbers; it is possible to obtain a relation between the temperature rise and the color change in thermochromic liquid crystal films for a temperature qualitative measurement (Lopez et al., 2008).

This chapter describes the acquisition of a sequence of thermal images of physiotherapy ultrasound equipment with a thermographical system; a chromatic modeling of TLC films in the temperature range from 35°C to 40°C; DIP for extracting a quantity that represents color and obtaining a mathematical relation between color-temperature for a quantitative evaluation of temperature rise in ultrasonic thermal images and finally, a tree-dimensional reconstruction for a future medical parameters evaluation.

2. Methodology

A. Thermocromic Liquid Crystal Films

These materials reflect incident white light selectively due to their layered-molecular structure to show bright iridescent color. Their color change, due to temperature increases, usually from colorless to red at low temperature, through the colors of the visible spectrum to blue and colorless again (Hallcrest,2008). They are viewed normally against a black background, and the materials have good long-term stability.

Fig. 2. Color changes in TLC films

The TLC film used in this work (R35C5W, Hallcrest®) has a start temperature of 35°C and a bandwidth of 5°C (Hallcrest, 2008).

Red Start (black to red)	Green Start	Blue Start	Clear Point (blue to black)
°C	°C	°C	°C
35±0.5	36±0.5	40±0.5	49±0.5

Table 1. Color-temperature relation for the TLC film

2.1 Practical implementation

A. Chromatic Modelling

The experimental setup for obtaining the sequence of thermal images is depicted in Figure 3. A TLC film was immersed in a black painted water tank (dimensions: 70 cm x 20 cm x 20 cm). In order to avoid image distortion due to bubble formation, degasified distilled water was used. The water temperature was increased up to 45°C by using an aluminum water heater. A water pump was used to avoid the temperature gradients formation.
To get thermal images, the heater was removed and the temperature was monitored with an Hg thermometer (Brannan®, 0.1°C resolution). The sequence of images started at 40°C, with a temperature increase between acquisitions of 0.5°C, and finished at 35°C. The thermal images were viewed by using a metallic reflecting film at 45° from the TLC film and recorded by a commercial camera (Sony®, DSC-W55). A white light illumination system (color temperature of 6500° K) was used as floodlighting to control the white light intensity.

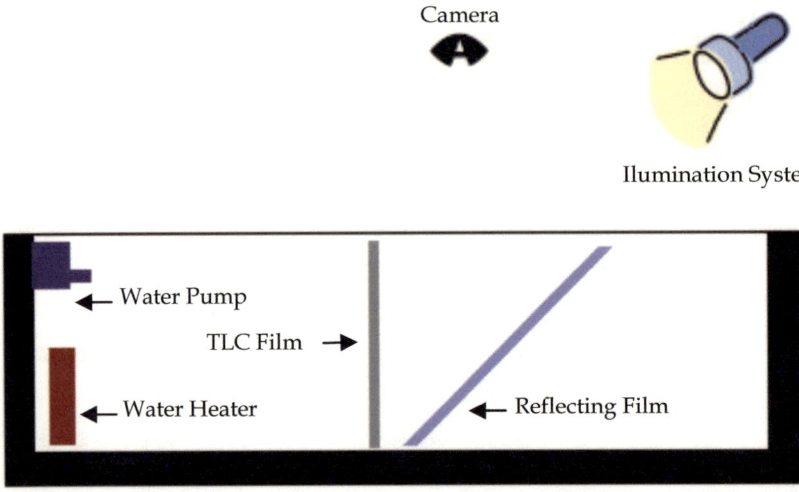

Fig. 3. Experimental setup

Numerical methods were applied to each thermal image to obtain the chromatic components in the RGB color model. The values of the components were between 0 and 255. The color was digitally represented with 1 byte. The algorithm is illustrated in Figure 4.

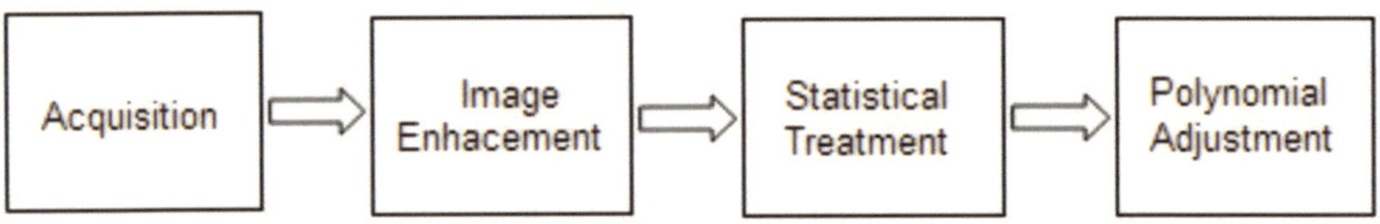

Fig. 4. Flow diagram of the algorithm for image processing

The image enhancement consists of spatial filtering to reduce noise of the images originated by electronic noise in the input device (digital camera) sensor and circuitry, or particles in the medium. For doing that a median filter was implemented.

The median filter, instead of the replacement of the pixel value with the mean values of neighboring pixels, replaces the pixel with the median of the neighboring values. The median is calculated by first sorting all the pixel values from the surrounding neighborhood into numerical order and then replacing the pixel being considered with the average pixel value. This filter is applied to each matrix (3) that represents the image in the RGB color model. The arithmetic average of the pixel values in filtered images is obtained for each chromatic component.

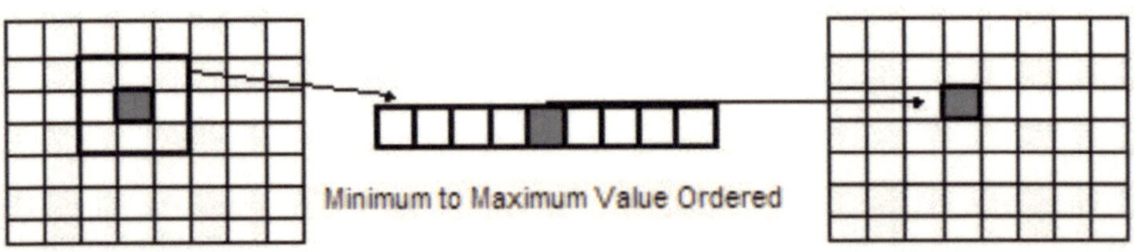

Fig. 5. Median Filter Implementation

Statistical treatment consists of obtaining the median value of the chromatic components on different thermal image sequences to describe the tendency in the color components with the change of temperature.

To obtain the polynomials that relate the change of the chromatic components and the temperature, the minimum quadratics adjustment is applied to the median of the pixel values in the color components.

B. Thermal Mapping

The experimental setup for exposing and viewing the thermal imaging system is illustrated in Figure 6.

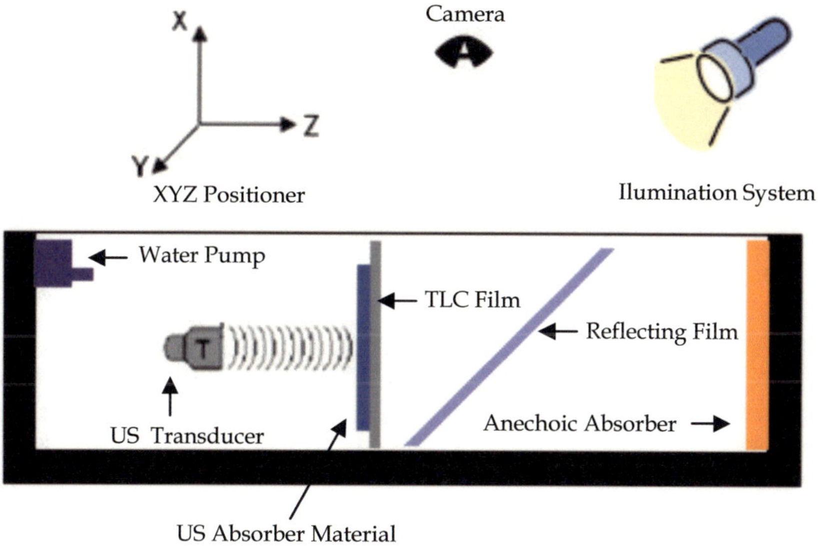

Fig. 6. Schematic of the experimental setup

Thermography system sensibility is related with the temperature rise induced by the ultrasound intensity and is determined by the absorbed ultrasonic energy (Martin & Fernandez, 1997).

The use of low reflection and high absorption attenuation coefficient materials enhances the heating of the TLC film. Polyvinyl chloride and polyurethane copolymer film with a water sonic wave reflection coefficient of 0,06 ±0,01 and a sonic attenuation coefficient of 23,5 ±1,02 dB•cm^{-1}•MHz was used (Macedo et al., 2003). These values agree with the IEC 61161 regulations for absorber materials (IEC, 1992).

A physiotherapy transducer (1MHz) and TLC film coupled to copolymer film with acoustic gel; were immersed in the black painted water tank filled with degasified distilled water. To minimize standing wave effects, ultrasound transmitted through the imaging system was absorbed and scattered by an anechoic rubber absorber positioned at the end of the system (Ham A, National Physical Laboratory) (Zequiri & Bickley, 2000). For mapping the temperature distribution on the copolymer film, the temperature of water was fixed at 34.5°C, below the start point of the TLC film. Circulation from water pump was added to system for a homogeneous water temperature.

Transducer was excited with a physiotherapy equipment (Ibramed®, Sonopulse) in continuous mode at 1 MHz and nominal ultrasonic intensity was set up to 2W•cm^{-2}.
An algorithm for processing the bi-dimensional thermal images sequence and for enhancing the visualization of the registered thermal phenomenon was developed. Figure 7 illustrates the processes of the digital image treatment.

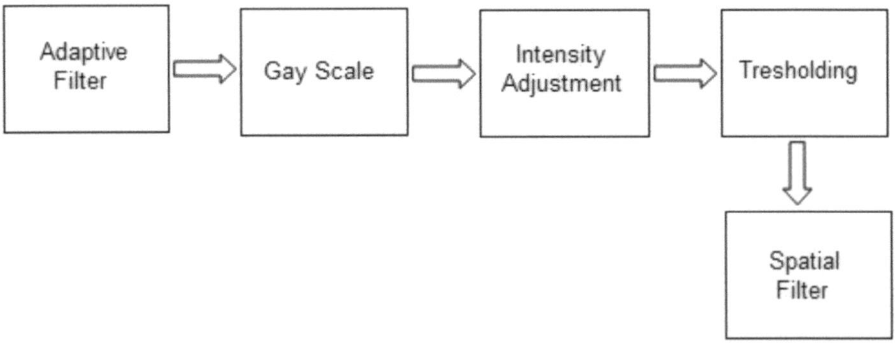

Fig. 7. Algorithm for image processing

Wiener adaptive filter implementation allows preserving edges and reducing noise in images facilitating three-dimensional reconstruction and temperature detection. Pixelwise adaptive Wiener method based on statistics estimated from a local neighborhood of each pixel was applied to the three layers that represent the image considering that the software uses RGB model for representing an image.
Adaptive filter estimates the local mean and variance around each pixel by using the following equations (Math Works, 2006):

$$\mu = \frac{1}{NM} \sum_{n_1, n_2 \in \eta} a(n_1, n_2), \tag{2}$$

$$\sigma^2 = \frac{1}{NM} \sum_{n_1, n_2 \in \eta} a^2(n_1, n_2) - \mu^2, \tag{3}$$

where μ is the local mean and σ^2 the variance in a *N*-by-*M* local neighborhood matrix for each pixel in the evaluated image. Then a pixelwise is created using the following expression:

$$b(n_1, n_2) = \mu + \frac{\sigma^2 - v^2}{\sigma^2}(a(n_1, n_2) - \mu), \tag{4}$$

where v^2 is the noise variance; the algorithm uses the average of all the local estimated noise variances.
Once the images are filtered, the image sequence is converted from color to gray-scale; three-dimensional reconstruction cannot be used with images represented with more than one matrix. The expression for the conversion is given by (Math Works, 2006):

$$I = 0.2989R + 0.5870G + 0.1140B, \tag{5}$$

where *I* is the gray intensity, *R,G,B* represent red, green and blue intensities for the transformed pixel.

Afterward, intensity adjustment for improving image contrast based on gamma correction was carried out. Gamma correction modifies middle values. Contrast in clear or dark areas is enhanced without affect neither white (255) nor black (0). Figure 8 illustrates image contrast improvement by applying gamma correction.

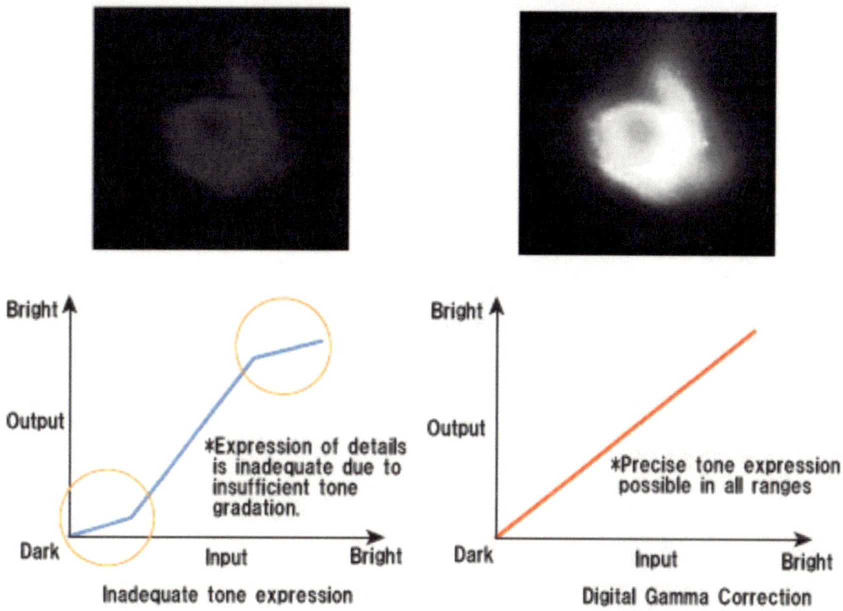

Fig. 8. Image intensity adjustment

Finally, the spatial filter consists of image thresholding and the resulting binary mask multiplied with the intensity adjusted image. Spatial filter eliminates environmental noise that distorts temperature distribution in thermal images enhancing visualization and three dimensional reconstruction.

C. Temperature Detection

Euclidean distance definition for two-point separation was used for the temperature detection taking into account color spatial point representation in the RGB color model. A numerical method for calculating Euclidean distance between evaluated pixel chromatic components in filtered thermal image and chromatic components for each one of the obtained temperatures in TLC characterization was developed. Distance is given by the following expression:

$$D(t) = \sqrt{(Rp - Rt)^2 + (Gp - Gt)^2 + (Bp - Bt)^2}\,, \qquad (6)$$

where D is the Euclidean distance, t the evaluated temperature, Rp, Gp, Bp represent chromatic component values for the evaluated pixel and Rt, Gt, Bt chromatic component values for the evaluated temperature.

Subsequently, the minimum distance was calculated and was related with the temperature for the evaluated pixel.

D. Three-Dimensional Thermal Pattern Reconstruction

Gray scale filtered thermal images for volumetric thermal distribution reconstruction are used. The goal is to create a three dimensional matrix with all the processed images and to interpolate similar gray intensity values.

Three dimensional reconstruction is divided in the following four steps:

- Volumetric data organization and processing.
- Isosurfaces creation (interpolation).
- Reconstruction configuration: color, illumination and graphic texture.
- Vision angle and perspective.

First, images in a three dimensional matrix *V(x,y,z)* were organized, where *x,y* represent the coordinate and gray intensity level for the image and *z* represents the image number of the sequence. Manipulating the matrix for just taking a part of it was possible; that implies that it is possible to obtain cuttings of the temperature pattern in any transversal section for viewing the thermal distribution inside the volume.

Thermal image edge filtering was applied for a smooth volume isosurface. Then, image outline pixel values are interpolated with their respective gray intensity level. This procedure allows creating the volume geometry and giving a specific color to each of the interpolated gray levels. Finally, perspective, texture and illumination are configured. Figure 9 illustrates edge interpolation carried out by the algorithm for reconstructing the thermal distribution.

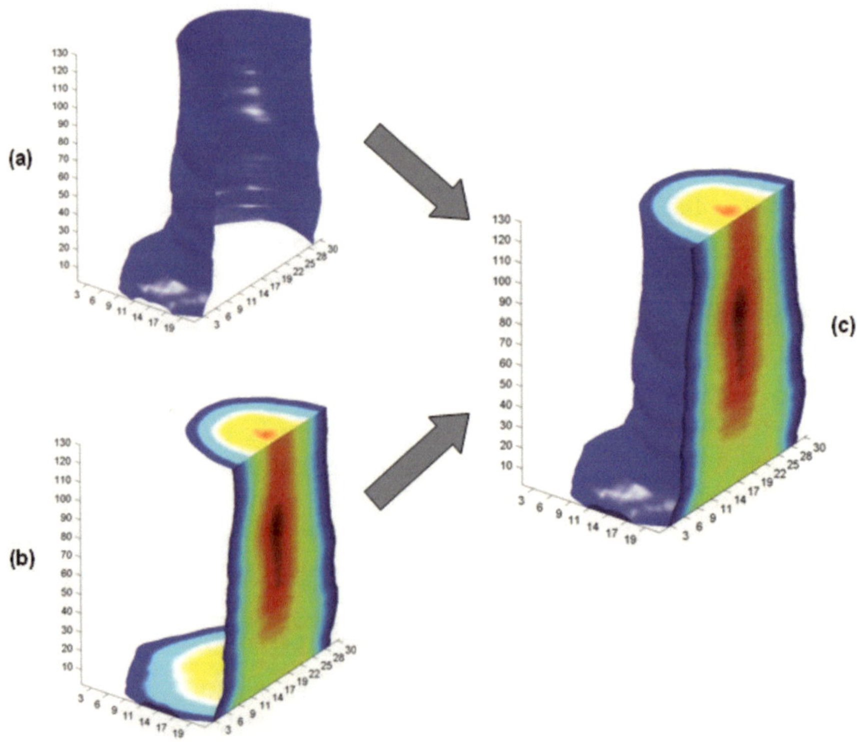

Fig. 9. (a) External edges interpolated. (b) Internal edges interpolated. (c) Reconstructed volume (Gómez et al.,2006).

3. Results

A. Chromatic Modelling

The images were taken at different times. Nine sequences of thermal images were taken. A median filter with a 15 x 15 kernel was implemented to improve the images. The arithmetic mean was applied to get the tuples of values for the chromatic components in the images. The tuples for the nine measurements were treated statistically with the median and the standard deviation.

For the polynomial adjustment, the minimum quadratics adjustment and the median were used. The polynomial degree was adjusted to the best fitting and the lowest degree considering the behavior tendency of the color component values.

Red components are higher in the range from 35°C to 36°C. Red components have a non-lineal behavior tendency (Figure 10).

The mathematical model for the chromatic red components is defined by the equation:

$$R=-4.30t^5+11.05t^4-4.75t^3-5.03t^2+2.66t+0.69, \quad (7)$$

where R is the intensity of the red components and t is the temperature defined between 35°C and 40°C.

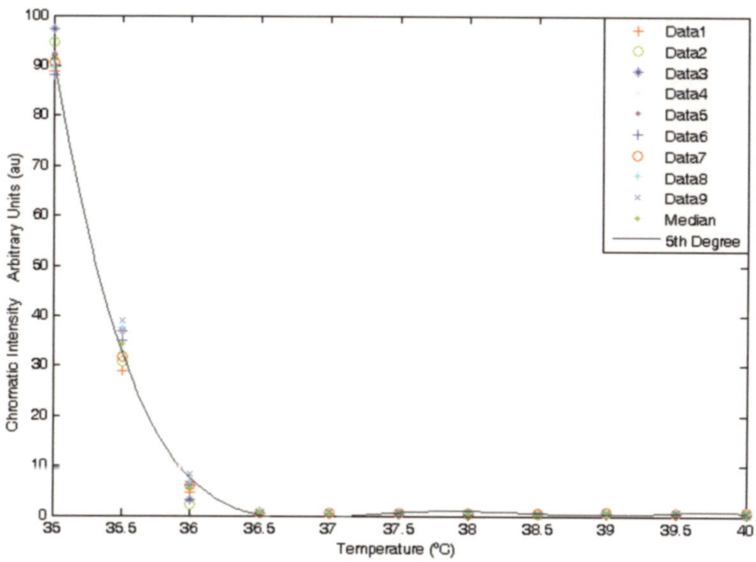

Fig. 10. Polynomial adjustment of the chromatic red component

Green components are present in the range from 35.5°C to 37°C. Green components have also a marked non-lineal behavior tendency (Figure 11).

The mathematical model for the chromatic green components is defined by the algebraic expression:

$$G=-3.12t^6+5.55t^5+1.43t^4-5.44t^3+7.37t^2-21.50t+77.3, \quad (8)$$

where G is the intensity of the green components and t is the temperature defined between 35°C and 40°C.

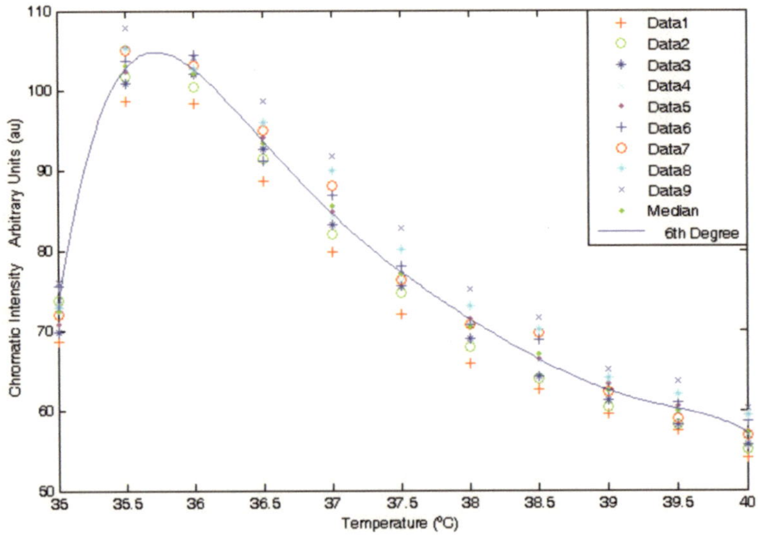

Fig. 11. Polynomial adjustment of the chromatic green component

Blue components are present in the range from 37.5°C to 40°C. Blue component has an almost-lineal behavior tendency (Figure 12).

The mathematical model for the chromatic blue component is defined by the expression:

$$B=-1.63t^4+0.41t^3-0.82t^2+31.81t+127.72, \qquad (9)$$

where B is the intensity of the blue component and t is the temperature defined between 35°C and 40°C.

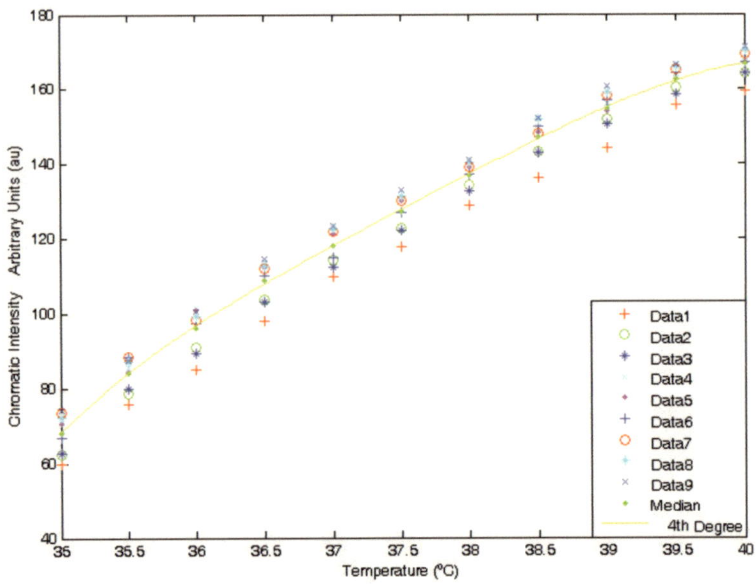

Fig. 12. Polynomial adjustment of the chromatic blue component

Temp.	Median			Standard Deviation		
°C	Red	Blue	Green	Red	Blue	Green
40	0.56	167.20	56.92	0.19	3.60	2.03
39.5	0.63	163.92	59.85	0.19	3.60	2.04
39	0.54	155.87	62.68	0.16	5.19	1.78
38.5	0.47	148.34	66.37	0.10	5.13	3.15
38	0.44	139.03	70.67	0.14	4.03	2.75
37.5	0.52	130.34	76.33	0.13	5.20	3.15
37	0.51	121.16	84.83	0.14	5.14	3.87
36.5	0.64	11.20	93.11	0.13	5.82	2.93
36	6.06	98.25	102.11	1.90	5.88	1.86
35.5	35.10	85.76	102.85	3.46	4.78	2.69
35	91.75	70.76	72.90	2.87	5.39	2.50

Table 2. Statistical treatment for the chromatic components measurements

B. Thermal Mapping

Once the ultrasound beam was switched on, the formation of a stable thermal image on the TLC film took approximately 30 s. Measurements were made at 2 mm intervals up to 80 mm. Thermal images were taken every 30 s so that the image was steady. At the end a sequence of 40 images was obtained (Figure 13).

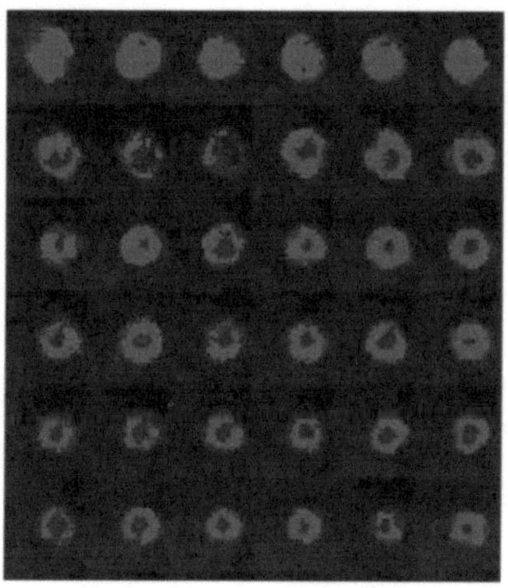

Fig. 13. Thermal image sequence

An adaptive Wiener filter with a 25 x 25 kernel was implemented to improve the images; binarization was applied with a 0.5 luminance level (Figure 14).

C. Temperature Detection

A graphic interface was developed. This graphic interface facilitates user interaction and avoids the necessity of working with the programming code. The algorithm is semiautomatic. Users have to take part in some image treatment process and parameters configuration (Figure 15).

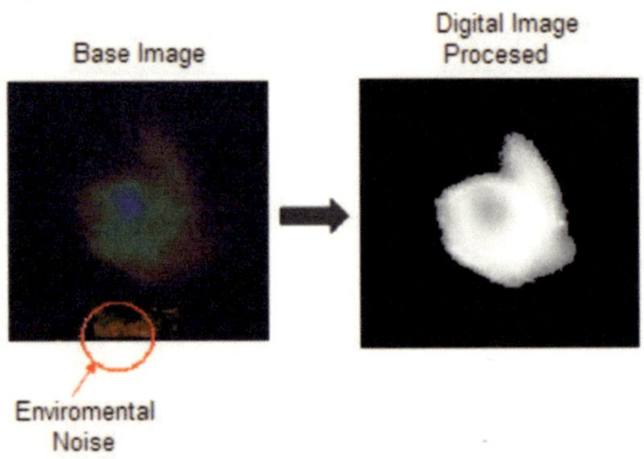

Fig. 14. Thermal image sequence

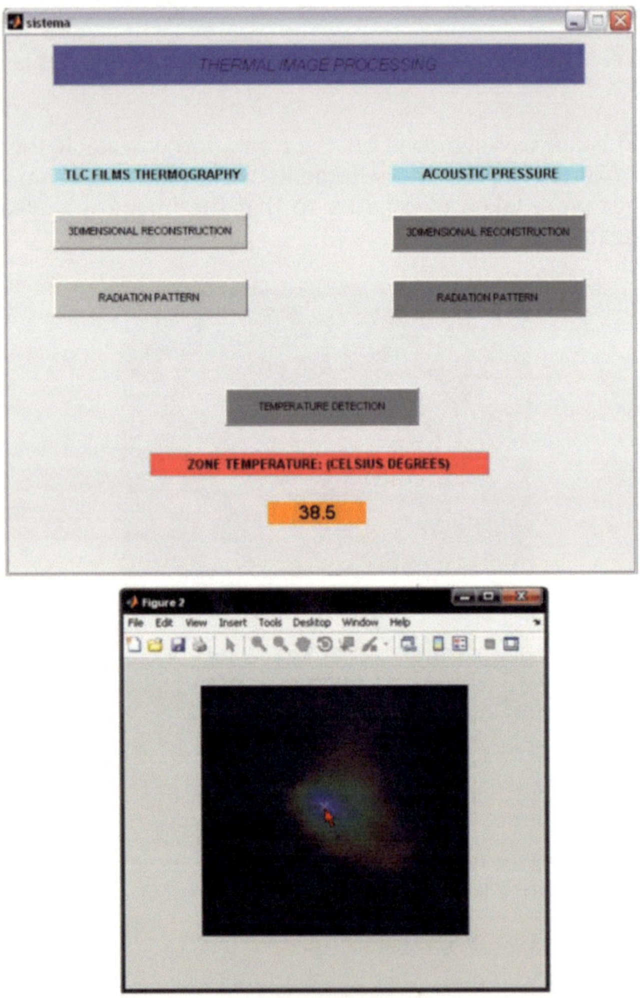

Fig. 15. Temperature detection with graphic interface

D. Three-Dimensional Thermal Pattern Reconstruction

For making three-dimensional reconstruction, each image of the sequence was superposed one behind the other in a three-dimensional array to interpolate pixel values corresponding to the edge of each image with the same level of intensity to create an isosurface. Color map, color bar, illumination and material were configured for the scene.
Radiation pattern gives information about the non-homogeneities of acoustic intensity when they are related to the gradients of temperature in thermal images (Figure 16).

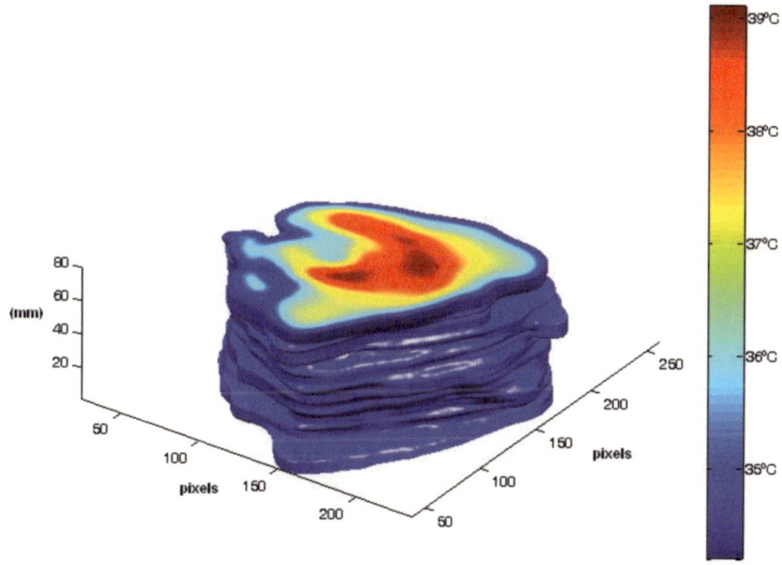

Fig. 16. Three-Dimensional temperature pattern reconstruction

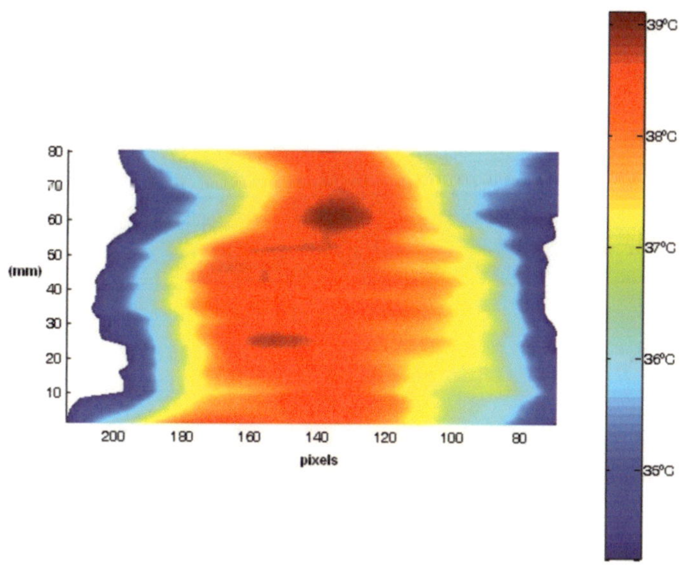

Fig. 17. Temperature pattern reconstruction.

Data of the three-dimensional (3-D) array was used for taking a transversal section for viewing thermal distribution inside the volume.
The thermal distribution pattern in the radiated copolymer shows a high temperature region distributed throughout z axis with two regions where the temperature reaches its maximum point, between 25 - 60 mm of depth (Figure 17).

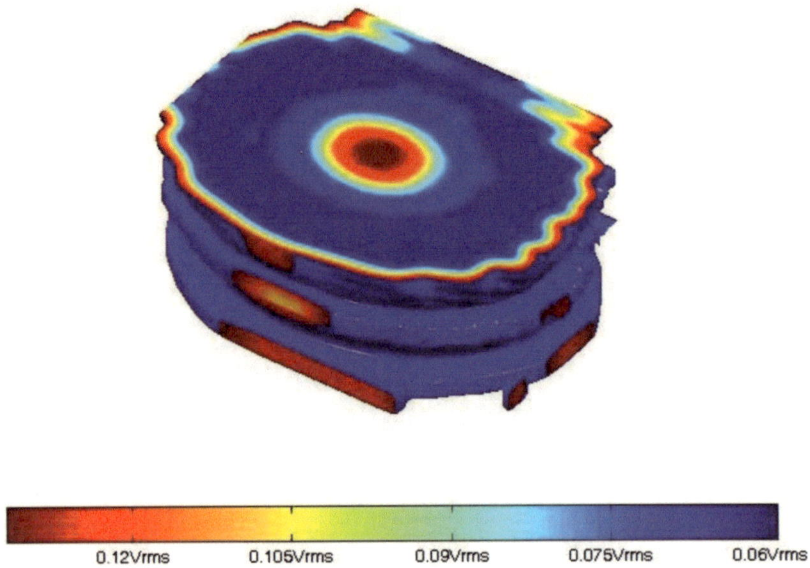

Fig. 18. Three-Dimensional acoustic pattern reconstruction

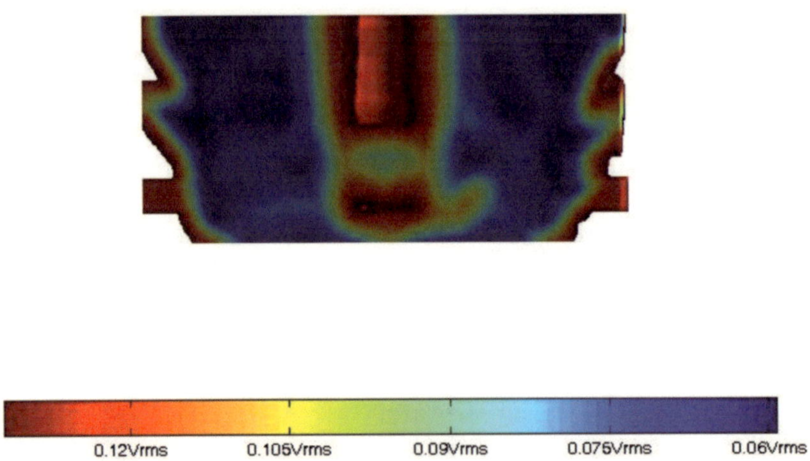

Fig. 19. Acoustic pattern reconstruction

System Validation wad done by comparing the radiation pattern obtained with the thermocromic liquid crystal films and the pressure acoustic mapping technique.This comparison was done using the acoustic field data obtained by Bazán (Bazán, 2005) for the theraphy transducer (Ibramed®, Sonopulse) @ 1MHz, along Z axis with 2mm separation between each measurement starting at Z = 0 mm and finishing at Z = 80mm.

4. Discussion

One of the main advantages of the technique based on TLC sheets is that the reflected light is within the visible spectrum allowing viewing and photography by using a rather conventional camera.

Chromatic modeling. DIP was used for a qualitative measurement of thermal distribution in gray-scale images (Gómez et al., 2006). In this work a sequence of thermal images between 30-35 Celsius degrees were obtained and a DIP algorithm was implemented for the mathematically modeling in the TLC film using the RGB color model. The use of a color model allows obtaining a mathematic relation between temperature and an abstract representation, as color is considered, to get a temperature quantitative measurement.

Thermal mapping. This method applied to ultrasound fields was proposed by Cook and Werchan (Cook & Werchan, 1971) improved by Macedo (Macedo et al., 2003) and Gómez (Gómez et al., 2006). Coupling polyvinyl chloride and polyurethane copolymer material to TLC film improves ultrasound absorption, delays thermal equilibrium and enhances system sensibility.

These techniques are dependent of the following factors:

Surrounding illumination - Uncontrolled light sources and possible reflections from surfaces in the TLC may cause errors in the measured temperatures; an illumination analysis is required in other applications of the method.

Viewing angle - The TLC colors depend on the angles between the camera and the TLC surface.

Controlled temperature conditions - It is important to control the temperature to achieve desired accuracy and to avoid the temperature gradients formation. Hysteresis is one of the problems when the crystals are heated above their clearing point temperature (Baknaria & Anderson, 2002).

Digital resolution - The minimal number of pixels needed per unit length of the TLC surface depends on the magnitude of the temperature gradient: locations with large gradients needing more pixels to achieve a desired accuracy. There is a possible algorithm failure when the thermal image resolution is increased. Time of processing also increases when resolution increases.

Temperature detection. A classifier algorithm for temperature detection was developed by using the RGB color model and the chromatic modeling by relating TLC film temperatures with colors and calculating distance separation between each temperature and the evaluated pixel in thermal image. Minimum separation distance was taken for estimating temperature in the evaluated pixel.

Three-dimensional thermal pattern reconstruction. The three-dimensional reconstruction of ultrasound thermal beamshape gives the possibility to know the radiation pattern (temperature distribution) and penetration depth which are important medical parameters for ultrasound physiotherapy treatment.

5. Conclusion

A thermographic technique based on TLC films to obtain two dimensional thermal images was developed; the practical implementation of the method is simple and low cost, it is a suitable way for chromatic and ultrasound therapy transducers characterization with TLC films.

DIP allows image improvement by reducing noise in thermal images and the possibility to manipulate images as a mathematical representation for an efficient obtainment of the required information in an image. Adaptive filter Wiener implementation allows reducing noise selectively enhancing definition in thermal images and tree-dimensional reconstruction.

The color change in TLC films is related with the chromic contribution of the components in the RGB color model and related with the changes in temperature. The statistical analysis shows the non-linear but regular change in the tendency for the RGB components. Tests with different TLC films in controlled environmental conditions are necessary trying to find a general mathematic model that describes thermochromic behavior allowing a higher resolution and extending measurement range.

An algorithm for chromatic modeling in TLC films based on RGB color model was developed obtaining a mathematical relation between color and temperature allowing a quantitative temperature measurement by using a classifier algorithm based on Euclidean distance. Neural networks are proposed for temperature detection algorithm alternative technique.

The future work is to develop tissue equivalent materials that mimic thermal and acoustic proprieties for obtaining the thermal distribution and to detect changes in beam uniformity for ultrasonic therapy equipment.

6. Acknowledgment

Authors acknowledge *the National Polytechnic Institute* UPIITA-IPN, *CONACyT (projects 68799,* 45041 and 60903), ICyTDF for their financial support. Authors also thank to *Ph.D. Arturo Vera Hernández* and *M. Elsa Veronica López Muñoz* for their technical support.

7. References

Artho, P.A.; Thyne, J.G.; Warring, B.P.; Willis, C.D.; Brisme´e, J.M.; Latman, N.S. (2002). "A Calibration Study of Therapeutic Ultrasound Units", *Physical Therapy*, v. 82, n. 3, p. 257-263.

Baknaria, S.; Anderson, A.M. (2002). "A Transient Technique for Calibrating Thermochromic Liquid Crystals: The Effects of surface Preparation, Lighting and Overheat", *ASME International Mechanical Engineering Congress ad Exposition*, Louisiana, USA.

Bazan, T.I. (2005). "Ultrasound Therapy Equipment Characterization by using Temperature and Acoustic Pressure Measurement Techniques", *Master's Degree Thesis*, CINVESTAV-IPN, Electrical Engineer, Mexico.

Castleman, K.(1996). *Digital Image Processing*, Eds. Prentice Hall, New Jersey, USA.

Cook, B.D. & Werchan, R.E. (1971) "Mapping Ultrasonic Fields with Cholesteric Liquid Crystals" *Ultrasonics*, v. 9, n. 2, pp. 101-102.

Cristoforetti, L.; Pontalti, R.; Cescatti, L. & Antolini, R.(1993). "Quantitative Colorimetric Analysis of Liquid Crystal Films (LCF) for Phantom Dosymetry in Microwave Hyperthermia," *IEEE Trans. Biomed. Eng.*, vol. 40, no. 11, pp. 1159-1165.

Fish, P. (1994). Physics and Instrumentation of Diagnostic Medical Ultrasound, Eds. John Wiley & Sons, New York, EUA.

Gómez, W.; Pereira, W.C.A.; Leija, L.; Krüger, M.A.V. & Vera, A. (2006). "3D Thermal Mapping of Ultrasonic Beam for Ultrasonic Transducers", *XXIX National Biomedical Engineering Congress*, Guerrero, México.

Haar, G. (1999). "Therapeutic ultrasound," *European Journal of Ultrasound*, vol. 9, pp. 3–9.

Hallcrest Data Sheet Listing: *"Handbook of Thermochromic Liquid Crystal Technology" Available* http://www.hallcrest.com/downloads/randtk_TLC_Handbook.pdf Access on 5 April 2008.

Hallcrest Data Sheet Listing: *"Thermacolor Thermochromic Sheets & Films" Available* http://www.hallcrest.com/downloads/ThermacolorThermochromicSheetsFilms_SS.pdf
Access on 10 February 2008.

Image Processing Toolbox (6.0) User's Guide; The Math Works, Natick, MA, 2008.

Ireland, P.T.; Jones, T.V. (2000). "Liquid Crystal Measurement of Heat Transfer and Surface Shear Stress"; *Measurement Science and Technology*, vol. 11, no. 7, pp. 969-986.

Jones, C.H.; Carnochan, P. (1986). "Infrared Thermography and Liquid Crystal Plate Thermography"; *Physical Techniques in Clinical Hyperthermia*; J. W. Hand, J. R. James, Eds. Research Studies Press, pp. 507–547.

López, G.A.; Valentino, A.; Vera, A.; Leija, L.(2008). "Chromatic Modelling in Liquid Crystal Films for Ultrasound Thermography", *International Conference on Advances in Electronics and Micro-Electronics* ENICS 2008, Valencia, Spain.

Macedo, A. R.; Alveranga, A. V.; Pereira, W. C. A.; Machado, J. C. (2003). "Ultrasonic Beam Mapping Using the Chromothermic Proprieties of Cholesteric Liquid Crystals", *Brazilian Biomedical Engineering Magazine* vol. 19, no. 2, pp. 61-68.

Martin, J. K.; Fernández, R. (1997). "A Thermal Beam-Shape Phantom for Ultrasound Physiotherapy Transducers," *Ultrasound in Medicine and Biology*, vol. 23, no. 8, pp. 1267-1274.

Stasiek, J. A. & Kowalewski, T. A.(2002) Thermochromic liquid crystals applied for heat transfer research. *Opto-Electron*, Rev. 10, pp. 1–10.

Ultrasonic Power Measurement in Liquids in the Frequency Range 0.5 to 25 MHz (1992). *International Electrotechnical Commision*, IEC 61161.

Zequiri, B.; Bickley, J. C. (2000). "A New Anechoic Material for Medical Ultrasonic Applications," *Ultrasound in Medicine and Biology*, vol. 26, no. 3. pp. 481–485.

6

Simple Optical Methods for Measuring Optical Nonlinearities and Rotational Viscosity in Nematic Liquid Crystals

Gun Yeup Kim, and Chong Hoon Kwak
Yeungnam University
South Korea

1. Introduction

Knowing the nonlinear optical properties such as nonlinear refraction, nonlinear absorption and rotational viscosity in dye-doped nematic liquid crystals (NLCs) is crucially important for photonic applications like liquid crystal display (LCD). Versatile optical methods have been developed for measuring the nonlinear refraction and nonlinear absorption including third harmonic generation (deSouza et al., 1999), degenerate four wave mixing (Liao et al., 1998), optical Kerr gate (Gomes et al., 1996), nonlinear interferometry (Yavrian et al., 1999), ellipse rotation (Lefkiry et al., 1998), phase distortion method (Watanabe et al., 1995), among others. Although these techniques are sensitive, they require complicated experimental setup to measure the optical nonlinearities. The Z-scan technique (Sheik-Bahae et al., 1990), is a simple and highly sensitive single beam method that uses the principle of spatial beam distortion to measure both the sign and the magnitude of the optical nonlinearity. Nowadays, the Z scan method has been most popularly used.

To determine the rotational viscosity coefficient, various experimental methods have been extensively investigated which include: the light scattering measurement (Durand et al., 1995), capillary flow measurement (Gähwiller, 1971), rotating magnetic field technique (Prost & Gasparoux, 1971), shear-waves reflectance technique (Kneppe et al., 1982; Martinoty & Candau, 1971), nuclear magnetic resonance technique (Martins et al., 1986), capacitance method (Leenhouts, 1985) and transient current method (Imai et al., 1995). Among them most widely used techniques are capacitance method and transient current method. Capacitance method measures the time constant of an exponentially decaying director relaxation from a non-equilibrium to an equilibrium state in a twisted nematic cell upon switching on or off the applied voltage. Transient current method measures the peak amplitude of the transient current induced by a direct-current voltage pulse application.

In this chapter, we propose simple and accurate optical methods for determining the nonlinear refractive coefficient, the nonlinear absorption and the rotational viscosity coefficient in dye-doped nematic liquid-crystal and also present the theories developed. This chapter is organized as follows. In section 2, the theoretical backgrounds are provided for closed- and open- aperture Z-scan transmittance, knife-edge X-scan and modified Z-scan by switching on or off the applied electric field, taking into account two photon absorption. In

Source: New Developments in Liquid Crystals, Book edited by: Georgiy V. Tkachenko,
ISBN 978-953-307-015-5, pp. 234, November 2009, I-Tech, Vienna, Austria

section 3, a series of experiments are performed and the experimental results are discussed, and finally, conclusions are drawn in section 4.

2. Theory

2.1 Closed and open aperture Z-scan theory for complex optical nonlinearity

In this section, we present the optical method to determine the nonlinear refraction and nonlinear absorption by Z-scan technique. The nonlinear medium is scanned along the z-axis in the back focal region of an external lens, and the far-field on-axis (i.e., closed aperture) transmittance and the whole (i.e., open aperture) transmittance are monitored as a function of the scan distance *z*. The open aperture Z-scan transmittance is insensitive to the nonlinear refraction and solely determines the nonlinear absorption, whilst the closed aperture Z-scan transmittance is coupled with both of the nonlinear effects. Actually, both the nonlinear refraction and the nonlinear absorption are often present simultaneously in nonlinear optical materials. Nonlinear absorption is inevitably present for resonant absorption wavelength ranges as well as for transparent regions owing to multi-photon absorption when the laser beam intensity is sufficiently high or because of other nonlinear processes. For simplicity, we only concentrate on two-photon absorption (TPA). Consider the fundamental Gaussian electric field (TEM_{00} mode) of travelling in the z direction as $E(r,z)=E_o\frac{w_o}{w(z)}\exp\left[-\left(\frac{r^2}{w^2}+i\frac{kr^2}{2R}\right)\right]\exp\left[-i\varphi(z)\right]=E(0,z)\exp\left[-i\frac{kr^2}{2\tilde{q}}\right]$ where E_0 is the amplitude of the electric field at the focus (i.e., z=0), $w(z)=w_o\left(1+\tilde{z}^2\right)^{1/2}$ is the beam radius at, is the beam at z, w_0 waist radius at focus, $R(z)=z\left(1+1/\tilde{z}^2\right)$ is the radius of curvature of the wave front at z, z , $\tilde{z}=z/z_o$ is the dimensionless propagation distance, $z_o=kw_o^2/2$ is the Rayleigh diffraction length of the beam, $k=2\pi/\lambda$ is the wave number and λ is the wavelength of laser beam. Here, $\tilde{q}$ denotes the complex beam parameter that contains all the information needed to specify its characteristics in the beam propagation and is defined by $1/\tilde{q}=1/R-i\lambda/\pi w^2$. For a Kerr medium with TPA process the nonlinear refractive index and the nonlinear absorption coefficient can be written as (Sheik-Bahae et al., 1990; Kwak et al., 1999)

$$n=n_o+n_2 I \tag{1a}$$

$$\alpha=\alpha_o+\beta I \tag{1b}$$

where n_0 is the linear refractive index, n_2 is the nonlinear refractive coefficient, $I(r,z)=\varepsilon_o c n_o\left|E(r,z)\right|^2/2$ denotes the incident Gaussian laser beam intensity, and α_0 is the linear absorption coefficient and β is the two photon absorption coefficient. Solving the wave equation that describes the propagation of a Gaussian laser beam through the medium and neglecting the transverse effect, the intensity variation for two photon absorption and the nonlinear phase shift of the beam at the exit surface of the sample are given by, respectively, (Sheik-Bahae et al., 1990)

$$I_e(r,z)=\frac{I(r,z)\exp\left[-\alpha_o L\right]}{1+q(r,z)} \tag{2a}$$

$$\Delta\phi(r,z)=\frac{kn_2}{\beta}\ln\left[1+q(r,z)\right] \tag{2b}$$

where $I(r,z)=I_o\exp\left[-2r^2/w^2\right]w_o^2/w^2$ is the incident beam intensity, I_0 is the on-axis intensity at focus, $q(r,z)=\beta I(r,z)L_{eff}$ is the nonlinear absorbance of the medium, $L_{eff}=(1-\exp[-\alpha_0 L])/\alpha_o$ is the effective length of the sample, and L is the sample thickness. By combining Eq.(2a) and Eq.(2b) we obtain the complex electric field at the exit surface of the sample:

$$\begin{aligned}E_e(r,z)&=\left[\frac{I(r,z)\exp[-\alpha_o L]}{1+q(r,z)}\right]^{1/2}\exp(i\Delta\phi)\\&=E(r,z)\exp\left[i\Delta\phi\left(1+i\frac{\beta}{2kn_2}\right)\right]\exp[-\alpha_o L/2].\end{aligned} \tag{3}$$

According to the aberration-free approximation of a Gaussian beam, which requires the Gaussian beam profile be approximated as being parabolic, by expanding the exponential in the intensity and retaining only the quadratic term, the nonlinear phase shift, Eq. (2b) can be approximated as:

$$\Delta\phi(r,z)\cong-\frac{kn_2}{\beta}\frac{q_o}{1+q_o}\frac{2r^2}{w^2} \tag{4}$$

where $q_o=q(0,z)=q_\infty/\left(1+\tilde{z}^2\right)$ and $q_\infty=\beta I_o L_{eff}$. It is noted that Eq.(4) is always valid whether $|q|\geq 1$ or not. By substituting Eq.(4) into Eq.(3) and employing the complex beam parameter formulation (Kwak et al., 1999) we have finally obtain the closed aperture Z-scan transmittance of the far-field at the aperture plane, including both of the effects of nonlinear absorption and nonlinear refraction as follows:

$$T_{close}(z)=\frac{1}{1-\dfrac{4(\tilde{z}-\eta)}{\left(1+\tilde{z}^2\right)^2(1+q_0)}\Delta\Phi_0+\dfrac{4\left(1+\eta^2\right)}{\left(1+\tilde{z}^2\right)^3(1+q_0)^2}\Delta\Phi_0^2} \tag{5}$$

where $\Delta\Phi_o=kn_2 I_o L_{eff}$ is the on-axis nonlinear phase at focus and $\eta=q_\infty/2\Delta\Phi_o=\beta/2kn_2$ is the ratio of the imaginary part to the real part of the complex nonlinearity and is inversely proportional to the figure of merit (FOM), defined as $\mathrm{FOM}=n_2/\beta\lambda$ (Lenz et al., 2000). The nonlinear absorptive and refractive contributions to the closed aperture Z-scan transmittance are coupled in terms of η or FOM. When the aperture is removed, however, the Z-scan is irrelevant to beam distortion caused by nonlinear refraction and is only a function of the nonlinear absorption, as mentioned above. Hence, the nonlinear absorption coefficient can readily be determined from the open aperture Z-scan transmittance. By spatially integrating Eq.(2a) at z over all r $(0\leq r\leq\infty)$ without having to include the free space propagation process, we have the CW open aperture Z-scan transmittance as:

$$T_{open}(z)=\frac{\ln\left[1+q_o(z)\right]}{q_o(z)} \tag{6}$$

where $q_o = q_{oo}/(1+\tilde{z}^2)$ and $q_{oo} = \beta I_o L_{eff}$. Once the nonlinear absorption coefficient β or q_{oo} is unambiguously extracted from an open aperture Z-scan, one can use the closed aperture Z-scan transmittance to determine the remaining unknown coefficient n_2 or $\Delta\Phi_o$ from Eq.(5).

2.2 Knife-edge X-scan theory for nonlinear absorption

In this section, we propose an alternative optical method for determining the nonlinear absorption coefficient, so-called knife-edge X-scan method. The knife-edge scanning technique is a simple single beam method for measuring a laser beam profile such as the beam radius and the radius of curvature of the wave front (Suzaki & Tachibana, 1975). Due to its high accuracy, simple apparatus and easy to data analysis, the knife-edge scanning method has been widely used. As the knife-edge along the x-axis moves across the beam propagation direction, the beam power at the far-field gradually decreases and eventually goes to zero. For a Gaussian beam distribution, the (measured) beam power is given by integrating the Gaussian function from negative infinity to present knife-edge position and becomes the error function.

Figure 1 represents schematic diagram for the knife-edge X-scan method proposed in this work to determine the nonlinear absorption coefficient.

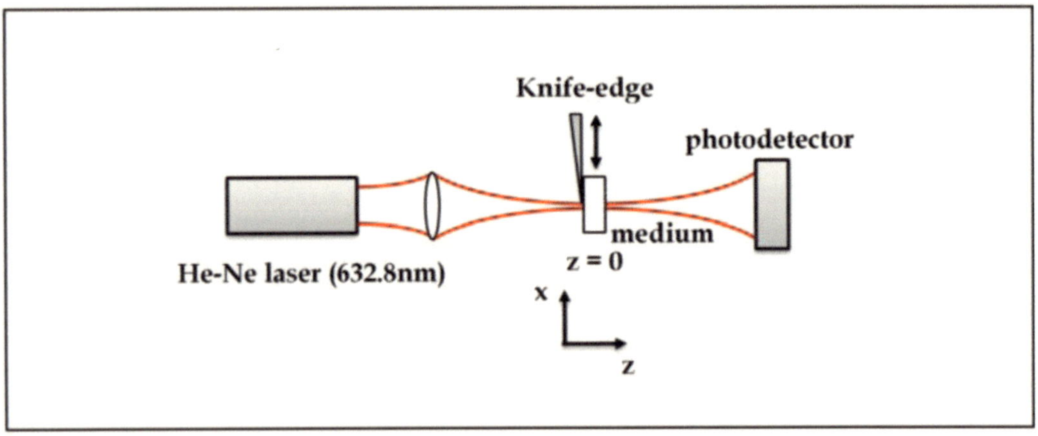

Fig. 1. Experimental setup for knife-edge X-scan technique for measuring nonlinear absorption.

The knife edge is positioned in front of a nonlinear optical medium placed at the focus (i.e., z=0) and is transversely scanned to the beam propagation axis from negative infinity to present knife-edge position. In case of two photon absorption process, the variation of beam power for a fundamental Gaussian laser beam passing through the medium can be written as

$$P(x',z) = \int_{-\infty}^{\infty}\int_{-\infty}^{x'} I_e(x,y,z)\,dx\,dy \qquad (7)$$

where $I_e(x,y,z)$ is the output beam intensity at the exit surface of the sample and is given by Eq.(2a). For a small nonlinear absorbance (i.e., $q(x,y,z)<1$), which is valid for our moderate experimental conditions, substituting Eq. (2a) into Eq.(7) with binomial expansion for $q(x,y,z)$ and integrating yields the transmitted power for knife-edge X-scan as

$$P(x',z) = P_o \sum_{m=0}^{\infty} \frac{(-q_o)^m}{2(m+1)} \left[erf\left(\sqrt{2(m+1)}\, x'/w(z)\right) + 1 \right] \tag{8}$$

where $q_o = q(r=0,z) = q_{oo}/(1+\tilde{z}^2)$, $q_{oo} = \beta I_o L_{eff}$, $P_o = \pi w_o^2 I_o / 2$ is the on-axis power at focus, and $erf(\bullet)$ is the error function. As is evident from Eq.(8), the first term (i.e., m=0) is exactly equivalent to the formula for conventional knife-edge scanning without nonlinear sample. The derivative of the transmitted power with respect to x' corresponds to a variation of incident Gaussian beam power (i.e., nonlinear Gaussian beam profile) caused by nonlinear absorption and is given by

$$\frac{dP(x',z)}{dx'} = \frac{P_o}{w(z)} \sqrt{\frac{2}{\pi}} \sum_{m=0}^{\infty} \frac{(-q_o)^m}{\sqrt{m+1}} \exp\left[-2(m+1) \frac{x'^2}{w^2(z)} \right]. \tag{9}$$

Figure 2 represents theoretical curves for normalized transmitted power and its derivative relative to knife-edge position x' for various nonlinear absorbance q_0=–0.5 and +0.5. Note that the first term (i.e., m=0) in Eq.(9) reveals one dimensional Gaussian beam power without nonlinear material (i.e., q_0=0) for knife-edge X-scan. For negative nonlinear absorption (i.e., q_0<0 or amplification), the beam radius or full width at half maximum (FWHM) decreases when compared with q_0=0, while for positive nonlinear absorption (i.e., q_0>0 or *real* absorption), the beam radius is much broaden than that of q_0=0.

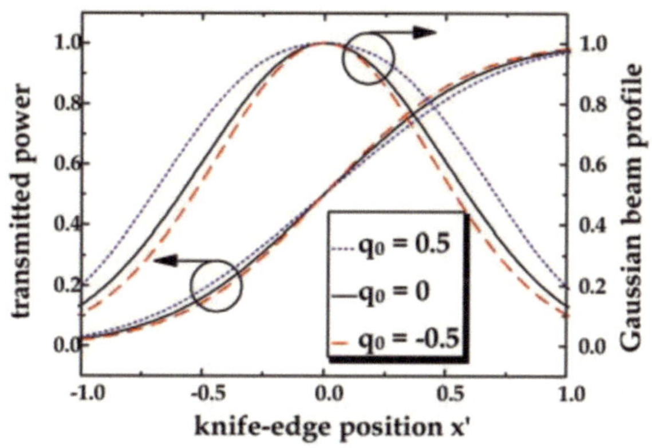

Fig. 2. Theoretical curves for normalized transmitted power and its derivative relative to knife-edge position x' (nonlinear Gaussian beam profile) for various nonlinear absorbance.

2.3 Orientational nonlinear refraction kinetics in nematic liquid crystals for rotational viscosity: Modified closed-aperture Z-scan

In this section, we will derive the kinetics of orientational refractive index change via director axis torque of nematic liquid crystals (NLCs), which is caused by a Gaussian optical field with/without an applied electric field. We also present a simple and accurate method to measure the rotational viscosity, the response time and the orientational nonlinear refraction in NLCs by modifying the closed Z-scan. Figure 3 shows the experimental setup. The optical method proposed in this work has basically the same experimental geometry used in closed aperture Z-scan. The sole distinction is that the NLC sample is placed at focus (i.e., z=0) of an external lens and is fixed at that place during the experiments, unlike Z-scan technique.

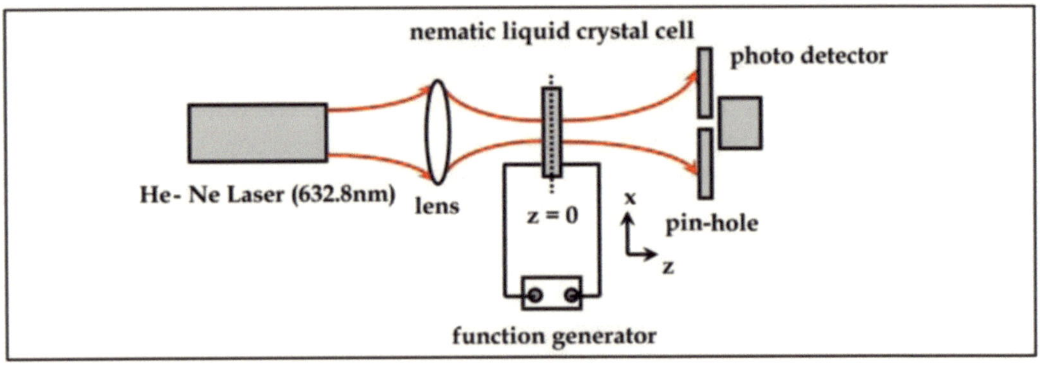

Fig. 3. Schematic diagram of the experimental setup for measuring rotational viscosity of nematic liquid crystal. A rectangular electric field with a pulse duration time t_0 is applied to the sample.

Before supplying an external electric field by a function generator, a focused optical beam is continuously illuminated to the sample, producing the optical field-induced director axis reorientation (Khoo, 1995), which gives rise to the orientational Kerr effect (OKE) and is given by $\Delta n_{OKE}(r,z=0)=n_{2,OKE}I_o\exp\left[-2r^2/w_o^2\right]$, where $n_{2,OKE}$ is the nonlinear refractive coefficient for OKE, I_0 is the on-axis intensity at focus and w_0 is the beam waist. The on-axis optical intensity of the far-field beam at the aperture plane is measured as a function of time. In this experimental situation, we adopt the closed aperture Z-scan formula, Eq.(5), just by taking z=0, which is given by

$$T_{OKE}(t)=\frac{1}{1+\dfrac{4\eta}{1+q_{\infty}}\Delta\Phi_o+\dfrac{4(1+\eta^2)}{(1+q_{\infty})^2}\Delta\Phi_o^2} \tag{10}$$

Where $\Delta\Phi_o=kn_{2,OKE}I_oL_{eff}$, $q_{\infty}=\beta I_oL_{eff}$ and $\eta=q_{\infty}/2\Delta\Phi_o$. When a rectangular electric field with a pulse duration time t_0 is applied to the sample, the field-induced director axis reorientation will be transient from a non-equilibrium state to an equilibrium state of OKE. In NLCs the field induced reorientation of the director axis is described by a torque balance equation (Khoo, 1995). We define an angle $\theta(r,t)$ as a (small) variation of the director axis orientation angle from stationary director axis angle induced by constant optical field, being spatially and temporally varying. Using the small reorientation angle approximation (i.e., $|\theta| << 1$) with the one elastic constant K, the torque balance equation is given by (Khoo, 1995; Kim et al., 2004; Kim et al., 2008)

$$\gamma_1\frac{\partial\theta(r,t)}{\partial t}=K\left(\frac{\partial^2\theta(r,t)}{\partial x^2}+\frac{\partial^2\theta(r,t)}{\partial z^2}\right)+\left|\vec{\Gamma}_E\right| \tag{11}$$

where γ_1 is the rotational viscosity coefficient, $\left|\vec{\Gamma}_E\right|=\Delta\varepsilon\varepsilon_o\left|\hat{n}\cdot\vec{E}(\hat{n}\times\vec{E})\right|/4\pi$ is the magnitude of the director axis torque, which is induced by the applied electric field $\vec{E}_{appl.}(t)$ and the optical electric field $\vec{E}_{opt.}(r,t)$, and $\hat{n}$ is a unit vector parallel to the reoriented director axis of NLCs, where $\Delta\varepsilon = \varepsilon_{\|} - \varepsilon_{\perp}$ is the dielectric anisotropy. The $\hat{n}$ and the total external electric field $\vec{E}(r,t) = \vec{E}_{appl.}(r,t) + \vec{E}_{opt.}(r)$ are expressed as $\hat{n} = (\sin\theta, 0, \cos\theta)$ and $\vec{E}(r,t) = (E_{opt.}, 0, E_{appl.})$. Then, the director axis torque is approximated as $|\vec{\Gamma}_E| = \Delta\varepsilon\varepsilon_o\left(E_{appl.}E_{opt.} - (E_{appl.}^2 - E_{opt.}^2)\theta\right)/4\pi$ for $|\theta| << 1$. In order to avoid the complexity for solving the equation, we assume that the NLC sample is placed at focus, so the Gaussian optical field can be considered as spatially uniform plane wave. The applied electric field is also spatially uniform and the variation of reorientation angle $\theta(r,t)$ is so small that $\theta(r,t)$ being considered as negligibly small for the second-order spatial derivatives, but temporally varying. Furthermore, when $E_{appl.} > E_{opt.}$, we take the direction of $\theta(t)$ is directed *toward* the direction of $\vec{E}_{appl.}(t)$, whereas for the case of when $E_{appl.} < E_{opt.}$ the direction of $\theta(t)$ is directed *away from* the direction of $\vec{E}_{appl.}(t)$. With this in mind, Eq.(11) can then be simplified as

$$\frac{d\theta(t)}{dt} + \frac{b^2-1}{\tau}\theta(t) = \frac{b}{\tau}, \quad for\ E_{appl.} > E_{opt.}\ or\ b > 1 \tag{12a}$$

$$\frac{d\theta(t)}{dt} + \frac{1-b^2}{\tau}\theta(t) = -\frac{b}{\tau}, \quad for\ E_{appl.} < E_{opt.}\ or\ b < 1 \tag{12b}$$

where $b = E_{appl.}/E_{opt.}$ and $\tau = 4\pi\gamma_1/\Delta\varepsilon\varepsilon_o E_{opt.}^2$ is a response time of NLCs, which is linearly proportional to the rotational viscosity coefficient and is inversely proportional to the optical beam intensity. Consider a rectangular electric pulse with a pulse width t_0 and an amplitude of $E_{appl.} = const.$is suddenly applied to the sample at time t=0 (i.e., $E_{appl.}(t) = E_{appl.} = const.$ for $0 < t < t_o$ otherwise $E_{appl.}(t) = 0$), while an optical field is continuously illuminated the sample from $t = -\infty$ to $t = \infty$. For the case of b>1 (i.e., $E_{appl.} > E_{opt.}$), Eqs.(12) becomes

$$\frac{d\theta(t)}{dt} + \frac{b^2-1}{\tau}\theta(t) = \frac{b}{\tau}, \quad for\ 0 < t < t_o \tag{13a}$$

$$\frac{d\theta(t)}{dt} + \frac{\theta(t)}{\tau} = 0, \quad for\ t > t_o. \tag{13b}$$

Eqs.(13) can be readily solved by using the boundary conditions of which $\theta(t)$ is continuous at $t = t_o$ and $\theta(t=0) = 0$, and the solution is given by

$$\theta(t < t_o) = \frac{b}{b^2-1}\left(1 - \exp\left[-\frac{b^2-1}{\tau}t\right]\right), \quad for\ 0 < t < t_o \tag{14a}$$

$$\theta(t \geq t_o) = \frac{b}{b^2-1}\left(1 - \exp\left[-\frac{b^2-1}{\tau}t_o\right]\right)\exp\left[-\frac{t-t_o}{\tau}\right] \quad for\ t > t_o. \tag{14b}$$

Similary, for the case of $b < 1$ (i.e., $E_{appl.} < E_{opt.}$), Eqs.(12) becomes

$$\frac{d\theta(t)}{dt}+\frac{1-b^2}{\tau}\theta(t)=-\frac{b}{\tau},\ for\ 0<t<t_o \tag{15a}$$

$$\frac{d\theta(t)}{dt}+\frac{\theta(t)}{\tau}=0,\quad for\ t>t_o. \tag{15b}$$

The solution to Eqs.(15) is given by

$$\theta(t<t_o)=-\frac{b}{1-b^2}\left(1-\exp\left[-\frac{1-b^2}{\tau}t\right]\right),\ for\ 0<t<t_o \tag{16a}$$

$$\theta(t\geq t_o)=-\frac{b}{1-b^2}\left(1-\exp\left[-\frac{1-b^2}{\tau}t_o\right]\right)\exp\left[-\frac{t-t_o}{\tau}\right]\quad for\ t>t_o. \tag{16b}$$

Since the orientational refractive index is proportional to $\theta(t)$ (Khoo, 1995), defining the transient orientational nonlinear refractive index as $\Delta n_{trans.}(t)\equiv\delta n_{trans.}\theta(t)$ where $\delta n_{trans.}$ is a proportional constant, mainly depending on the dielectric anisotropy, then the total orientational nonlinear refraction consists of the transient contribution, $\Delta n_{trans.}(t)$ owing to the transient electric field and the stationary contribution, Δn_{OKE} due to the constant optical field. Therefore, the total nonlinear phase shift $\Delta\Phi_o(t)$, experienced by the optical beam in travelling the NLC sample is given by $\Delta\Phi_o(t)=k\left[n_{2,OKE}I_o+\delta n_{trans.}\theta(t)\right]L_{eff}$, where $L_{eff}=(1-\exp[-\alpha_0 L])/\alpha_0$. Figure 4 represents the theoretical predictions of the transient optical transmittance of Eq.(10) and the reorientation angle of the director axis of Eqs.(14) in NLC sample. In this simulation, we use the following parameters: $\Delta\varepsilon=10$, $E_{opt}=0.1\text{V}/\mu\text{m}$ $E_{appl}=0.5\text{V}/\mu\text{m}$, $b=5$, $q_{\infty}=0.5$, $\eta=10^{-2}$, $\gamma_1=10^{-1}$ Pa·s and $t_o=100ms$, respectively.

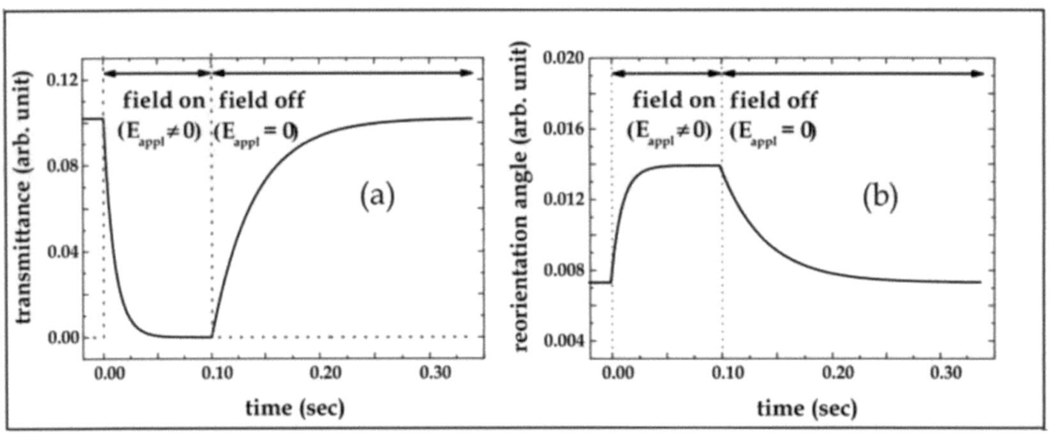

Fig. 4. Theoretical curves of (a) normalized transmittance and (b) reorientation angle as a function of time.

3. Experiments and discussions

3.1 Sample preparation of nematic liquid crystals cell

We fabricated porphyrin:Zn-doped nematic liquid crystal (NLC) cells filled by capillary phenomenon between two transparent indium-tin-oxide coated glass substrates with 20 μm thick beads as a spacer. Two glass substrates were assembled by UV bond and then filled inside of cells with porphyrin:Zn-doped nematic liquid crystal for various concentrations of dye (0, 0.006, 0.13, 0.50*wt%*) . The liquid crystal used was the eutectic liquid crystal mixture, commercially known as E7 (Merck Ltd.), which has a positive dielectric anisotropy $\Delta\varepsilon$=13.8, the elastic constants $K_{11} = 1.11\times10^{-11} N$ and $K_{33} = 1.71\times10^{-11} N$ at room temperature and wavelength λ=589nm. Zn-doped porphyrin dye [5, 10, 15, 20-tetraphenylporphyrinatozinc (ZnTPP)] was supplied by Busan National University. We made no surface treatments to NLC sample, so the director axis orientations are random before they are subject to any optical field or applied electric field.

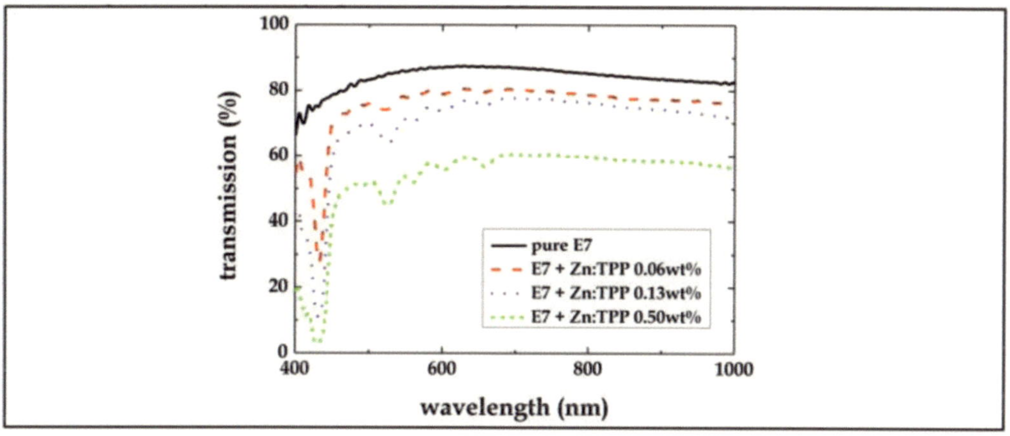

Fig. 5. Transmission spectra for various dye concentrations of porphyrin:Zn in nematic E7 liquid crystal.

The transmission spectrum for pure E7 NLC cell reveals nearly transparent of about 90 % in visible wavelength range, as shown in Fig. 5. As increasing the concentrations of dye the transmisstion spectrum is gradually decreased. It is also shown from Fig. 5 that Zn-doped porphyrin dye is photosensitive to blue-green wavelength region. The linear absorption coefficients for various dye concentrations at wavelength 632.8nm were estimated by using the Beer-Lambert law $\left(T = \exp[-\alpha_0 L]\right)$, neglecting the Fresnel reflection at surfaces of the sample as follows: $\alpha_{pure} = 0.676\times10^2\ \mathrm{cm}^{-1}$, $\alpha_{0.06wt\%} = 1.095\times10^2\ \mathrm{cm}^{-1}$, $\alpha_{0.13wt\%} = 1.315\times10^2\ \mathrm{cm}^{-1}$, and $\alpha_{0.50wt\%} = 2.607\times10^2\ \mathrm{cm}^{-1}$.

3.2 Determinations of nonlinear absorption coefficient by using knife-edge X-scan and open-aperture Z-scan

In this section, we determine the nonlinear absorption coefficients for various dye concentrations in NLC sample by means of knife-edge X-scan method and open-aperture Z-scan method and compare the experimental results quantitatively. Figure 1 represents the schematic diagram for the knife-edge X-scan method. The cw He-Ne laser of wavelength λ=632.8*nm* is used for experiments and the laser beam power is 3mW. The focal length of biconvex lens is 20cm. The whole transmitted power is measured by a photo detector during the knife-edge scan. Before conducting the knife-edge X-scan experiment, we have to determine the incident Gaussian laser beam profiles such as beam radius $w(z)$, beam waist w_0 and radius of curvature of the wave front $R(z)$ at z. Figure 6(a) shows the typical experimental results of normalized power for knife-edge scan against scan x' distance at several, which are well fitted with the theoretic formula as $P(x',z)/P_o = \left[erf\left(\sqrt{2(m+1)}\,x'/w(z)\right)+1 \right]/2$. Figure 6(b) represents the beam radius extracted from Fig.6(a) with theoretical curve, yielding the beam waist w_0=4.90μm, the on-axis intensity at focus I_0= 8.0kW/cm^2 and the optical field $E_{optc.}$=0.22V/μm.

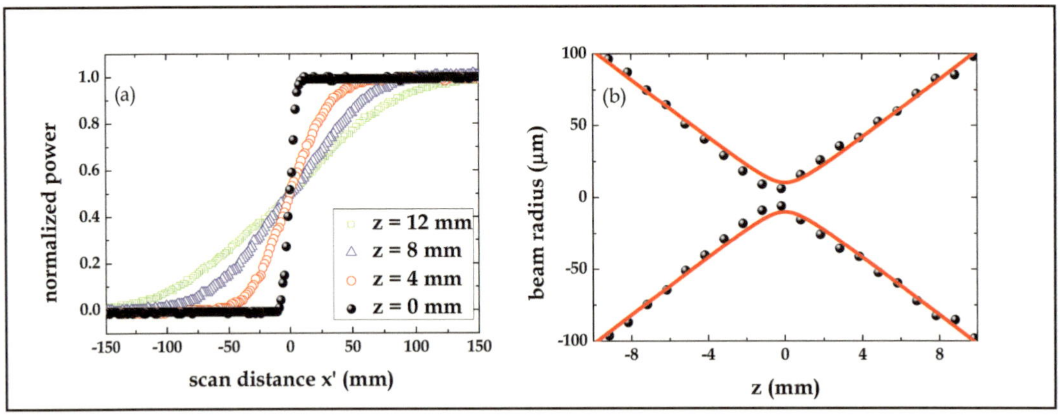

Fig. 6. (a) The measured laser beam power for knife-edge scanning vs. scan distance x', at several fixed z positions and (b) the measured Gaussian beam radius $w(z)$with theoretical curve.

To determine the nonlinear absorption coefficient of the sample we performed two kinds of experiments; one is the knife-edge X-scan in which the sample is placed at rear face of the knife-edge, as shown in Fig. 1, and the other is the conventional open-aperture Z-scan. Since the closed-aperture Z-scan transmittance is entangled with the nonlinear refraction and the nonlinear absorption, as described in Eq.(5), one should determine the nonlinear absorption coefficient before finding the nonlinear refractive coefficient. Once the nonlinear absorption coefficient β is extracted from the open aperture Z-scan or the knife-edge X-scan, one can extracts the remaining unknown nonlinear refractive coefficient n_2 from the closed aperture Z-scan transmittance. Figure 7 represents the typical experimental results of the knife-edge X-scan and the open aperture Z-scan for various dye concentrations with the theoretical predictions.

Table 1 compares the nonlinear absorption coefficient β for various dye concentrations, determined by the knife-edge X-scan method with the open aperture Z-scan method. Nonlinear absorption coefficients determined by two methods are in good agreement with each other.

Dye concentration [*wt*%]	Knife-edge X-scan β (cm/kW)	Open aperture Z-scan β (cm/kW)
0	1	3
0.06	38	33
0.13	72	75
0.50	286	300

Table 1. Comparison of knife-edge X-scan with open aperture Z-scan for determined nonlinear absorption coefficients for various dye concentrations of nematic liquid crystal.

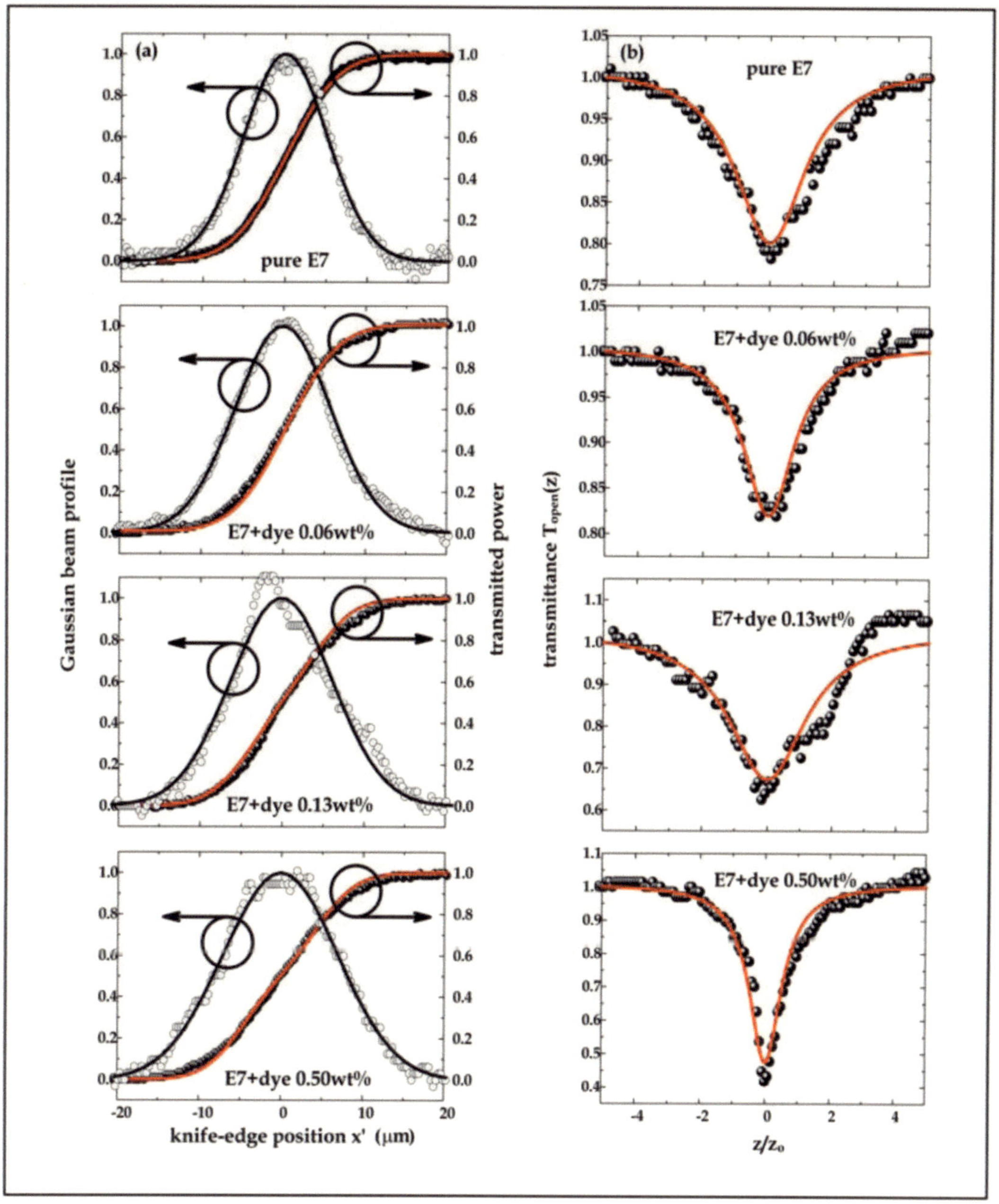

Fig. 7. (a) Knife-edge X-scan data and (b) open aperture Z-scan data for various dye concentrations of nematic liquid crystal with theoretical curves.

3.3 Determinations of nonlinear refractive coefficient by using closed aperture Z-scan

Figure 8 depicts the typical closed aperture Z-scan data, revealing a self-defocusing nature. The nonlinear refractive coefficients are determined from the best curve fitting using Eq.(5) with the known nonlinear absorption coefficients obtained from preceding subsection.

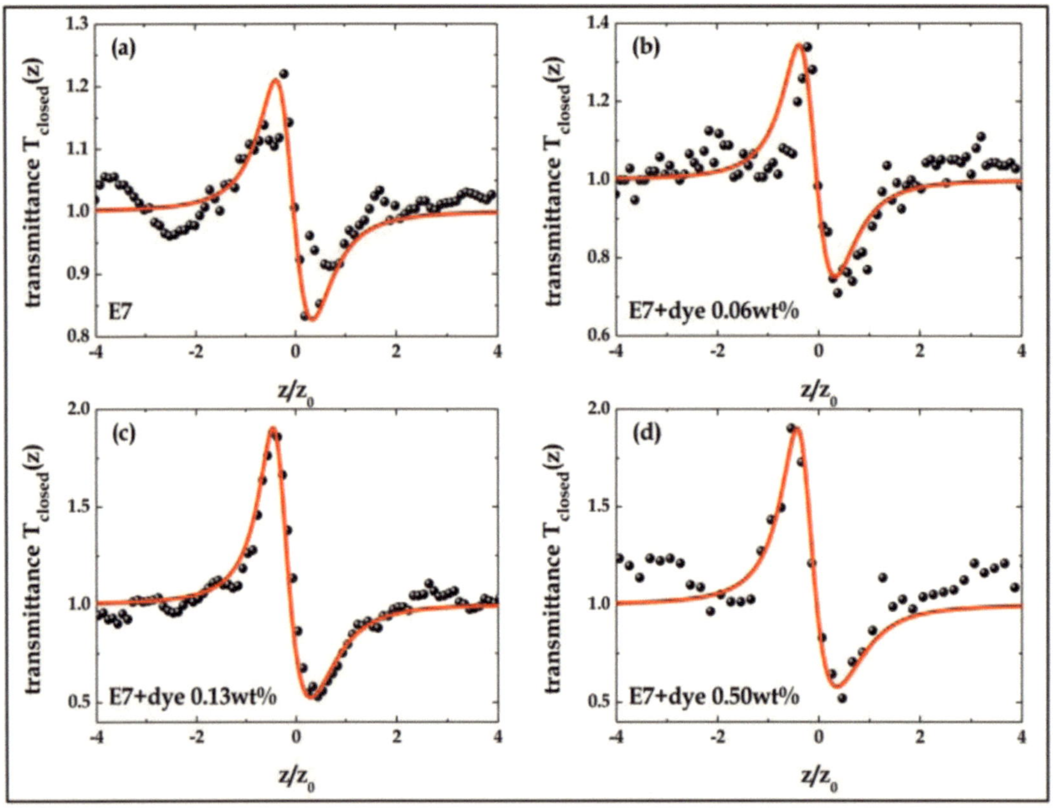

Fig. 8. The closed-aperture Z-scan transmittance data for various dye concentrations of nematic liquid crystal with theoretic curves of Eq.(5).

3.4 Determinations of rotational viscosity by modified closed-aperture Z-scan

Following the method described in subsection 2.3, we conducted the transient optical transmittance experiments by applying the rectangular electric field with the pulse duration time of $t_o = 30\ ms$ and the amplitude of $E_{appl.} = 0.5\ V/\mu m$. The NLC sample is placed at focus (i.e., z=0) of an external lens and is fixed at that place during the experiments. The optical field is $E_{opt.} = 0.22\ \text{V}/\mu\text{m}$ at focus (i.e., $z = 0$), so $b = E_{appl.} / E_{opt.} = 2.3$. Before applying the rectangular pulse field, the focused optical beam which is continuously illuminated produces a bias nonlinear refraction, which is called the optical-field induced orientational

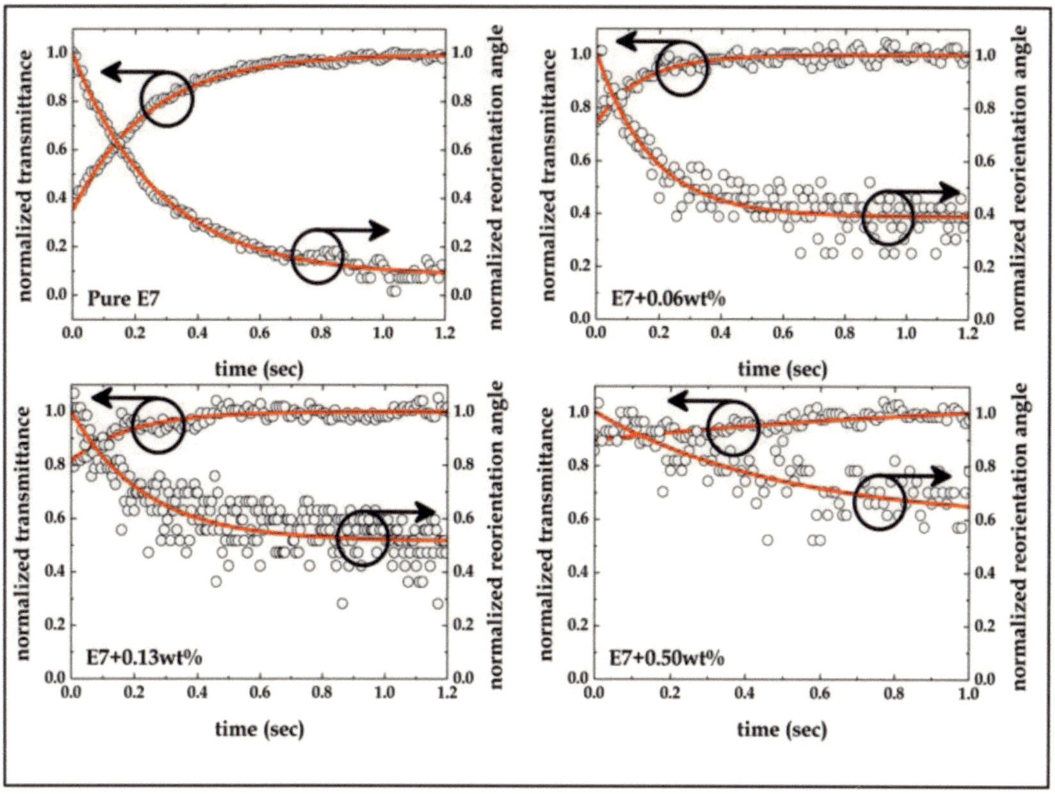

Fig. 9. The experimental data of normalized transient optical transmittance and normalized reorientation angle against time for various dye concentrations with theoretical curves.

Kerr effect (OKE), $\Delta n_{OKE} = n_{2,OKE} I_o$. Figure 9 represents the transient optical transmittances and the corresponding reorientation angles for various dye concentrations after the electric field is switched off. It should be noted that the steady state value of the reorientation angles do not go to zero, strongly implying the existence of the bias OKE caused by constant optical field, as theoretically predicted.

Table 2 compares the rotational viscosity coefficient and nonlinear refractive index coefficient for various dye concentrations, determined by the transient optical transmittance (or modified Z-scan) method with the closed aperture Z-scan method. Nonlinear refractive index coefficient, which is determined by two methods are in good agreement with each other. It reveals that the physical mechanism of the Kerr effect in NLC is caused by the optical field-induced director axis reorientation. It is also noted that the measured value of the rotational viscosity coefficient of 0.23Pa·s for pure E7 is almost the same value of 0.224Pa·s at 25°C by means of transient current method (Chen & Lee, 2007).

Dye concentration [wt%]	Rotational viscosity coefficient $\gamma_1(\mathrm{Pa \cdot s})$	Modified closed aperture Z-scan $n_{2,\,\mathrm{OKE}}(\mathrm{cm^2/MW})$	Closed aperture Z-scan $n_2(\mathrm{cm^2/MW})$
0	0.23	–0.40	–0.42
0.06	0.16	–0.85	–0.94
0.13	0.20	–1.60	–1.47
0.50	0.39	–4.27	–4.46

Table 2. Rotational viscosity coefficient and nonlinear refractive coefficient for various dye concentrations.

4. Conclusion

In this chapter, we propose simple and accurate optical methods to determine the nonlinear refraction, the nonlinear absorption and the rotational viscosity coefficient in dye-doped nematic liquid crystals, and also develop the corresponding theories. The versatile optical methods presented are as follows: (i) closed aperture Z-scan for measuring both the sign and the magnitude of the optical nonlinear refraction, taking into accounting two photon absorption, (ii) open aperture Z-scan and knife-edge X-scan for measuring the nonlinear absorption coefficient unambiguously, and (iii) modified closed aperture Z-scan (or transient optical transmittance) method by applying a rectangular electric field to measure the orientational Kerr effect (OKE) and the rotational viscosity coefficient. The measured values of optical nonlinearities and the rotational viscosity by an optical method are cross-checked by another method, showing excellent agreement with each other.

5. References

Chen, H. Y.; Lee, W. & Clark N. A. (2007). Faster electro-optical response characteristics of a carbon-nanotube-nematic suspension. *Appl. Phys. Lett.*, Vol 90, Issue 3, 033510, ISSN: 0003-6951

deSouza, P. C.; Nader, G.; Catunda, T.; Muramatsu, M. & Horowicz, R. J. (1999). Application of the Z-scan technique to a saturable photorefractive medium with the overlapped ground and excited state absorption. *Opt. Comm.*, Vol. 177, Issue 1-6, 417-423, ISSN: 0030-4018

Durand, G.; Leger, L.; Rondelez, F. & Veyssie, M. (Orsay Liquid Crystal Group) (1995). Quasielastic Rayleigh scattering in nematic liquid crystals. *Phys. Rev. Lett.*, Vol. 22, Issue 25, 1361-1363, ISSN: 0031-9007

Gomes, A. S. L.; Demenicis, L.; Petrov, D. V.; de Araujo, C. B.; de Melo, C. P. & Souto-Maior, R. (1996). Time-resolved picosecond optical nonlinearity and all-optical Kerr gate in poly (3-hexadecylthiophene). *Appl. Phys. Lett.*, Vol. 69, Issue 15, 2166-2168, ISSN : 0003-6951

Imai, M.; Naito, H.; Okuda, M. & Sugimura, A. (1995). Determination of rotational viscosity and pretilt angle in nematic liquid crystals from transient current: influence of ionic conduction. *Mol. Cryst. Liq. Cryst.*, Vol. 259, 37-46, ISSN: 1542-1406

Khoo, I. C. (1995). *Liquid crystals: Physical properties and nonlinear optical phenomena.* 121-150, John Wiley & Sons, Inc., ISBN: 0-471-30362-3, New York

Kneppe, H.; Schneider, F. & Sharma, N. K. (1982). Rotational viscosity of nematic liquid crystals. *J. Chem. Phys.* Vol. 77, Issue 6, 3203-3208, ISSN: 0021-9606 1

Kim, E. J.; Yang, H. R.; Lee, S. J.; Kim, G. Y. & Kwak, C. H. (2008). Orientational photorefractive holograms in porphyrin:Zn-doped nematic liquid crystals. *Opt. Express*, Vol. 16, Issue 22, 17329-17341, ISSN: 1094-4087

Kim, K. H.; Kim, E. J.; Lee, S. J.; Lee, J. H.; Kim, J. E. & Kwak, C. H. (2004). Effects of applied field on orientational photorefraction in porphyrin:Zn-doped nematic liquid crystals. *Appl. Phys. Lett.*, Vol. 85, Issue 3, 366-368, ISSN: 0003-6951

Kwak, C. H.; Lee, Y. L. & Kim, S. G. (1999). Analysis of asymmetric Z-scan measurement for large optical nonlinearities in an amorphous As_2S_3 thin film. *J. Opt. Soc. Am. B,* Vol. 16, Issue 4, 600-604, ISSN: 0740-3224

Leenhouts, F. (1985). Determination of the rotational viscosity from the director pattern relaxation in twisted nematic cells. *J. Appl. Phys.*, Vol. 58, Issue 6, 2180-2183, ISSN: 0021-8979

Lenz, G.; Zimmermann, J.; Katsufuji, T.; Lines, M. E.; Hwang, H. Y.; Spalter, S.; Slusher, R. E.; Cheong, S. W.; Sangher, J. S. & Aggarwal, I. D. (2000). Large Kerr effect in bulk Se-based chalcogenide glasses. *Opt. Lett.*, Vol. 25, Issue 4, 254-256, ISSN: 0146-9592

Lefkiry, M.; Phu, X. N. and Rivoire, G. (1998). Existence of a bistable polarization state in a Kerr medium in the presence of two-photon absorption. *Quantum Semiclass. Opt.*, Vol. 10, No. 1, 283-292 ISSN: 1464-4266

Liao, H. B.; Xiao, R. F.; Wang, H.; Wong, K. S. & Wong, G. K. L. (1998). Large third-order optical nonlinearity in Au:TiO_2 composite films measured on a femtosecond time scale. *Appl. Phys. Lett.*, Vol. 72, Issue 15, 1817-1819, ISSN: 0003-6951

Martinoty, P. & Candau, S. (1971). Determination of viscosity coefficents of a nematic liqid crystal using a shear waves relectance technique. *Mol. Cryst. Liq. Cryst.*, Vol. 14 243-271, ISSN: 1542-1406

Martins, A. F.; Esnault, P. & Volino, F. (1986). Measurement of the viscoelastic coefficients of main-chain nematic polymers by an NMR technique. *Phys. Rev. Lett.*, Vol. 57, Issue 14, 1745-1748, ISSN: 0031-9007

Prost, J. & Gasparoux, H. (1971). Determination of twist viscosity coefficient in the nematic mesophases *Phys. Lett. A*, Vol. 36, Issue 3, 245-256, ISSN: 0375-9601

Sheik-Bahae, M.; Said, A. A.; Wei, T. H.; Hagan, D. J. & Stryland, E. V. (1990). Sensitive measurement of optical nonlinearities using a single beam. *IEEE J. Quantum Electron.*, Vol. 26, Issue 4, 760-769, ISSN: 0018-9197

Suzaki, Y. & Tachibana, A. (1975). Measurement of the μm sized radius of Gaussian laser beam using the scanning knife-edge. *Appl. Opt.*, Vol. 14, Issue 12, 2809-2810, ISSN: 0003-6935

Watanabe, Y.; Ohnishi, M. & Tsuchiya, T. (1995). Measurement of nonlinear absorption and refraction in titanium dioxide single crystal by using a phase distortion method. *Appl. Phys. Lett.*, Vol 66, Issue 25, 3431-3432, ISSN: 0003-6951

Yavrian, A.; Galstyan, T. V. & Pich´e, M. (1999). Single-mirror interferometer for nonlinear optical characterization. *IEEE J. Quantum Electron.*, Vol. 35, 1430-1433, ISSN: 0018-9197

7

A Polarizer-free Liquid Crystal Display using Dye-doped Liquid Crystal Gels

Yi-Hsin Lin, Jhih-Ming Yang, Hung-Chun Lin, and Jing-Nuo Wu
Department of Photonics, National Chiao Tung University
Taiwan, R. O. C.

1. Introduction

Liquid crystal (LC) based photonic devices are important in light modulated applications including amplitude modulation and phase modulation. [Yang *et al.* (2006)] The anisotropic properties of LC result in employing two polarizers in most of LC devices. The optical efficiency (~3%) and the viewing angle are limited. Therefore, it is highly desirable to develop polarizer-free LC devices. [Yang *et al.* (2006); Lin *et al.* (2008)] In polarizer-free liquid crystal displays (LCDs), two types are demonstrated. One is polarizer-free Guest-Host LCD which obtains dark state by doping small amount of dichroic dye molecules into LC host. [White *et al.* (1974); Cole *et al.* (1977); Bahadur (1992); Wu *et al.* (2001); Yang (2008)] However, the contrast ratio and reflectance are low due to the dichroic ratio (~10:1) of dyes. The other is scattering-absorption type, which combines light scattering and absorption. The second type is to mimic the display shown in white paper. In a printed paper, the printed areas turn out dark because the ink absorbs light and light is scattered by the fibers in the white paper. In order to obtain scattering effect, the polymer and liquid crystal complex system is used, including polymer dispersed liquid crystals (PDLCs), polymer networks liquid crystals or liquid crystal gels. [Drzaic (1995)] The scattering mainly results from the mismatch of refractive indices of polymer networks and liquid crystal molecules. To further increase light absorption, dye molecules are doped into the polymer and liquid crystal complex system, for example, dye-doped PDLCs[Drzaic (1995); Lin *et al.* (2004)], and dye-doped LC gels. [Lin *et al.* (2005); Lin *et al.* (2006); Lin *et al.* (2008); Lin *et al.* (2009)] Contrast ratio (CR) of dye-doped PDLC is still not good enough because the dye solubility with polymer matrix, the order parameter of dye and dichroic ratio (typically ~10:1) of dye. [Drzaic (1995); Lin *et al.* (2004)] In 2005, we have developed a polarizer-free LCD using a dye-doped dualfrequency liquid crystal (DFLC) gel on the ITO-only glass substrates [Lin *et al.* (2005); Lin *et al.* (2006)]. Although its contrast ratio reaches ~150:1 and response time ~6 ms under frequency modulation and the laser-based measurement, the frequency driving scheme, high driving voltage (~30 V_{rms}) and unavoidable dielectric heating effect. [Wen *et al.* (2005); Yin *et al.* (2006)] need to be overcome for TFT-LCDs and flexible displays applications.

To avoid the dielectric heating effect of DFLC, negative LC within vertical alignment layer is a good alternative and it is suitable for making a transflective LCD [Lin *et al.* (2006)]. The

Source: New Developments in Liquid Crystals, Book edited by: Georgiy V. Tkachenko,
ISBN 978-953-307-015-5, pp. 234, November 2009, I-Tech, Vienna, Austria

gel-like feature of materials, vertically aligned polymer network and low temperature processes drives us to realize a trim-able and bendable polarizer-free flexible display in reflective mode. In 2008, we demonstrated a polarizer-free flexible electro-optical switch using dye-doped LC gels which is polarizer-free, fast response, high contrast. [Lin *et al.* (2008)] Many parameters affect the phase separation process [Yang *et al.* (2006); Ren *et al.* (2008)]and then have influence on the electro-optical properties of dye-doped LC gels, such as curing temperature effect. The normally white gels exhibit ~55% reflectance, ~450:1 contrast ratio, ~6.4 ms response time, and ~30 V_{rms} at f=1 kHz driving voltage at curing temperature 10 °C. A single pixel flexible reflective display using such dye-doped LC gels are also demonstrated under bending and trimming. To further realize a display with multi-pixels, the substrate is patterned by pixilated indium-tin-oxide (ITO) in general. However, the involved fabrication and driving are complicated. It also causes problems especially in the fabrication process of flexible displays, such as chemical stability of plastic substrates, failure of ITO under tension and so on. [Crawford (2005)] In order to achieve a simple and easy process for flexible displays, we developed a multiple-step switch using distinct dye-doped LC gels without patterning ITO layers in 2009. [Lin *et al.* (2009)] Moreover, the switch should provide extra information states besides voltage-on and voltage-off states for the applications of the decorative displays within a simple driving and manufacturing process. The distinct dye-doped LC gels can display information by the spatial distribution of polymer network density without patterned ITO layers. The different regions of polymer network densities have different threshold voltages, but the similar bright and dark states. The distinct dye-doped LC gel is transparent (or in bright state) at 0 V_{rms} and opaque (or in dark state) at 30 V_{rms}. At 9 V_{rms}, it shows the colored pattern (or information) because of distinct polymer networks. The response time is ~10 ms and contrast ratio is ~200:1. We can also extend the concept for a polarizer-free four step switch. The potential applications are for decorative displays, electrically tunable diaphragm, and electrically tunable low pass or high pass filter.

In this book chapter, we introduce a polarizer-free LCD using dye-doped LC gels whose the physical mechanism is mainly the combination of both light scattering and absorption. In the beginning, we introduce the structure and mechanism of our dye-doped liquid crystal gels. Second, the experimental setups, results (morphologies, electro-optical properties, response time) and mathematical model are also discussed. Third, the performance of such a polarizer-free liquid crystal display is shown. Finally, we will introduce multiple step switches using distinct dye-doped LC gels for the application of decorative displays. The potential applications are flexible displays, electrically tunable light shutters, and decorative displays.

2. Polarizer-free dye-doped liquid crystal gels

2.1 Structures and operating principles

The structure and operation principles of the dye-doped LC gel are schematically depicted in Fig. 1. At V=0, the cell does not scatter light and the absorption is rather weak due to the vertically aligned polymer networks, liquid crystal directors and dye molecules. Therefore, the display has the highest reflectance. When the applied voltage (V_1) is larger than the threshold voltage (V_{th}), the LC directors with a negative dielectric anisotropy are tilted away from the electric field and LC molecules bring dye molecules to rotate with them, as shown in Fig. 1. The cell is then switched into a multi-domain structure. As a result, both of scattering and absorption increase. Then the reflectance decreases. In addition, all the LC directors and dye molecules have the same tilt angle but random orientations. Therefore, the reflectance is polarization-independent. When we further increase the voltage ($V_2 >> V_{th}$), the LC directors and dye molecules are aligned along x-y plane randomly, as Fig. 1 depicts. The scattering and absorption are maximal. Besides, all the polarizations of incident light experience the same averaged scattering and absorption. As a result, the dye-doped LC gel appears black and is still polarization independent. The appearance of color is mainly because of the light absorption of dye. The scattering and reflection assist the multiple absorption due to the elongate the paths of light propagation.

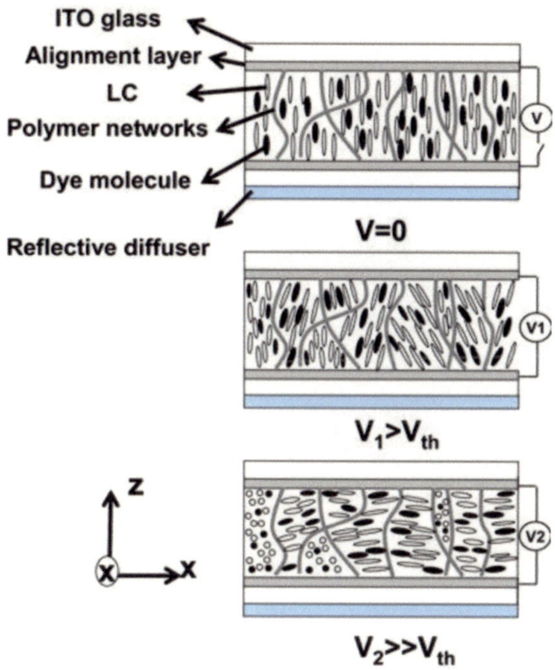

Fig. 1. Schematic operating principle of dye-doped liquid crystal display at V=0, $V_1 > V_{th}$, and $V_2 >> V_{th}$. The alignment layer has no rubbing treatment.

2.2 Fabrications

The dye-doped LC gel we employed is a mixture of negative nematic liquid crystal ZLI-4788 (Merck, n_e= 1.6567, Δn=0.1647 at λ=589 nm; $\Delta\varepsilon$= -5.7 at f= 1 kHz) and a diacrylate monomer (bisphenol-A-dimethacrylate) with a dichroic dye S428 (Mitsui, Japan) at 90:5:5 wt% ratios. The structure of the diacrylate monomer is shown as follows:

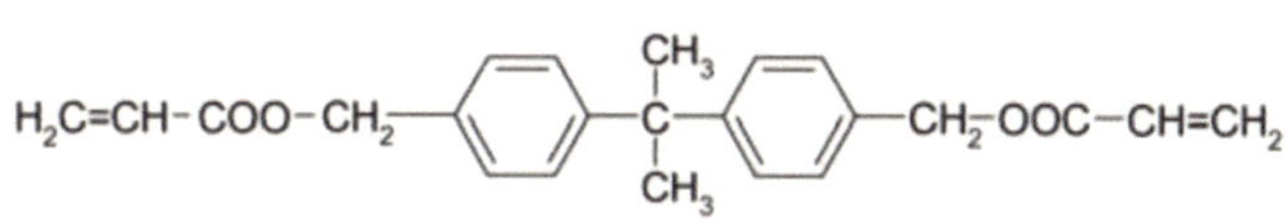

The dye-doped LC mixture was then injected into an empty cell consisting of two glass substrates whose inner surfaces were coated with a thin conductive layer, indium-tin-oxide (ITO) and polyimide (PI) layer *without* rubbing treatment. The PI layer provides vertical alignment for the LC directors. The cell gap was 5 μm. The filled cell was irradiated by a UV light (λ~365 nm, I~3 mW/cm^2). The cell was cured at a fixed temperature for 1.5 hr. Such a fixed temperature is called curing temperature. After photo-polymerization, the formed chainlike polymer networks are along the z direction because the LC directors are aligned perpendicular to the glass substrates during the UV curing process, as shown in Fig. 1.

2.3 Morphologies

Fig. 2 show the morphologies observing under an optical microscope with a single polarizer only. The top region of the two regions in Fig. 2 is the patterned ITO area. The bright region represents the state of V=0. The dark area represents the ITO electrodes applied 30 V_{rms} at *f*=1 kHz. At the voltage-off state, the cell shows good bright state because of the vertically aligned polymer networks, LC and dye molecules. At 30 V_{rms}, it shows the fine domain textures of the polymer networks, and red color because of dye molecules, as shown in Fig. 2. Our LC cell shows good dark and bright states although the dark state up to now is redish, not truly black.

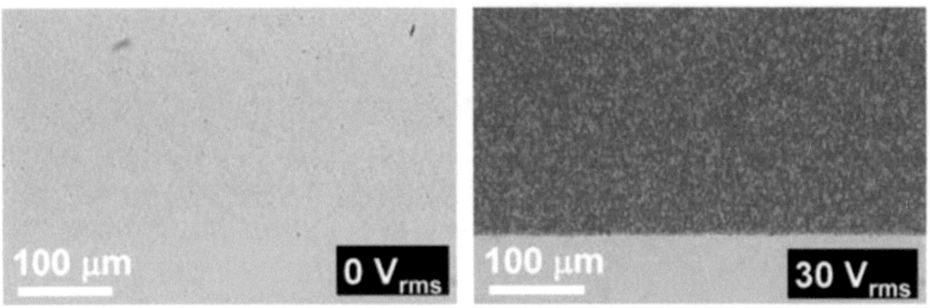

Fig. 2. The microscopic images of dye-doped LC gels at 0 V_{rms} and 30 V_{rms}.

The photographs of Scanning Electron Microscopy (SEM) are shown in Fig. 3 (a)-(h). Fig. 3 (a), (c), (e) and (g) are the top views of the cells at different curing temperatures (T). Fig. 3 (b), (d), (f), and (h) are the side views of the cells at different curing temperatures. In Fig. 3 (b), (d), (f), and (h), the polymer networks are perpendicular to the glass substrates. The polymer networks of dye-doped LC gels consist of chain-linked polymer grains. The averaged sizes of polymer grains measured from Fig. 3 (a), (c), (e) and (g) are around 68 nm at T=10°C, 94nm at T=20°C, 125nm at T=30°C, 132nm at T=40°C. The averaged domain sizes of polymer networks measured from Fig. 3 (b), (d), (f), and (h) are around 3.25 μm at T=10°C, 4.62 μm at T=20°C, 4.78 μm at T=30°C, 6.12 μm at T=40°C. Both of the domain sizes and the size of polymer grains increase with curing temperatures. The scattering is mainly because of the domain sizes of polymer networks because the domain sizes of polymer networks are near wavelength of incident light while the sizes of polymer grains are smaller than the wavelength.

2.4 Electro-optical properties

We adopt the typical reflectance measurement to measure the electro-optical properties of dye-doped LC gels. Because the guest-host system we employed appears dark red rather than black, we used an unpolarized green He-Ne laser (λ=543.5 nm, Melles Griot, Model 05-LGR-173) instead of a white light source for characterizing the device performances. A dielectric mirror was placed behind the cell so that the laser beam passed through the cell twice. A large area photodiode detector (New Focus, Model 2031) was placed at ~25 cm (the

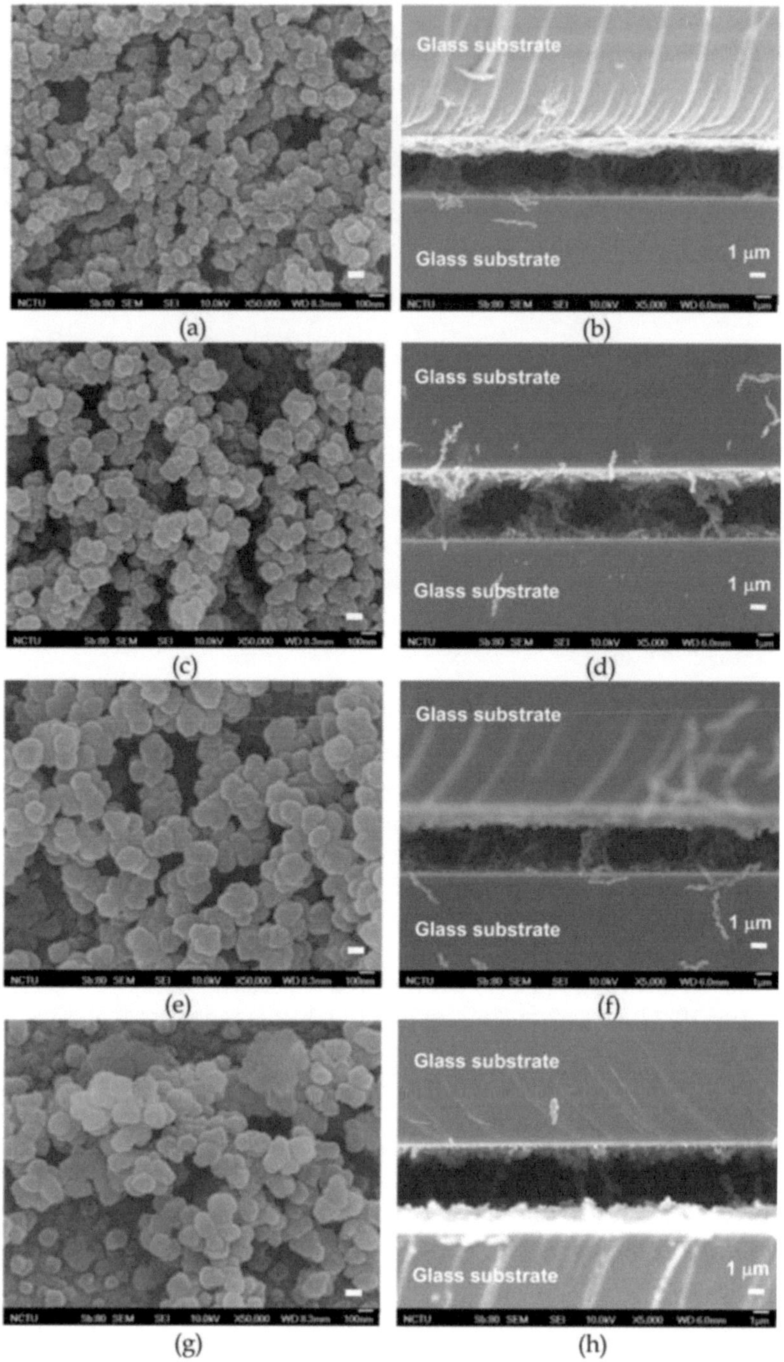

Fig. 3. SEM photographs of dye-doped LC gels at curing temperatures 10°C (a, b), 20°C (c, d), 30°C (e, f), and 40°C (g, h). The LC and dye were extracted. (a), (c), (e) and (g) are the top views of the cells. (b), (d), (f), and (h) are the side views of the cells. The white-indicated bars in (a), (c), (e) and (g) are 100 nm.

normal distance for viewing a mobile display) behind the sample which corresponds to ~2° collection angle. A computer controlled LabVIEW data acquisition system was used for driving the sample and recording the light reflectance. In order to prove our dye-doped LC gel is polarization independent, we placed a polarizer between the laser and the LC cell. The reflectance as a function of an angle of the polarizer at different applied voltages is shown in Fig. 4. The variation of reflectance is less than 5% when we rotated the polarizer. It indicates the dye-doped LC gels are indeed polarization independent at all applied voltages. The reflectance at 0 V_{rms} is around 50%.

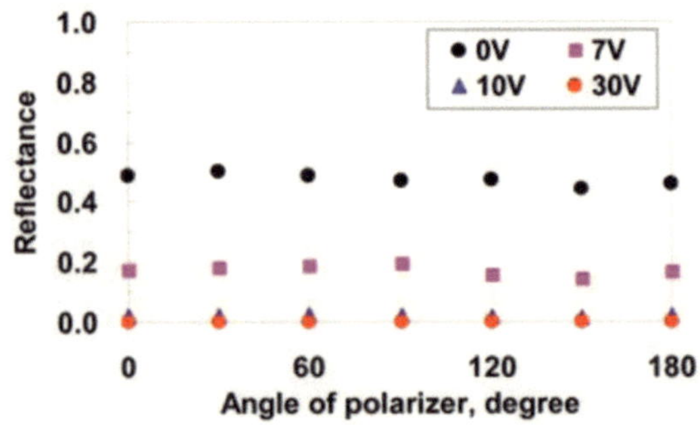

Fig. 4. Reflectance as a function of angle of the polarizer at different applied voltages. The curing temperature was 20 °C.

2.4.1 Curing temperature effect

After we removed the polarizer, the incident light then was unpolarized green laser beam. Fig. 5 (a) is the measured voltage-dependant reflectance of the dye-doped LC gels at various curing temperatures. The reflectance was normalized to that of a pure LC cell with the same cell gap. The reflectance decreases gradually with the applied voltage $V>V_{th}$ due to the increases of the scattering and the absorption. As curing temperature decreases, the increases (~40% to ~55%) of maximum reflectance at V=0 and that is because of the better vertical alignment of LC directors, dye molecules and polymer networks at a low curing temperature. The contrast ratio (CR) is defined as a reflectance ratio of 0 V_{rms} to 30 V_{rms}. The CRs are ~450: 1 at 10 °C, ~250: 1 at 20 °C, ~200: 1 at 30 °C, and ~300: 1 at 40 °C. The contrast ratio decreases as T< 30°C and then increases as T>30°C. That is because the increase of a curing temperature results in larger polydomains; therefore, the CR and threshold voltage decrease. Moreover, the decay time increases, as shown in Fig. 5(b). When the temperature is higher than 30°C, we found the cell has dynamic scattering, a fluctuation of liquid crystal directors in polymer domains, to help rebooting the contrast ratio in spite of the larger domain size. To lower the driving voltage, a high birefringence and high absolute value of dielectric anisotropy (Δε) of a negative LC and slightly lower polymer concentration could be considered.

Response time is also an important parameter for Guest-Host LCDs. The response time of the dye-doped LC gels was measured using 30 V_{rms} squared pulses with time duration 500ms at *f*=1 kHz. The curing temperature-dependent rise times and decay times are shown in Fig. 5(b). The rise times are about 0.4 ms and the decay time decreases with decreasing curing temperatures. A typical response time of a guest-host display is around 50 ms. The response time of our dye-doped LC gel (~6.4ms) is faster because polymer network helps LC directors to relax back. The rise time is ~0.4 ms and decay time is ~6 ms at 10 °C.

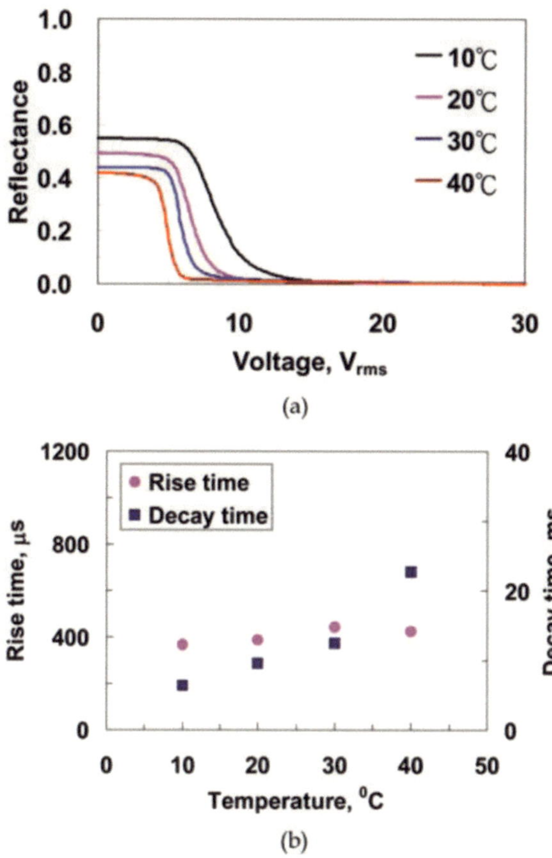

Fig. 5. (a) Voltage-dependent reflectance at various curing temperature. (b) Curing temperature-dependent response time

Besides the curing temperature, the UV curing intensities, monomer concentrations and dye concentrations also affect the performance of the dye-doped LC gels. We discuss in the following sections.

2.4.2 Curing UV intensity effect

To examine the effect of UV curing intensity (I), we prepared four samples with the same mixtures at the curing temperature 10 °C, but at different UV curing intensities which were 2.6, 1.37, 0.733, 0.354 mW/cm^2 respectively. The cell gaps were still 5 μm. The measured reflectance as a function of voltage is shown in Fig. 6(a). With the increases of UV curing intensity, threshold voltage (V_{th}) increases from 4.82 V_{rms} (I ~ 0.354 mW/cm^2) to 5.92 V_{rms} (I~2.6 mW/cm^2). The maximal reflectance at V=0 increases from 44% (I ~ 0.354 mW/cm^2) to ~57% (I~2.6 mW/cm^2). In Fig. 6(b), rise time is around 0.4 ms, but decay time decreases from 9 ms (I ~ 0.354 mW/cm^2) to 6.88 ms (I ~ 2.6 mW/cm^2). The larger UV curing intensity results in smaller domain size of polymer networks. That causes a stronger anchoring energy and then enlarges the threshold voltage. Therefore, it boosts the response. The reflectance at V=0 decreases under higher UV curing intensity owing to better vertically alignment at V=0. Generally speaking, UV curing intensity does not affect the electrooptical properties of dye-doped LC gels dramatically. However, weak UV curing intensity can result in the longer curing time in order to complete the polymerization.

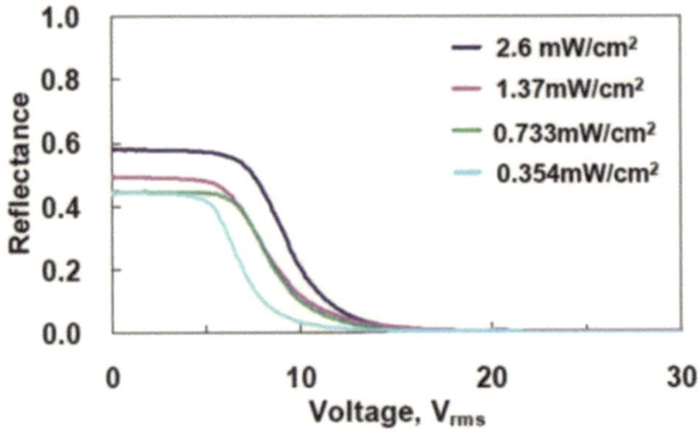

(a)

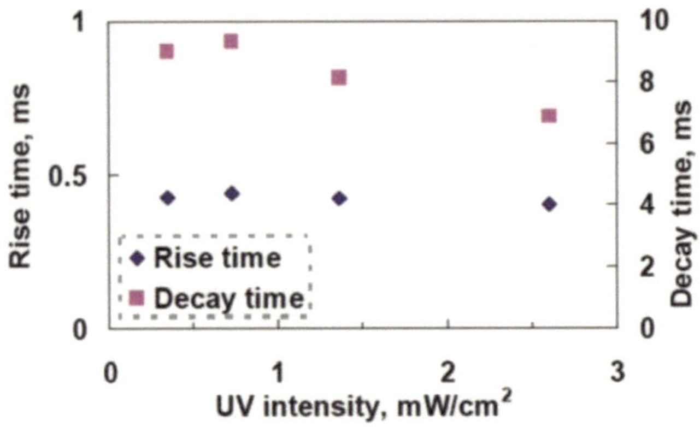

(b)

Fig. 6. (a)Voltage-dependent reflectance at various curing intensity and (b) response time as a function of UV curing intensity. The curing temperature was 10 °C.

2.4.3 Monomer concentration

In this section, the monomer concentration effect is discussed. We prepared three cells at different monomer concentrations which were 3 wt%, 5 wt%, and 7 wt%. The fabrication processes of the cells were at the same UV curing intensity 2.6 mW/cm^2 and also at the same curing temperature 20 °C. The cell gaps were still 5μm. The voltage-dependent reflectance at different monomer concentrations is shown in Fig. 7(a). The threshold voltage increases from 2.1 V_{rms} (at 3 wt% M1) to 6.52 V_{rms} (at 7 wt% M1) due to the denser polymer networks. The reflectance at V=0 decreases slightly (from 51% at 3 wt% M1 to 46% at 7 wt% M1). That is because denser polymer networks affect the vertical alignment of LC directors and also increases the scattering slightly. CR increases from 222 : 1 (at 3 wt% M1) to 486 : 1 (at 7 wt% M1) owning to better scattering of higher monomer concentration at the high driving voltage. Fig. 7(b) shows the measured response time as a function of monomer concentration. Rise time is around 0.2 ms-0.4 ms, and decay time increases from 52 ms (at 3 wt% M1) to 7.3 ms (at 7 wt%M1).

The higher monomer concentration has smaller domain sizes; therefore, the LC directors are relaxed back faster due to stronger anchoring energy after turning off the applied voltage. To lower driving voltage, we can reduce the monomer concentration; however, the polymer network of dye-doped LC gels is not stable as the monomer concentration is too low.

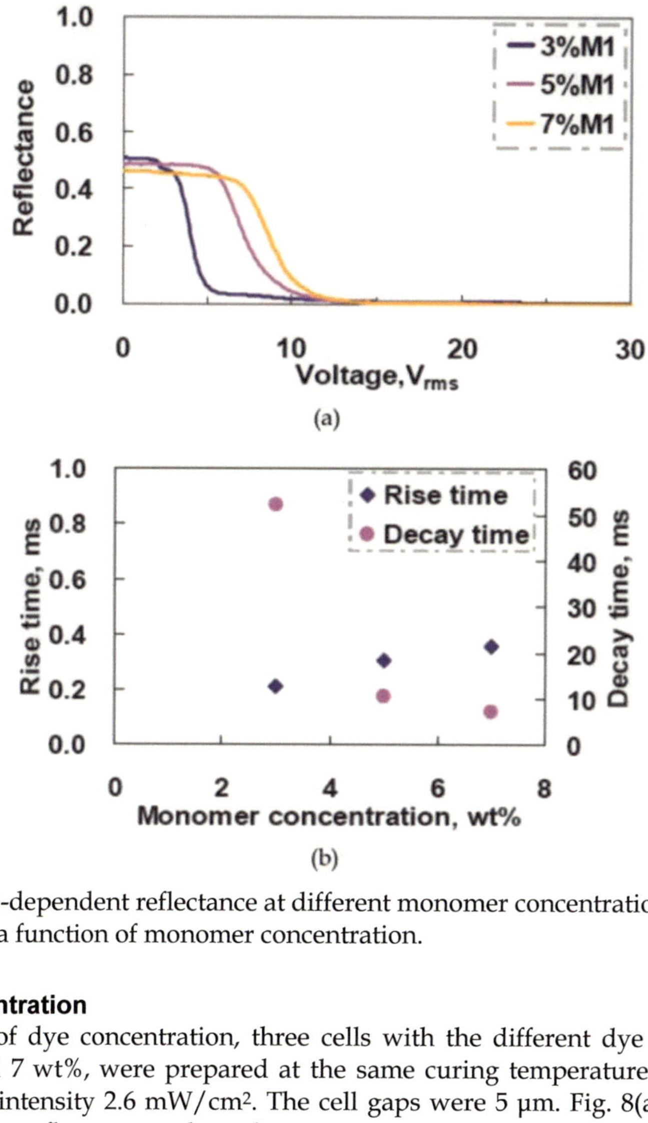

Fig. 7. (a) Voltage-dependent reflectance at different monomer concentrations. (b) Measured response time as a function of monomer concentration.

2.4.4 Dye concentration

As to the effect of dye concentration, three cells with the different dye concentrations, 3 wt%, 5 wt%, and 7 wt%, were prepared at the same curing temperature 20 °C and at the same UV curing intensity 2.6 mW/cm^2. The cell gaps were 5 μm. Fig. 8(a) plots measured voltage-dependent reflectance at three dye concentrations. The reflectance at V=0 decreases from ~ 63% (at 3 wt% S428) to ~ 42.5% (7 wt% S428) as the dye concentration increases due to the increases of absorption, even at the same concentration of monomer. The threshold voltages of cells with 3 wt%, 5 wt%, and 7 wt% S428 are almost the same around 5 V_{rms}. That is because the monomer concentrations are the same. The CRs of the cells with 3 wt%, 5 wt%, and 7 wt% S428 are 668:1, 342:1, and 31:1, respectively. Higher dye monomer concentration reduces not only reflectance at V=0, but also degrades the dark state at V=30.

In Fig. 8(b), rise times are around 0.3 ms and the decay time increases 2 times from 9.8 ms at 3wt% S428 to 20.96 ms at 7wt% S428) with the dye concentration. That is because dye molecules hinder LC directors relax back at higher concentration of dye under same monomer concentration.

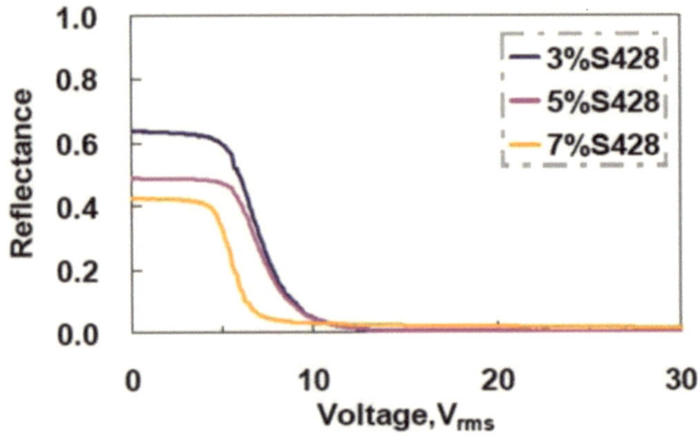

(a)

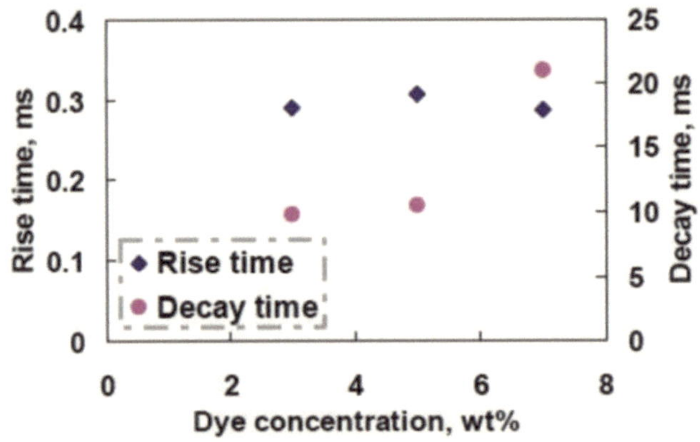

(b)

Fig. 8. (a) Voltage-dependent reflectance at different dye concentrations.(b) Measured response time as a function of dye concentration.

3. Mathematical models

By considering the scattering and absorption, the reflectance (R(θ)) as a function of tilt angle (θ) of LC directors with respect to x-axis can be expressed as:

$$R(\theta) \approx e^{-\alpha_{ave}(\theta)\cdot 2d} \cdot e^{-\beta_{ave}(\theta)\cdot 2d} \tag{1}$$

where d is cell gap, $\alpha_{ave}(\theta)$ is the average absorption coefficient, and $\beta_{ave}(\theta)$ is the average scattering coefficient. $\alpha_{ave}(\theta)$ and $\beta_{ave}(\theta)$ satisfy the following equations.

$$\alpha_{ave}(\theta) = \rho_1 \cdot \frac{\alpha_{eff}(\theta) + \alpha_{\perp}}{2} \tag{2}$$

$$\beta_{ave}(\theta) = \rho_0 \cdot \frac{\sigma_{eff}(\theta)}{V} \tag{3}$$

$\alpha_{eff}(\theta)$ can be expressed as:

$$\alpha_{eff}(\theta) = \frac{\alpha_{\parallel} \cdot \alpha_{\perp}}{\sqrt{\alpha_{\parallel} \cdot Cos^2\theta + \alpha_{\perp} \cdot Sin^2\theta}} \tag{4}$$

where ρ_1 is the dye concentration, $\alpha_{//}$ and $\alpha_{\perp}$ are the absorption coefficients when the polarization of incident light is parallel or perpendicular to the principal axis of dye molecule. ρ_0 is the LC concentration, V is the average volume of a domain. σ_{eff} in Eq. (4) is the effective scattering cross section from all liquid crystal droplets and can be expressed as:

$$\sigma_{eff}(\theta) = \frac{1}{\pi}\int_0^{\pi} \sigma_s(\theta,\alpha_o)\cdot d\alpha_o \tag{5}$$

Based on anomalous diffraction approach [Drzaic (1995)], scattering cross section results from a single LC domain is:

$$\sigma_s(\theta,\delta) = 2\sigma_o[H_{ve}(\theta)\cdot Cos^2\delta + H_{vo}\cdot Sin^2\delta] \tag{6}$$

where σ_0 is the geometrical optics cross section related to the domain size, δ is the polarization angle; $H_{ve}(\theta)$ and $H_{vo}(\theta)$ stand for phase shift induced by e-ray and o-ray respectively. The averaged domain sizes of polymer networks are measured around 1.5 μm~3μm. For numerical calculations we have chosen the following parameters according to the experiments: ρ_1=0.05 g/cm^3, and ρ_0=0.89 g/cm^3. $\alpha_{//}$= 11.83 μm^{-1} and $\alpha_{\perp}$= 0.926 μm^{-1}. The simulation result is shown in Fig. 9. In Fig. 9, the reflectance decreases with the tilt angle. The R-θ curve shifts to right as the domain size is smaller. The simulation results agree with the experimental results in Fig. 5. The smaller domain size or larger density of polymer networks can result in the larger operating voltage and better dark state. We can adjust UV intensity, curing temperature, the controlled temperature under UV illumination, and concentration of LC, dye or monomer to change the domain sizes of polymer networks.

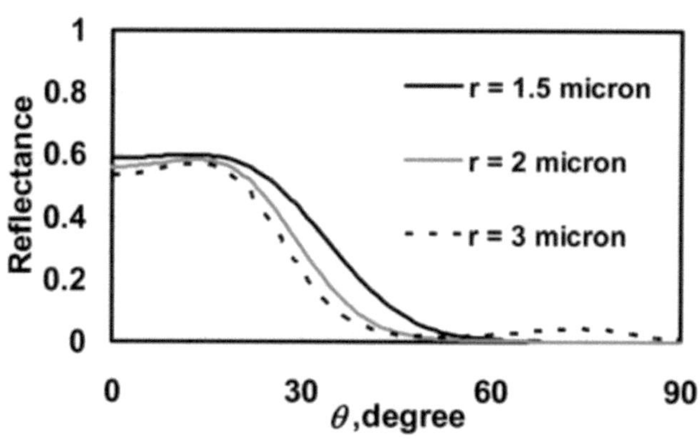

Fig. 9. Calculated reflectance as a function of tilt angle in dye-doped LC gels at different domain sizes.

4. Polarizer-free flexible LCDs using dye-doped LC gels

Roll-able, bendable, trim-able, and conformable paper-like flexible displays are useful for electronic paper, electronic tag, and decorative displays. [Crawford (2005)] Many liquid crystal (LC) technologies, such as polymer-dispersed liquid crystals (PDLC) [Mach *et al.* (2001); Sheraw *et al.* (2002); Hohnholz et al. (2005); Buyuktanir *et al.* (2006); Wang *et al.* (2007)], cholesteric liquid crystals [Wu *et al.* (2001); Yang *et al.* (1994); Chari *et al.* (2006); Khan *et al.* (2005); Khan *et al.* (2007)], and single-substrate LCDs using photoenforced stratification [Penterman *et al.* (2002); Raynes (2002); Vogels *et al.* (2004)] or using LC/polymer composites [Kim *et al.* (2002); Kim *et al.* (2004); Lin *et al.* (2006); Ren *et al.* (2007)], and non-liquid crystal technologies, such as electrophoretic imaging [Comiskey *et al.* (1998); Gelinck *et al.* (2004); Daniel *et al.* (2007)], Gyricon [Crowley *et al.* (2002)], and organic light-emitting diode (OLED) [Gu *et al.* (1997); Burrows *et al.* (1997); Krasnov (2002); Sugimoto *et al.* (2004); Zhou *et al.* (2006)], have been carried out to achieve transmissive type or reflective type flexible displays. In liquid crystal-based flexible displays, bistability and colors of cholesteric liquid crystals limits the application due to the complexity of driving and color shift at off angle. Instead of cholesteric liquid crystals, dye-doped LC gels can be used in flexible displays because the dye-doped LC gels is gel-like and the polymer networks of dye-doped LC gels are perpendicular to the glass substrates.

The images of a single pixel of the polarizer-free LCD using dye-doped LC gels at V=0 and V=30 V_{rms} are shown in Fig. 10(a). By replacing glass substrates with flexible substrate, dye-doped LC gel is not only bendable but also trim-able because our material is gel-like, as shown in Fig. 10(b). The flexible substrates are provided by EOL/ITRI (Electronics& Optoelectronics Research Laboratories, Industrial Technology Research Institute, Taiwan). IZO was over coated on the top of flexible substrates made by polycarbonate with thickness 120 μm. The cross shaped microstructures made by photo-spacers, resins, were developed on the flexible substrates by photolithography process. The width of photo-spacers is 10 μm and the pitch of photo-spacers is 430 μm. The ambient white light was used to illuminate the cells. However, the CR is degraded. The CR is higher under laser-based measurement because of the collimation of the laser beam. Figure 10(c) is the transmission as a function of radius of curvature under bending at 0 and 30 V_{rms}. The measurement method is two-point bending technique. The transmission of dye-doped LC gels is almost the same as the radius of curvature larger than 21 mm. The dye-doped LC gel is trim-able as well because our material is gel-like, as shown in Fig. 10 (d) and (e). The flexible display performance remains almost the same after cutting by a scissor. Since no polarizer in needed, the residual birefringence of polycarbonate does not affect the performance of our flexible display.

5. Polarizer-free multiple step switches using distinct dye-doped LC gels

In order to realize a display with multi-pixels, we usually pattern the substrate by pixilated indium-tin-oxide (ITO). However, the involved fabrication and driving are complicated. Especially in the fabrication process of flexible displays, chemical stability of plastic substrates and failure of ITO under tension degrade the performance of flexible displays. In order to achieve a simple and easy process for flexible displays, we can develop a multiple step switch using dye-doped LC gels without patterning ITO layers. Moreover, the switch should provide extra information states besides voltage-on and voltage-off states for the applications of the decorative displays within a simple driving and manufacturing process. A polarizer-free multiple step switches using distinct dye-doped liquid crystal gels

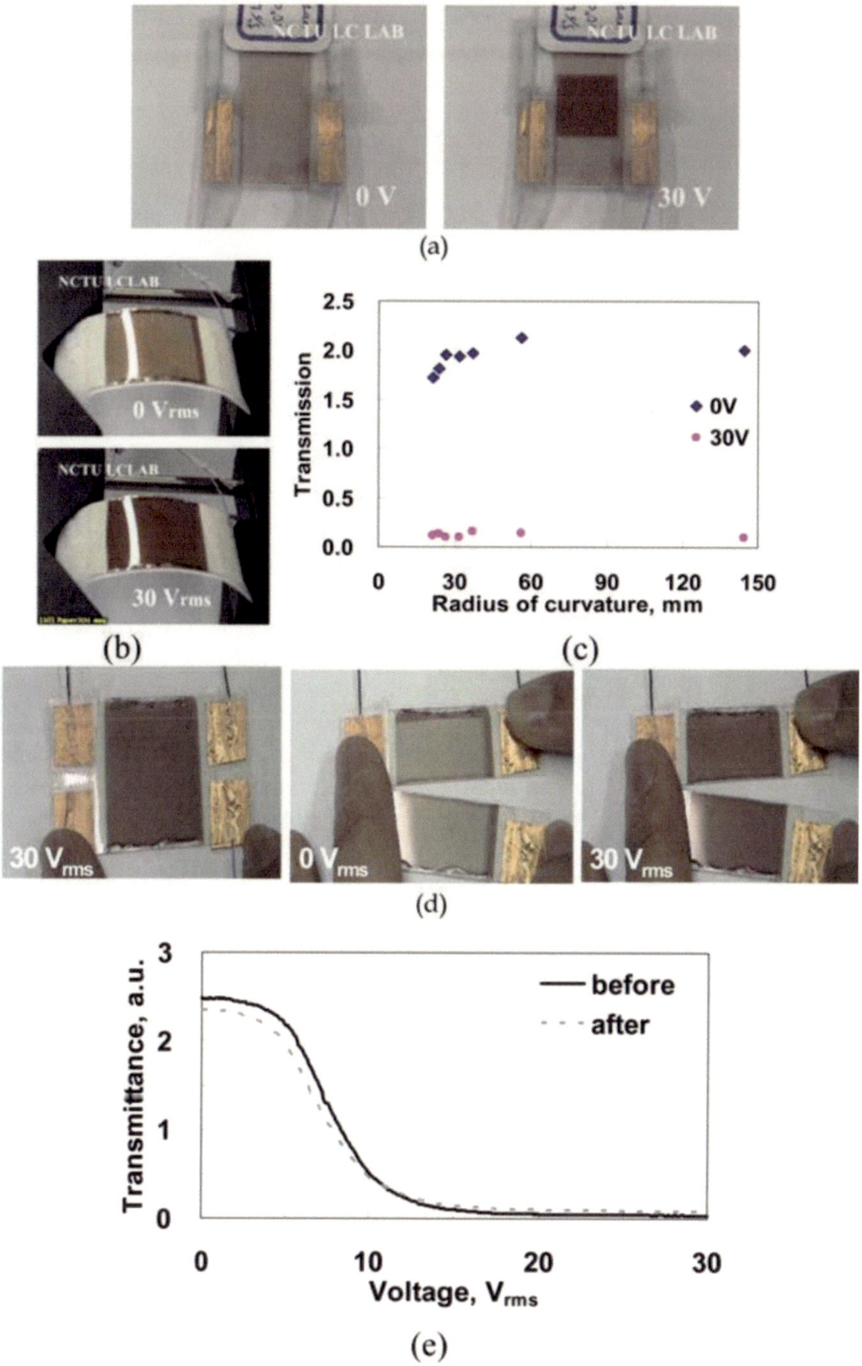

Fig. 10. (a) A single pixel polarizer-free reflective LCD using the dye-doped LC gels in glass substrates and (b) in flexible substrates. (c) The transmission as a function of bending radius of curvature. (d)The dye-doped LC gel is trim-able. (e) The voltage dependent transmission before and after trimming. A piece of white paper was used as a diffusive reflector.

is demonstrated in reflective mode. By controlling the spatial distribution of the density of polymer networks, the distinct dye-doped LC gels can produce multiple states: bright, dark and information states without patterning ITO. The multiple states are generated by adapting different polymer network density in different display regions. The dark state is due to the combination of scattering and absorption. In this section, we demonstrated a three step switch using distinct dye-doped LC gels. [Lin *et al.* (2009); Du *et al.* (2003)] The potential applications are decorative displays, electrically tunable iris, and electrically tunable low pass or high pass filter.

5.1 Operating principles of a polarizer-free three step switch

Fig. 11 illustrates the operating principles of a polarizer-free three step switch using distinct dye-doped LC gels. The structure consists of ITO glass substrates, vertical alignment layer without rubbing treatment, a diffusive reflector, negative liquid crystals, dichroic dye molecules, and the distinct distribution of polymer networks which divide the cell into two parts, the low density of polymer networks (LDPN) and the high density of polymer networks (HDPN), as shown in Fig. 11(a). Without an applied voltage (V=0), polymer networks, LC and dye molecules are aligned vertically; hence, the cell does not scatter light and the absorption is weak. The cell has high polarization-independent reflectance. When V_1 is larger than the threshold voltage in LDPN (V_{th1}), the LC directors are reoriented first and then bring dye molecules to rotate accordingly due to weak anchoring energy provided by LDPN; meanwhile, the LC directors remain vertically aligned in HDPN, as shown in Fig. 11(b). The reflectance then reduces in LDPN region due to the increases the scattering and absorption while the reflectance remains high in HDPN region. Furthermore, all the LC directors and dye molecules in LDPN have the same tilt angle within random orientations which originate from the alignment layer without rubbing treatment. The reflectance in LDPN is polarization-independent. When the applied voltage is larger than the threshold voltage V_{th2} (>V_1) of HDPN, the LC directors in HDPN begin to reorient by the electric fields and then the reflectance in this region decreases as well. When the applied voltage V_2 is much larger than V_{th2}, the negative liquid crystals and dye molecules in both regions are reoriented randomly in the x-y plane, as shown in Fig. 11(c). Thus, the both regions appear polarization-independent dark state.

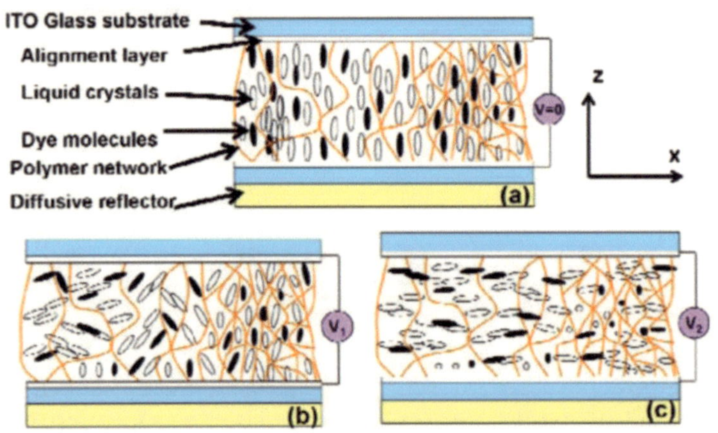

Fig. 11. Schematic structure and operating principles of distinct dye-doped LC gels at (a) V=0, (b)V_1, and (c)V_2 (V_2> V_1).

5.2 A polarizer-free three step switch using distinct dye-doped liquid crystal gels

The fabrication process is similar to dye-doped LC gels, but spatial distribution of polymer networks. To prepare a cell of distinct dye-doped LC gels, we mixed nematic LC ZLI-4788 with an UV-curable diacrylate monomer and a dichroic dye S428 (Mitsui, Japan) at 90: 5: 5 wt% ratios. In order to generate distinct distribution of polymer networks, two-step UV curing process was applied. First, one region of the cell with filled mixtures was exposed to a UV light (λ~ 365 nm) with UV intensity 1.37 mW/cm^2 at 10°C for 1.5 hours to generate HDPN. After that, the other region of the cell was irradiated by the UV light with lower UV intensity 0.73 mW/cm^2 at 20°C for another 1.5 hours to develop LDPN. The higher curing intensity of UV light and lower curing temperature results in HDPN. [Lin *et al.* (2008); Du *et al.* (2003)] The cell gap was 5 μm. The three step switch can also be a flexible switch by replacing substrates.

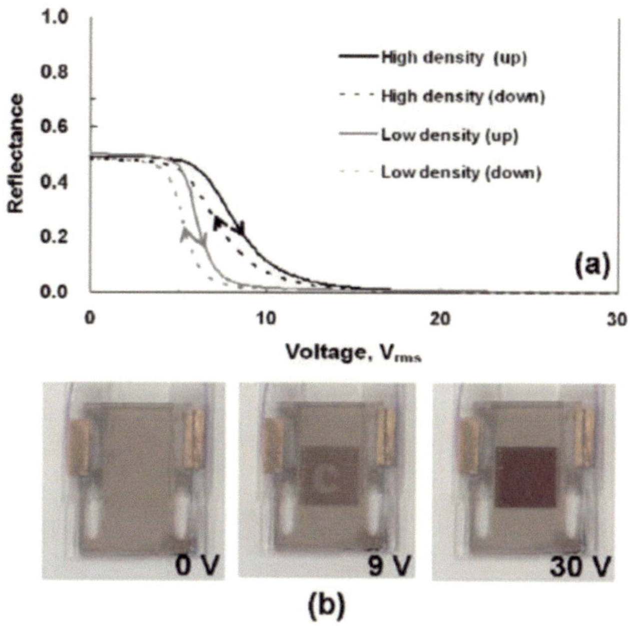

Fig. 12. (a) Voltage-dependent reflectance of the three step switch when voltage ia ramped up (solid line) and down(dotted line) (b) Images of the three step switch. A white paper was used as a diffusive reflector.

Fig. 12(a) plots the measured voltage-dependant reflectance of the sample for HDPN and LDPN when the voltage is ramped up (solid line) and down (dotted line). The threshold voltage in HDPN (5.42 V_{rms}) is larger than that in LDPN (4.62 V_{rms}). At V=0, the reflectance in HDPN and LDPN are 50.2% and 49.2%, respectively. The CR is still defined as a reflectance ratio of 0 V_{rms} to 30 V_{rms}. The CRs in both regions are ~ 200:1 under laser-based measurement. When we use an integrating sphere (DMS803, product of the AUTRONICMELCHERS GmbH) under a white light source, the CRs in both regions are ~3:1. Typically CR for newspaper under the same measurement is ~5:1. Improving the scattering state and dichroic ratio of dye can further improve the CR. The response time (rise time plus decay time) is 11 ms in LDPN and ~ 9 ms in HDPN by using 30 V_{rms} squared pulses with time duration 500 ms at *f*=1 kHz. The HDPN results in the stronger anchoring force between the LC and polymer networks; therefore, the threshold voltage is higher and response time is faster. In order to demonstrate the image of the 3-step switch, we patterned the distinct dyedoped LC gels with a photomask of a "C" pattern without patterning the ITO layer, as shown in Fig. 12(b). The cell can show three switching states: bright state at V=0, information state at 9 V_{rms}, and dark state at 30 V_{rms}.

The distinct dye-doped LC gels show the hysteresis when the voltage is ramped up and down. Without patterning ITO layer, the voltage is applied to different density regions simultaneously. Thus, the sample still can be operated as a three step or a four step switch as long as the electro-optical curves in different density regions are distinguishable no matter ramped-up voltages or ramped-down voltages.

The reflectance in HDPN (R_{high}) and in LDPN (R_{low}) can be expressed as a function of tilt angle θ of LC directors with respect to x-axis (in Fig. 11): [Drzaic (1995); Khoo *et al.* (1993)]

$$R_{high}(\theta) \approx e^{-2d\times(\alpha_1(\theta)+\beta_1(\theta))} \tag{7}$$

$$R_{low}(\theta) \approx e^{-2d\times(\alpha_2(\theta)+\beta_2(\theta))} \tag{8}$$

where d is the cell gap, $\alpha_1(\theta)$ and $\alpha_2(\theta)$ are the averaged absorption coefficients. $\beta_1(\theta)$ and $\beta_2(\theta)$ are the average scattering coefficients. The averaged absorption can be expressed in Eq. (9).

$$\alpha_i(\theta) = \rho \times \frac{\alpha_{eff}(\theta) + \alpha_\perp}{2}, i = 1, 2 \tag{9}$$

where ρ is dye concentration, and $\alpha_\perp$ is the absorption coefficient for the polarization of incident light perpendicular to the principal axis of the dye molecule. The effective absorption coefficient $\alpha_{eff}(\theta)$ of the dye molecules can be expressed as:

$$\alpha_{eff}(\theta) = \frac{\alpha_{//} \times \alpha_\perp}{\sqrt{\alpha_{//} \times \sin^2\theta + \alpha_\perp \times \cos^2\theta}} \tag{10}$$

Where $\alpha_{//}$ is the parallel absorption coefficient. At V=0, θ is near π/2; consequently, $\alpha_1 \approx \alpha_2$ approximately equals to $\rho\times\alpha_\perp$ and $\beta_1(\theta) \approx \beta_2(\theta)$ equals to 0 because of the vertically alignment of LC directors, dye molecules and polymer networks. Therefore, $R_{high}(\theta) \approx R_{low}(\theta) \sim 63\%$ after we calculated by ρ=0.05 and $\alpha_\perp$=0.926 m^{-1}. However, the measured reflectance (~50%) is lower than 63%. That is because of the imperfection of vertical alignment of LC, dye and polymer networks which is limited by the order parameter of dye molecules and the structure of the monomer without a rod-like central core. At V_2 (>> V_{th1} and V_{th2}) in Fig. 11(c), α_1 $(\theta = 0) \approx \alpha_2$ $(\theta = 0) \approx \rho \times (\alpha_\perp + \alpha_{//})$ is around 0.319 m^{-1} since measured $\alpha_{//}$ is 11.83 m^{-1}. Moreover, the scattering is stronger in HDPN due to smaller domains and the mismatch of refractive indices, i.e. β_1 $(\theta = 0) > \beta_2$ $(\theta = 0)$; therefore, theoretically R_{high} $(\theta = 0)$ is lower than R_{low} $(\theta = 0)$ which also means higher CR in HDPN. However, CRs in both regions are similar in our experiment. Owning to the strong anchoring energy in HDPN, LC directors and dye molecules are not totally aligned along x-y plane which means $\theta \neq 0°$ at 30 V_{rms}; hence, absorption and scattering is not as high as we expect in HDPN. At V_1(in Fig. 11(b)), $\theta_1 > \theta_2$ due to the different anchoring strengths in two regions. Thus, α_1 $(\theta_1) < \alpha_2$ (θ_2). So does β_1 $(\theta_1) < \beta_2$ (θ_2). Therefore, $R_{high}(\theta_1)$ is larger than $R_{low}(\theta_2)$ at V_1.

6. Conclusion

In conclusion, we have introduced and demonstrated polarizer-free LCDs using dye-doped LC gels. These polarizer-free dye-doped LC gels exhibit high reflectance, high contrast ratio, wide viewing angle, and fast response time. Especially the low temperature process is favorable for flexible displays. The gel-like materials assist stabilizing the flexible display under trimming. Our dye-doped LC gels provide a stable LC mode and open a new window in paper-like flexible displays. A polarizer-free three step switch using distinct dye doped LC gels is also demonstrated. The distinct dye-doped LC gels can display information by the spatial distribution of polymer network densities without patterned ITO layer. The advantages of such polarizer-free switches are 1) polarizer-free, 2) simple fabrication without ITO patterning, and 3) simple driving. However, the issues we have to overcome are high driving voltages, low resolution, non-black colors and long term stability of dye molecules due to photobleach. This concept can also be extended for making a polarizer-free multiple-switch. The potential applications are decorative displays, electrically tunable iris, and electrically tunable low pass or high pass filter.

7. Acknowledgement

The authors are indebted to Dr. Yan-Rung Lin, Dr. Shie-Chang Jeng, and Dr. Chi-Chang Liao (EOL/ITRI, Taiwan), and Dr. Yung-Hsun Wu (Innolux, Taiwan) for technical discussions. Thank Mr. Chun-Hsiang Lo, and Mr. Hung-Shan Chen for the technical assistance. This work is supported by National Science Council (NSC) in Taiwan under project number: 96-2112-M-009-019-MY2.

8. References

Bahadur, B. (1992). *Liquid Crystals Applications and Uses,* World Scientific, 9810201117, Singapore

Burrows, P. E.; Gu, G.; Bulovic, V.; Shen , Z.; Forrest, S. R. & Thompson, M. E. (1997). Achieving full-color organic light-emitting devices for lightweight, flat-panel displays. *IEEE T Electron Dev.*, 44, 1188-1203, 0018-9383

Buyuktanir, E. A.; Mitrokhin, M.; Holter, B.; Glushchenk, A. & West, J. L. (2006). Flexible bistable smectic-A polymer dispersed liquid crystal display. *Jpn. J. Appl. Phys. Part 1*, 45, 4146-4151, 0021-4922

Chari, K.; Rankin, C. M.; Johnson, D. M.; Blanton, T. N. & Capurso, R. G. (2006). Singlesubstrate cholesteric liquid crystal displays by colloidal self-assembly. *Appl. Phys. Lett.*, 88, 043502, 0003-6951

Cole, H. S. & Kashnow, R. A. (1977). A new reflective dichroic liquid-crystal display device. *Appl. Phys. Lett.*, 30, 619-621, 0003-6951

Comiskey, B.; Albert, J. D.; Yoshizawa, H. & Jacobson, J. (1998). An electrophoretic ink for all-printed reflective electronic displays. *Nature*, 394, 253-255, 0028-0836

Crawford, G. P. (2005). *Flexible Flat Panel Displays,* Wiley, 9780470870488, England

Crowley, J. M.; Sheridon, N. K. & Romano, L. (2002). Dipole moments of gyricon balls. *J. Electrostat.*, 55, 247-259, 0304-3886

Daniel, J.; Arias, A. C.; Wong, W.; Lujan, R.; Ready, S.; Krusor, B. & Street, R. (2007). Jetprinted active-matrix backplanes and electrophoretic displays. *Jpn. J. Appl. Phys. Part 1,* 46, 363-1369, 0021-4922

Drzaic, P. S. (1995). *Liquid Crystal Dispersions,* World Scientific, 9810217455, Singapore

Du, F. & Wu, S. T. (2003). Curing temperature effects on liquid crystal gels. *Appl. Phys. Lett.,* 83, 1310-1312, 0003-6951

Fan, Y. H.; Ren, H. W.; Liang, X.; Lin, Y. H. & Wu, S. T. (2004). Dual-frequency liquid crystal gels with submillisecond response time. *Appl. Phys. Lett.,* 85, 2451-2453, 0003-6951

Gelinck, G. H.; Huitema, H. E. A.; Veenendaal, E. Van; Cantatore, E.; Schrijnemakers, L.; H. Van der Putten, J. B. P.; Geuns, T. C. T.; Beenhakkers, M.; Giesbers, J. B.; Huisman, B. H.; Meijer, E. J.; Benito, E. M.; Touwslager, F. J.; Marsman, A. W.; Van Rens, B. J. E & De Leeuw, D. M. (2004). A Flexible active-matrix displays and shift registers based on solution-processed organic transistors. *Nature Mater,* 3, 106-110, 1476-1122

Gu, G.; Burrows, P. E.; Venkatesh, S.; Forrest, S. R. & Thompson, M. E. (1997). Vacuumdeposited, nonpolymeric flexible organic light-emitting devices. *Opt. Lett.,* 22, 172- 174, 0146-9592

Hohnholz, D.; Okuzaki, H. & MacDiarmid, A. G. (2005). Plastic electronic devices through line patterning of conducting polymers. *Adv. Funct. Mater.,* 15, 51-56, 1616-301X

Khan, A.; Shiyanovskaya, I.; Schneider, T.; Miller, N.; Ernst, T.; Marhefka, D.; Nicholson, F.;

Green, S.; Magyar, G. (2005). Reflective cholesteric displays: from rigid to flexible. *J. of Soc. Inform. Displays,* 13, 169-474, 1071-0922

Khan, A.; Shiyanovskaya, I.; Schneider, T.; Montbach, E.; Davis, D. J.; Miller, N.; Marhefka, D.; Ernst, T.; Nicholson, F. & Doane, J. W. (2007). Progress in flexible and drapable reflective cholesteric displays. *J. of Soc. Inform. Displays,* 15, 9-16, 1071-0922

Khoo, I. C. & Wu, S. T. (1993). *Optics and Nonlinear Optics of Liquid Crystals,* World Scientific, 9810209355, Singapore

Kim, I.; Kim, J. H.; Kang, D.; Agra-Kooijman, D. M. & Kumar, S. (2002). Fabrication of electro-optic devices using liquid crystals with a single glass substrate. *J. Appl. Phys.,* 92, 7699-7701, 0508-3443

Kim, J. H.; Vorflusev, V. & Kumar, S. (2004). Single glass substrate LCDs using a phase separated composite organic film method. *Displays,* 25, 207-213, 0141-9382

Krasnov, A. N. (2002). High-contrast organic light-emitting diodes on flexible substrates. *Appl. Phys. Lett.,* 80, 3853-3855, 0003-6951

Lin, Y. H.; Ren, H. & Wu, S. T. (2004). High contrast polymer-dispersed liquid crystal in a 90 Cell. *Appl. Phys. Lett.,* 84, 4083–4085, 0003-6951

Lin, Y. H.; Ren, H.; Gauza, S.; Wu, Y. H., Liang, X. & Wu, S. T. (2005). Reflective direct-view displays using a dye-doped dual-frequency liquid crystal gel. *J. Display Technology,* 1, 230-233, 1551-319X

Lin, Y. H.; Ren, R.; Gauza, S.; Wu, Y. H., Zhou, Y. & Wu. S. T. (2006). High contrast and fast response polarization-independent reflective display using a dye-doped dualfrequency liquid crystal gel. *Mol. Cryst. Liq. Cryst.,* 453, 371-378, 1542-1406

Lin, Y. H.; Ren, R.; Wu, Y. H.; Li, W. Y.; Liang, X. & Wu, S. T. (2006). High performance reflective and transflective displays using guest-host liquid crystal gels. *SID Tech. Digest,* 37, 780-782, 0003-966X

Lin, Y. H.; Ren, R.; Wu, Y. H.; Li, W. Y.; Liang, X. & Wu, S. T. (2006). High performance reflective and transflective displays using guest-host liquid crystal gels. *SID Tech. Digest*, 37, 780-782, 0003-966X

Lin, Y. H.; Ren, H.; Gauza, S.; Wu, Y. H., Zhao, Y.; Fang, J. & Wu, S. T. (2006). IPS-LCD using a glass substrate and an anisotropic polymer film. *J. Display Technology*, 2, 21- 25, 1551-319X

Lin, Y. H.; Ren, H. & Wu, S. T. (2008). Polarisation-independent liquid crystal devices. *Liquid Crystal Today*, 17, 2-8, 1358-314X

Lin, Y. H.; Yang, J. M.; Lin, Y. R.; Jeng, S. C. & Liao. C. C. (2008). A polarizer-free flexible and reflective electrooptical switch using dye-doped liquid crystal gels. *Opt. Express*, 16, 1777-1785, 1094-4087

Lin, Y. H.; Yang, J. M. ; Jeng, S. C.; Lin, Y. R. & Liao. C. C. (2008). Flexible and reflective polarizer-free liquid crystal displays using dye-doped liquid crystal gels, *Proc. SPIE*, pp. 691108, San Jose, Ca, USA, January 2008

Lin, Y. H. & Yang, C. M. (2009). A polarizer-free three step switch using distinct dye-doped liquid crystal gels. *Appl. Phys. Lett.*, 94, 143504, 0003-6951

Liu, K. H.; Chou, W. Y.; Liao, C. C.; Ho, C. T. & Shieh, H. P. (2006). Microcell liquid crystal film for high-contrast flexible display applications. *Jpn. J. Appl. Phys.* 45, 7761-7765, 0021-4922

Liu, K. H. ; Lee, C. Y. ; Ho, C. T. ; HCheng, H. L. ; Lin, S. T. ; Tang, H. C. ; Kuo, C. W. ; Liao, C. C.; Shieh, H. P. & Chou, W. Y. (2007). Innovative plasma alignment method in flexible liquid crystal display films. *Electrochem. Solid State Lett.*, 10, J132-J135, 1099-0062

Mach, P.; Rodriguez, S. J.; Nortrup, R.; Wiltzius, P. & Rogers, J. A. (2001). Monolithically integrated, flexible display of polymer-dispersed liquid crystal driven by rubberstamped organic thin-film transistors. *Appl. Phys. Lett.*, 78, 3592-3594, 0003-6951

Penterman, R.; Klink, S. L.; Koning, H. de; Nisato, G. & Broer, D. J. (2002). Single-substrate liquid-crystal displays by photo-enforced stratification. *Nature*, 417, 55-58, 0028-0836

Raynes, P. (2002). Liquid crystal painting. *Nature*, 417, 28-29, 0028-0836

Ren, H. W. & Wu, S. T. (2003). Tunable electronic lens using a gradient polymer network liquid crystal. *Appl. Phys. Lett.*, 82, 22-24, 0003-6951

Ren, H. W.; Wu, S. T. & Lin, Y. H. (2007). Single glass substrate liquid crystal device using electric field-enforced phase separation and photoinduced polymerization. *Appl. Phys. Lett.*, 90, 191105, 0003-6951

Ren, H.; Wu, S. T. & Lin, Y. H. (2008). In-situ observation of fringing field-induced phase separation in a liquid crystal and monomer mixture. *Phys. Rev. Lett.*, 100, 117801, 0031-9007

Sheraw, C. D.; Zhou, L.; Huang, J. R.; Gundlach D. J.; Jackson, T. N.; Kane, M. G.; Hill, I. G.; Hammond, M. S.; Campi, J.; Greening, B. K.; Francl, J. & West, J. (2002). Organic thin-film transistor-driven polymer-dispersed liquid crystal displays on flexible polymeric substrates. *Appl. Phys. Lett.*, 80, 1088-1090, 0003-6951

Sugimoto, A.; Ochi, H.; Fujimura, S.; Yoshida, A.; Miyadera, T. & Tsuchida, M. (2004). Flexible OLED displays using plastic substrates. *IEEE J. Sel. Top. Quantum Electron*, 10, 107-114, 1077-260X

Vogels, J. P. A.; Klink, S. I.; Penterman, R.; Koning, H. D.; Huitema, E. E. A. & Broer, D. J. (2004). Robust flexible LCDs with paintable technology. *Soc. Inform. Display Tech. Dig.*, 35, 767-769, 0003-966X

Wang , P. C. & MacDiarmid, A. G. (2007). Integration of polymer-dispersed liquid crystal composites with conducting polymer thin films toward the fabrication of flexible display devices. *Displays*, 28, 101-104, 0141-9382

Wen, C. H. & Wu, S. T. (2005). Dielectric heating effects of dual-frequency liquid crystals. *Appl. Phys. Lett.*, 86, 231104, 0003-6951

White, D. L. & Taylor, G. N. (1974). New absorptive mode reflective liquid-crystal display device. *J. Appl. Phys.*, 45, 4718-4723, 0508-3443

Wu, S. T. & Yang, D. K. (2001). *Reflective Liquid Crystal Displays,* Wiley, 0471496111, England

Yang, D. K.; West, J. L.; Chien, L. C. & Doane, J. W. (1994). Control of reflectivity and bistability in displays using cholesteric liquid crystals. *J. Appl. Phys.*, 76, 1331-1333, 0508-3443

Yang, D. K. & Wu, S. T. (2006). *Fundamentals of Liquid Crystal Devices,* Wiley, 047001542X, England

Yang, D. K. (2008). Review of operating principle and performance of polarizer-free reflective liquid-crystal displays. *J. Soc. Info. Display,* 16, 117-124, 1071-0922

Yin, Y.; Shiyanovskii, S. V. & Lavrentovich, O. D. (2006). Electric heating effects in nematic liquid crystals. *J. Appl. Phys.*, 100, 024906, 0508-3443

Zhou, L. S.; Wanga, A.; Wu, S. C.; Sun, J.; Park, S. & Jackson, T. N. (2006). All-organic active matrix flexible display. *Appl. Phys. Lett.*, 88, 083502, 0003-6951

8

Active-Matrix Liquid Crystal Displays - Operation, Electronics and Analog Circuits Design

Ilias Pappas, Stylianos Siskos and Charalambos A. Dimitriadis
Physics Department
Aristotle University of Thessaloniki
Thessaloniki,
Greece

1. Introduction

For more than four decades, Cathode Ray Tube (CRT) Displays have been the dominant display technology providing very attractive performance. Brightness, contrast ratio, high image quality, speed and resolution were the main high standard specifications that CRTs were satisfied.

The last two decades, there was a tremendous growth in small portable applications which required the necessary adjustment of the display technology to them. The large depth of the CRTs was the main disadvantage for preventing them to be used in these kinds of applications. Flat Panel Displays seem to be the most attractive solution to this problem. Displays engineers searched for many years in order to find the suitable flat panel display technologies that could replace CRT displays. The first successfully established flat panel technology was the plasma displays, which demonstrated to be of larger size and higher image quality compared to the CRT technology. However, the problem with the integration of plasma displays in small portable applications still exists. Finally, the inroad of the thin-film transistors liquid crystal displays (TFT-LCD), in late 1990's, was a milestone in the displays industry and technology.

The successful development of the TFT-LCDs was achieved not accidentally. It was the sequence of the liquid crystal cell technology development, in combination with the development of semiconductors technologies for large-area microelectronics on glass, like thin-film transistors. Although, both technologies were very-well known before the 90's, an extended research for establishing compatible fabrication processes for the materials and the manufacturing equipment has led to the TFT-LCDs realization.

TFT-LCDs were rapidly grown and dominated the displays industry, especially in small portable applications. The implementation of the TFT-LCD panel peripheral driving components with low-power CMOS blocks and, therefore, the compatibility with battery operation was the main reason for the ascendance of the TFT-LCD technology in small portable applications. Today, the TFT-LCD market has been expanded. They can be used in an extremely wide range of our everyday life products, like mobile phone applications, ATMs, PDAs, navigation systems, notebook PCs and home applications, such as wide screen TVs.

Source: New Developments in Liquid Crystals, Book edited by: Georgiy V. Tkachenko,
ISBN 978-953-307-015-5, pp. 234, November 2009, I-Tech, Vienna, Austria

At the beginning, when LCDs were used in calculators, watches and small sized displays, direct and passive matrix addressing were the applicable addressing methods. As the size, resolution and information content of the displays were increased, the number of the pixels array was, also, increased leading the existing addressing methods to become non-applicable. A solution to this problem was proposed by Lencher (Lechner et al., 1971) and by Marlone (Marlowe & Nester, 1972). A switch TFT was added at each pixel of the display matrix and in this way the pixels were controlled independently with the use of the external driving voltages. The first external voltage controls whether the switch TFT will be turned "ON" or "OFF" and the second voltage is the necessary, to the liquid crystal, bias voltage which is stored to a capacitor. The storage capacitor is placed in parallel to the liquid crystal and ensures that the necessary liquid crystal voltage remains constant during the frame time. This addressing method is called Active Matrix and the displays that used this method are called Active Matrix Liquid Crystal Displays (AMLCDs).

In this chapter, the fundamentals of an AMLCD will be presented. The fundamentals include the operation description, the driving methods and circuitry and finally the analog circuits design by using polycrystalline silicon TFTs. At the beginning of the chapter, a short presentation of the thin-film transistors technology, including structures and operations modes, will be given. More attention will be given to the analog circuit design due to the difficulties that arise by implementing TFTs in circuits. Moreover, design techniques for overcoming these difficulties will be described in more details.

2. Thin-Film Transistors

2.1 Historical overview

In the past twenty years, the development of Thin-Film Transistors (TFTs) has become the spearhead of the electronic flat panel displays industry. However, the generation of TFTs is originated many years before, back to the earliest days of semiconductors physics. The TFTs principals and their potential utilities were settled nearly seventy years ago, but the remarkable development of the bipolar transistors and their technological cousin, the metal oxide semiconductor field-effective transistors (MOSFETs), has overshadowed the TFTs concept.

The first attempt for implementing a TFT device is traced back to the 40's and it was a thin-film field effect device used a germanium film (Bardeen & Brattain, 1948). The history and the structure of the TFTs, as it is known today, began with the work of Weimer (P. K. Weimer, 1962) at RCA Labs. in 1962. Polycrystalline cadmium sulphide (CdS) was the material that used for the thin-film and silicon monoxide was the insulator. Source and drain contacts were placed on the opposite side of the gate. This structure is called "staggered" and it is shown in fig. 1. Glass was used as the insulating substrate, forming in this way a three-terminal device. On the theoretical part, the analysis of the device characteristics was realized by Borkan and Weimer (Borkan and Weimer, 1963), based on Shockley's JFET analysis.

In the 70's, two very important events changed dramatically the prospects for TFTs. The first was the implementation of a thin-film semiconductor instead of crystalline bulk silicon material, like Cadmium Selenide (CdSe) (Broody et al., 1973). The impact of this implementation was the reduction of the fabrication cost and the decrease of the transistor size. The second landmark was the Active Matrix addressing method proposed by (Lechner et al., 1971) and the fact that the switch device needed in each pixel of the matrix can be materialized with the use of a TFT device.

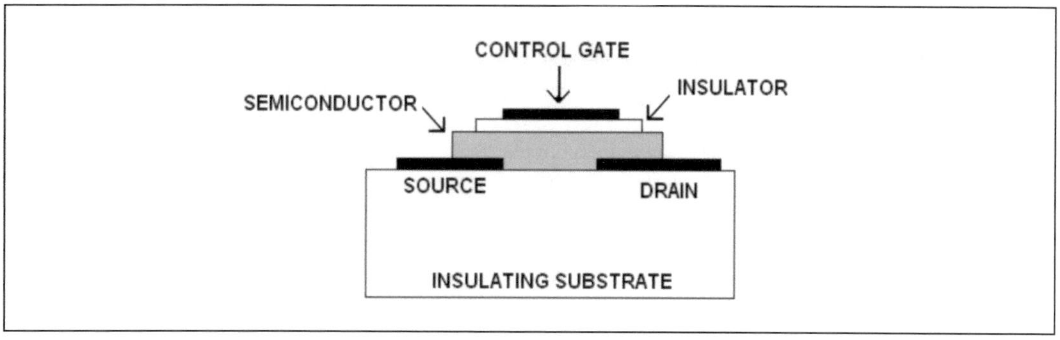

Fig. 1. Schematic diagram of Weimer's staggered CdS TFT structure.

The period from 1979 to 1981 was revolutionary for the TFT technology. In 1979, a new material was introduced for the implementation of the thin-film. (LeComber et. al., 1979) proposed a TFT using amorphous hydrogenated silicon (a-Si:H) as the active layer material. Amorphous hydrogenated silicon was preferred over the pure amorphous silicon due to the fact that it can be doped with both donors and acceptors forming n-type and p-type TFT devices. Thin-film, formed with a-Si:H, have no grain boundaries and it was the most cost-effective technology. However, the main disadvantage of this technology was the low mobility of the carriers and the instability of the device electrical characteristics.

In 1980, Lueder (Lueder, 1980) introduced a new approach in the TFT fabrication process. Photolithography was used instead of shadow masks that were being used since then. Their devices were optimized for liquid crystal displays applications, since the photolithography process made source-drain contacts that were self-aligned to the gate, causing the gate-source parasitic capacitance to be reduced. The compensation of the parasitic effects has improved the performance of the TFT-LCDs in the terms of the response time and the image quality.

In 1981, Depp (Depp et. al., 1981) from IBM proposed a polycrystalline silicon active layer. The use of chemical vapour-deposited polycrystalline silicon has improved the characteristics of the TFTs devices since the mobility has shown a significant increase. However, the chemical vapour-deposition was a high temperature process making the glass substrate in-appropriate. This was the reason for the replacement of the glass substrate with a high temperature substrate such as quartz.

In the forcoming years until now, the research of the TFT technology was focused on the improvement of the TFT devices characteristics and their fabrication process. Amorphous and polycrystalline silicon TFTs were the two main technologies that were used in the flat panel display industry. Recently, new TFTs technologies have been proposed, like nano-crystalline silicon (Pappas et. al, 2007.b), metamorphous silicon and organic TFTs (Lin et. al, 1997). These technologies have been emerged due to the constantly increasing performance of the flat panel displays which requires improved TFTs technologies. Furthermore, the introduction of new flat panel displays, like Organic Light Emitting Diode (OLED) displays, and the flexible substrate displays have created the need for new TFT technologies, compatible with the new types of flat panel displays.

2.2 Amorphous and polycrystalline silicon TFTs structures

Polycrystalline and amorphous silicon are the two most commonly used types of thin-film transistors in the display industry. Fig. 2 shows four basic structures of a-Si:H TFTs according to the positions of the electrodes, insulator and the active layer. Staggered structures have already being described, whereas coplanar TFTs have all electrodes on the same side of the active layer. For better control of the active layer, the inverted structures of the above have been proposed. The inverted structures are also called bottom-gated structures. Amorphous silicon TFT technology can provide only n-channel devices because their electronic properties do not allow high-quality p-channel devices, since the ON-current is significant low. This is caused due to the low mobility (less than unity) of the holes along the active layer. The amorphous silicon TFTs market includes notebooks, desktop monitors and home LCDs televisions. The main advantage of the a-Si:H technology is the easy fabrication process with limited number of process steps making this technology cost efficient.

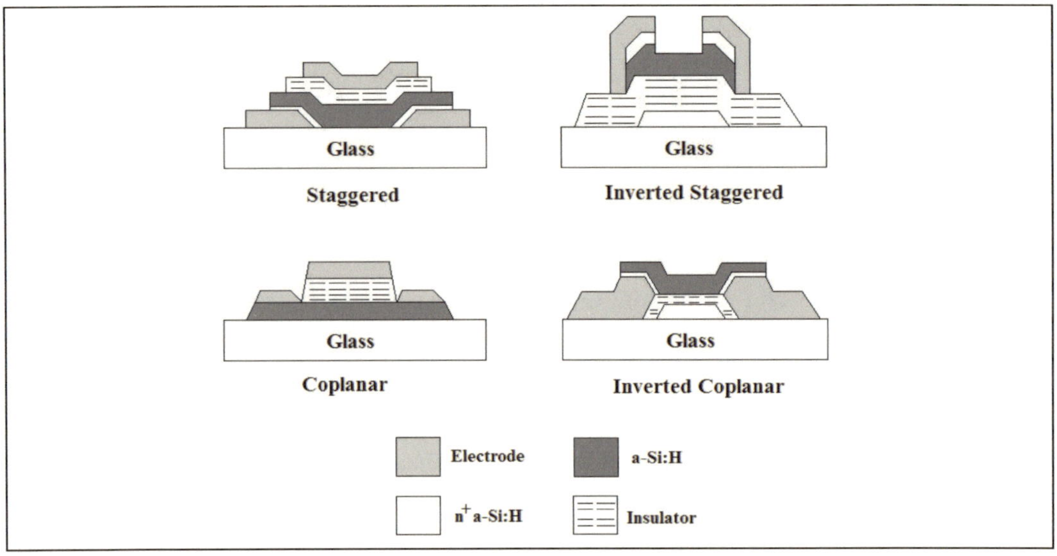

Fig. 2. Basic a-Si:H TFT structures.

On the other hand, polycrystalline silicon (poly-Si) technology can provide both n-type and p-type devices. Depending on the fabrication process, poly-Si TFT can be divided into high-temperature TFTs with fabrication processes similar to the ordinary silicon MOSFET devices and the low-temperature TFTs where the deposited amorphous silicon film is turned into polycrystalline silicon either with excimer laser annealing (ELA) or with solid state phase crystallization combined with ELA process. The basic top-gated poly-Si TFT structure is shown in fig. 3.

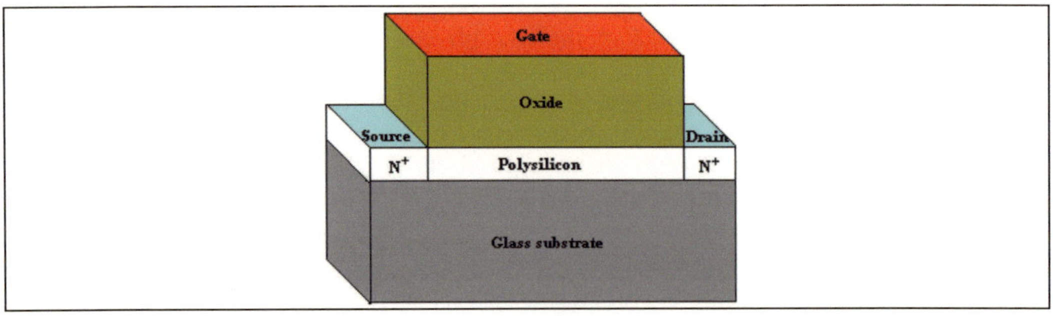

Fig. 3. Poly-Si top-gated TFT structure.

The main advantage of the poly-Si TFTs technology is the high carrier mobility (more than 100 cm^2 / Vs), meaning that poly-Si TFTs can provide high driving current. The high driving current has led to the implementation of the peripheral driving circuits of an AMLCD with poly-Si TFTs, causing the reduction of the total fabrication cost and the realization of Systems on Glass (SoG).

2.3 Polycrystalline and amorphous silicon TFT operation

The operation of both poly-Si and a-Si TFTs is similar to the ordinary silicon MOSFETs operation. Typical transfer characteristics of n-type and p-type poly-Si and n-type a-Si TFTs, with gate dimensions W / L = 100 μm / 10 μm, are shown in fig. 4. As it can be seen from fig. 4, poly-Si TFTs have higher ON-current due to the higher carrier mobility. The TFTs operation can be divided into three working modes: cut-off, linear and saturation modes. The drain current of an n-type TFT can be presented with the following expressions for each mode of operation:

Cut-off mode: $I_{DS} = 0$, when $V_{gs} < V_{thn}$
Linear mode: $I_{DS} = \mu C_{ox} W [V_{eff} V_{ds} - V_{ds}^2/2]/L$, when $0 < V_{ds} < V_{eff}$
Saturation mode: $I_{DS} = [\mu C_{ox} W V_{eff}^2]/2L$, when $V_{eff} < V_{ds}$

where I_{DS} is the drain current, μ is the effective surface mobility of the carriers, C_{OX} is the gate oxide capacitance per unit area, W is the effective gate mask width, L is the effective gate mask length, V_{ds} is the drain-to-source voltage and V_{eff} is the effective gate voltage, equal to the different between the gate to source voltage V_{gs} and the transistor threshold voltage V_{thn}, $V_{eff} = V_{gs} - V_{thn}$.

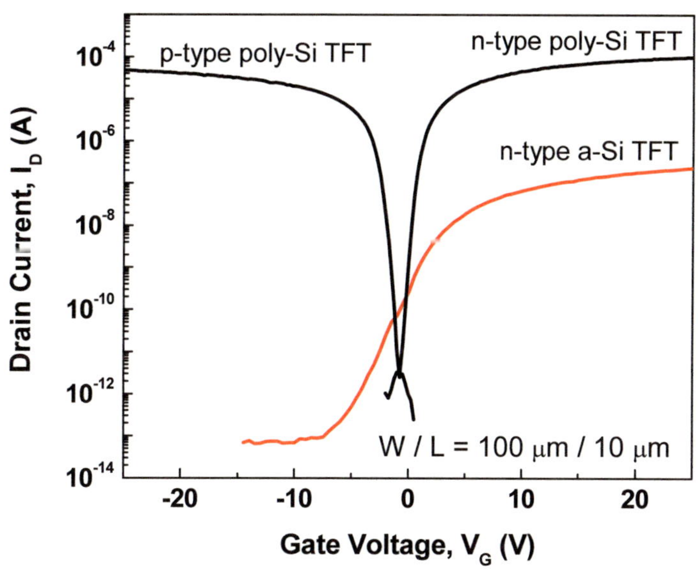

Fig. 4. Transfer characteristics of polysilicon and amorphous TFTs.

2.4 Instabilities of polycrystalline and amorphous TFTs

The expressions of the drain current for each working mode indicate the important role that the threshold voltage and mobility play in the TFTs operation. However, measurements in fabricated TFTs have shown that threshold voltage and mobility variations exist from device to device, even if the devices are implemented on the same wafer and with the same fabrication process.

For the poly-Si TFT technology, the threshold voltage and mobility variations are caused due the random distribution of the grain boundaries along the channel (Jagar et. al., 2003). The grain boundaries are produced when the amorphous silicon is turned into polycrystalline during the fabrication process and they can modeled as discontinues for the carriers along the active layer. For small sized display, the threshold voltage variation is about ± 300 mV (Zhang et. al., 2000), while for large substrate area it can be up to ± 1 V (Lee et. al., 1999).

Amorphous silicon TFT technology exhibits good uniformity in the electrical characteristics of the TFT devices over a large area even if some of the characteristics are rather poor, like the mobility. However, protracted bias stress will cause threshold voltage shift due to the inherent instability of the amorphous material (Powell et. al., 1989). The threshold voltage shift can be modeled by the equation (Den Boer, 2005)

$$\Delta V_{thn} = (V_{gs} - V_{thn}) \bullet \left(\frac{t}{\tau} \right)^{\beta} \tag{1}$$

where ΔV_{thn} is the threshold voltage shift in the case of an n-type TFT and β and τ are constants that depend on the temperature and the quality of the active layer-insulator interface. Threshold voltage shift can be more than 3 V (Arokia et.al., 2005) and even if it is temporal in nature, it can be considered to have the same impact on the circuits design as the threshold voltage variations of the poly-Si TFTs.

Another second order effect that produces instabilities in the TFTs operation is the impact ionization of the carriers which is known as "kink effect" (Valdinoci et. Al., 1997)). Kink effect appears when the device is operating in the saturation mode and for high values of the drain voltage. The result of the kink effect is an abrupt increase of the drain current caused by the impact ionization of the carriers near the drain electrode due to the high electric field. Kink effect can not be controlled, causing difficulties in the analog circuits design especially when high supply voltages are required. Among the variations that produce instabilities in the TFTs operation, the threshold voltage variations have the highest impact in the analog circuits design (Vaidya et. al., 2008) and compensation methods will be described in a next paragraph.

3. Addressing methods

There are three different addressing methods in the display technology, direct, passive matrix and active matrix. The addressing method that will be used in a LCD design is a very important choice, because the peripheral driving circuits of the pixel array depend on it. The addressing method can be selected by taking into account the specifications of the information content, the fabrication cost, the available TFT technology, the response time, the area and the power consumption, with the first three parameters being the most important.

3.1 Direct addressing method

The first used addressing method was the direct method. According to this method, its segment is directly connected and controlled individually by the peripheral electronics. The segments are arranged in such way so that they can produce the desired icon. The most common arrangement is the 7-segments, shown in fig. 5, used in simple alphanumerical displays, such as calculators and watches. In this method, no multiplexing is available and this is the reason for only being used in low information content applications. In the direct method, the smallest controllable component for the image production is called segment instead of pixel, which will be used in the two other addressing methods.

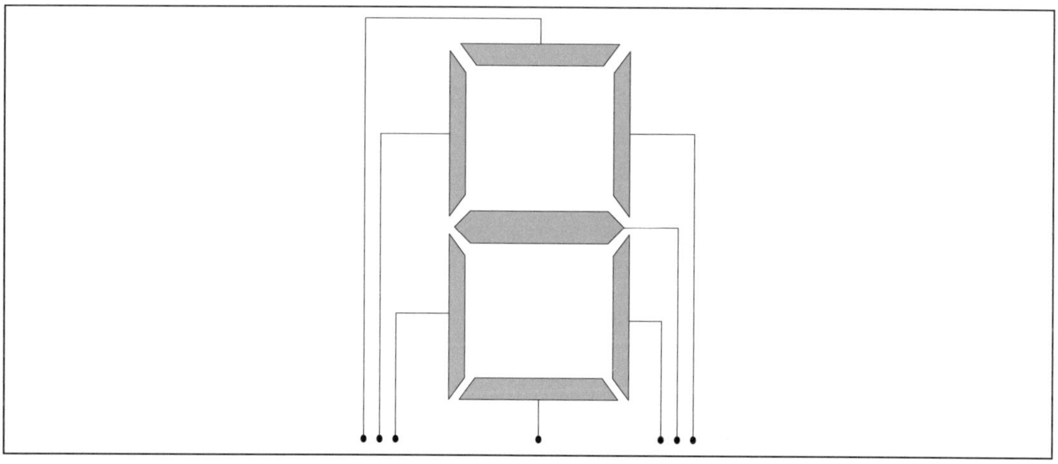

Fig. 5. 7-Segment arrangement with direct addressing method.

3.2 Passive matrix addressing method

As the displays and the information content were getting larger, the need for more picture elements was arisen. The solution to this problem was the modification of the segments arrangement into a pixels matrix with M rows and N columns. In this method, each pixel can not be controlled individually and a multiplexing addressing approach has to be realized. The new multiplexed addressing method was the passive matrix (PM). The configuration of the passive matrix addressing method is shown in fig. 6.

The passive matrix is a one-line-at-a-time driving method. During the programming time, a pulse from the row peripheral driver activates all the pixels of the programming line and at the same time the data voltage is delivered to the storage capacitor and the liquid crystal through the peripheral column driver. Passive matrix is the addressing method with the minimum number of interconnections. For example, for an M rows and N columns array, the direct method needs MxN interconnections while the passive matrix method needs M+N interconnections. Furthermore, passive matrix is the simplest and the most cost-efficient method. However, the disadvantages of the passive matrix method are the low multiplexing capability and the crosstalk effect between the pixels. Crosstalk effect is caused because all the row-pixels are electrically related and a small dc voltage can be added to the pixel liquid crystal voltage from its neighbour pixels. The results of the crosstalk are the poor contrast ratio and the small active region of operation for the displays.

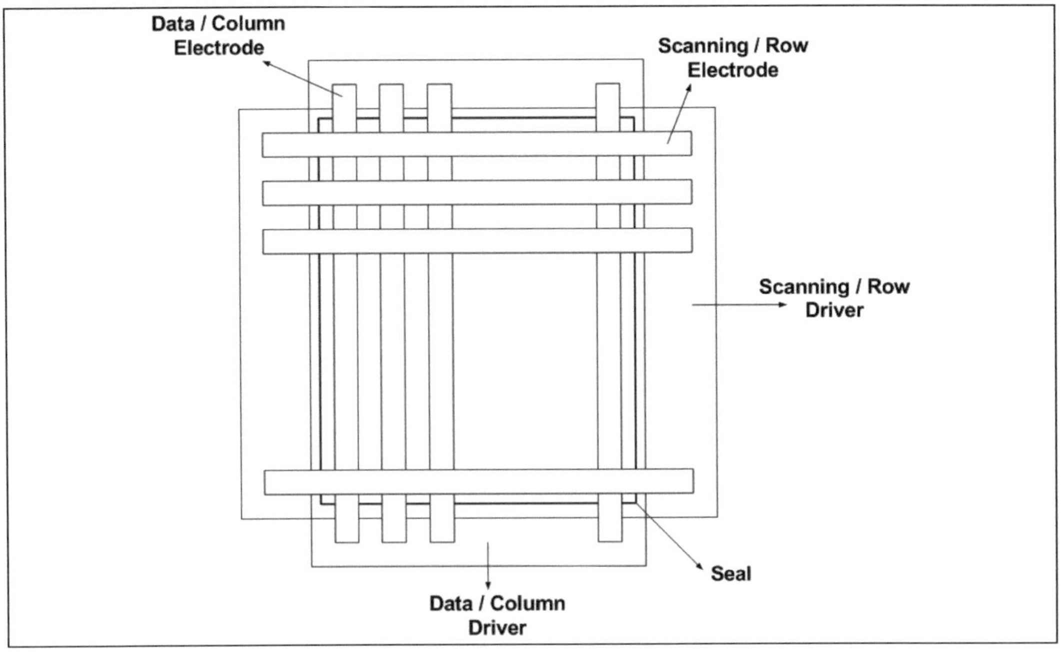

Fig. 6. Configuration of the passive matrix addressing method.

3.3 Active Matrix addressing method

The Active Matrix (AM) addressing method overcomes the multiplexing limitation of the PM method and the crosstalk effect. This can be achieved by incorporating a nonlinear control element, like a switch, in the cross point of the row and column lines (in series connection) of each pixel. The use of a switch will provide a 100 % duty ratio for the pixel by using the charge stored at the pixel during the row addressing time. Figure 7 illustrates the configuration of an AMLCD. The switch is controlled by two pulse signals which are produced by external driving circuits, the row and column drivers. As in most of the matrix addressed displays with line-at-a-time programming, the rows are scanned by a select gate pulse. The gate pulse of the selected row will turn "ON" the switch TFT of each pixel and simultaneously, the storage capacitor will be charged with the data voltage provided from the column driver. After the row time, the switch TFT will turn "OFF" as soon as the negative edge of the row pulse is delivered and the pixel will be isolated from all its neighbour pixels until the next frame time. In this way, the crosstalk effect and the multiplexing limitation are eliminated.

4. Active Matrix liquid crystal display electronics

The design of the displays electronics modules is an essential task due to the fact that the design defines the image quality of the display. The functionality of the electronics involves the acquisition of the incoming video signal, the signal processing and finally, the representation of the signal image on the pixels matrix. Therefore, the main objection of the electronics modules is the accurate supply of the data signal to each pixel with the right timing.

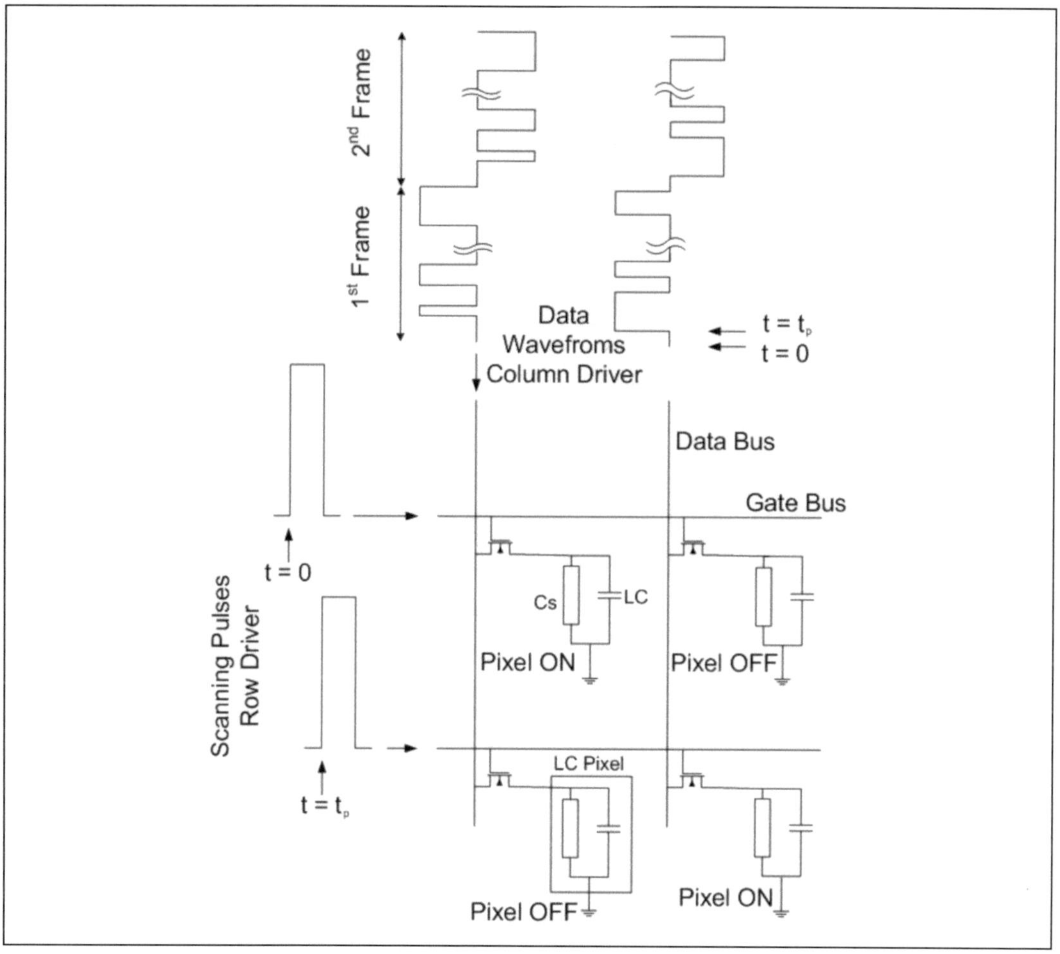

Fig. 7. Configuration of the active matrix addressing method.

4.1 Electronics modules overview

Two of the most important parameters in the AMLCD design are the definition of the addressing method and the size of the pixels matrix. These two parameters will determine the peripheral driving electronics modules and the timing and programming issues. Furthermore, the types of displays format are separated based on the size of the pixels matrix, proving the significance of this parameter. The various displays format with their timing specifications are shown in Table 1.

The resolution of the display is referred to the number of the active pixels in each dimension, horizontal (rows) and vertical (columns). The first given number is the number of the columns and the second is the rows number. Higher resolution results in better image quality of the display. However, higher resolution for a standard matrix size requires additional hardware for its implementation; higher resolution is achieved by decreasing the size of the pixels. Furthermore, the ratio of the column number to the row number will determine the aspect ratio of the display.

Video Format	Resolution	H-Sync. @60Hz	System Clock	Panel Size
VGA	640x480	31 μs	28 MHz	≤ 10.4"
SVGA	800x600	26 μs	40 MHz	10.4", 12.1"
XGA	1024x768	20 μs	65 MHz	12.1", 13.3", 14
QVGA	1280x960	16 μs	104 MHz	13.3", 14.1"
SXGA	1280x1024	15 μs	108 MHz	14.1", 15"
SXGA+	1400x1050	15 μs	124 MHz	17", 18.1"
HDTV	1920x1080	14 μs	144 MHz	≥ 24"
UXGA	1600x1200	13 μs	184 MHz	20.1", 23"
QXGA	2048x1536	10 μs	195 MHz	≥ 30"
QSXGA	2560X2048	8 μs	256 MHz	≥ 32"
QUXGA	3200X2400	6 μs	400 MHz	≥ 36"

Table 1. Displays format (VGA: Video Graphic Array, S : Super, X : Extended, Q : Quarter, U : Ultra, HDTV : High Definition Television)

$$aspect\,ratio = \frac{columns\,number}{rows\,number} \tag{2}$$

For example, in the case of a VGA display the aspect ratio is 4:3 while in HDTV is 16:9.
Another important specification is the refresh rate of the display. Refresh rate is the number of times that an image is refreshed per second. A very usual mistake is the confusion between the refresh and frame rate. The refresh rate is the repeated illumination of identical frames, while the frame rate measures how often a display image can change into another. The refresh rate is calculated by dividing the horizontal scan rate by the number of horizontal pixels and muliplying the result by 0.95. The refresh rate is measured in hertz and a typical bandwidth is starting from 60 Hz for small displays and it can be up to 100 Hz for high motion displays. The refresh rate depends upon the monitor resolution and its maximum horizontal scan rate, as higher resolution necessitates more scan lines per second. Increased refresh rates reduce flickering and thereby reduce eye strain for a viewer. Because of this, it is advisable to purchase monitors that have a refresh rate between 75 and 85 Hz.
For the timing (clock) issues, the horizontal synchronization (H-sync) period of each display format (i.e. the programming time of all the pixels of a row for a given refresh rate of 60 Hz) are shown in Table 1. The H-sync period is given by the following expression.

$$H - Sync. = \frac{1}{refresh\,rate} \div rows\,number \tag{3}$$

The system clock frequency is, also, depending strongly on the matrix size since it is given by the expression.

$$System\,clock = \frac{1}{H - Sync. \div columns\,number} \tag{4}$$

Figure 8 shows a block diagram of AMLCD electronics modules. An AMLCD consists of the pixels matrix, the peripheral column and row driving modules, the backlight, a timing - control unit and a booster module for producing DC voltages higher than the power supply voltage.
The pixels array and the row / column drivers are implemented using TFT technology meaning that they are fabricated on the glass substrate. The control and power supply generation blocks are separated, mounted on a PC board and connected to the row and column drivers on the one side and the host controller on the other. The control block may include level shifter, timing and analog functions generators. The objective of the control unit is to generate timing and data signals for biasing the row and column drivers, by taking as input digital signals which contain all the video, synchronization and timing information from the host system, which is a graphics controller chip.

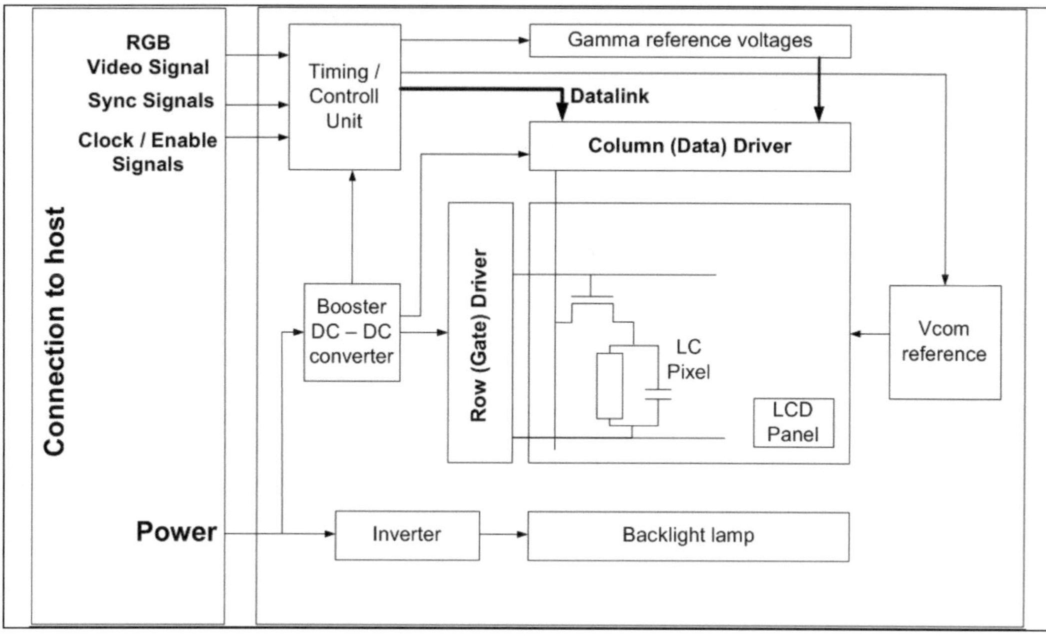

Fig. 8. Block diagram of an AMLCD architecture.

The architecture and design of the module electronics have a significant impact not only on the image quality of the display but also on the display system fabrication cost and power consumption. Figure 9 shows the contribution to the power consumption and the fabrication cost of each electronic module of a typical 10.4-in. backlit TFT-LCD. The typical power consumption of this display is 3 W at 150 cd/m^2 brightness.
Summarizing, an AMLCD designated for portable applications is expected to satisfy the specifications that are presented below:

- High Pixels Matrix (above QVGA)
- High Resolution (160 dots/inch)
- High Contrast Ratio (≥ 100 : 1)
- Full Color (8 bit/color)
- Full Motion Video (80 frames per second)
- Adequate Brightness (15-100 fl,)

- Wide Viewing Angle (≥ ± 45° in horizontal and vertical directions)
- Light Weight
- Small Volume (small depth)
- Low Power Consumption / High Luminous Efficiency
- Low Cost

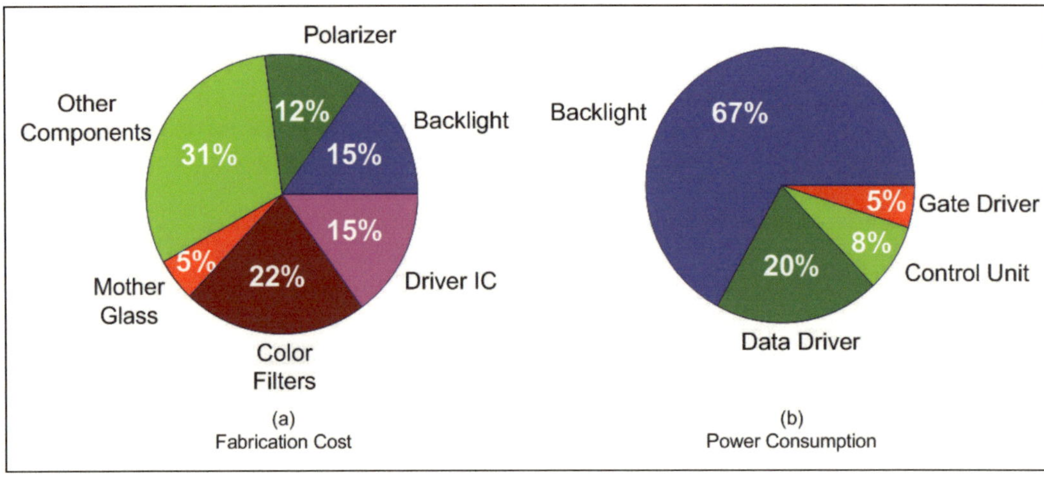

Fig. 9. (a) Fabrication cost contribution of each component. (b) Power consumption contribution of each electronic module.

4.2 Pixel structure and design

Liquid Crystal Pixels are transmission type pixels with a backlight, meaning that they are not emitting their own light. While there are many types of liquid crystal materials such as smectics, nematics and cholesterics, twisted nematic (TN) display mode is the most advanced and popular. A TN pixel cell consists of two glass substrates coated on their inner surfaces with transparent electrodes and separated by several millimeters from each other. A nematic liquid crystal material fills the space between the two substrates and two polarizers are attached on both sides of the pixel with their polarization axis crossed. The polarizer is a three-layer composite film with a stretched iodine doped polyvinyl alcohol (PVA) polarizing film in the center and two outer films for protecting the PVA film from the ambient. Since the two substrates, each having alignment layer, are oriented with their alignments perpendicular to each other, liquid molecule is twisted initially. In the voltage-off state, the polarizers are oriented perpendicular and the incoming light from a back light source, whose polarity is twisted by the liquid crystal, is transmitted through the output polarizer. When a voltage is applied to the electrodes, the director of the molecules tends to orient themselves parallel to the applied field, since liquid crystal materials have positive dielectric anisotropy. In this situation, the polarization of the light transmitted through liquid crystal is crossed to the output polarizer resulting in the cut off of the light and thus creating a black state for the display pixel. This operation is called normally white mode, while normally black mode can be achieved by changing the polarizers to a parallel orientation. Figure 10 shows the configuration a TN pixel cell in a normally white mode.
The transmission (luminance) versus the applied voltage characteristic is shown in Fig. 11. The shown characteristic is for normal viewing angle and indicates that grayscale levels can

be achieved by varying the voltage across the LCD. Unfortunately, the transmission - voltage curve is viewing angle dependent, leading to grayscale errors and color shift in a display when it is viewed from significant angles to the display normal.

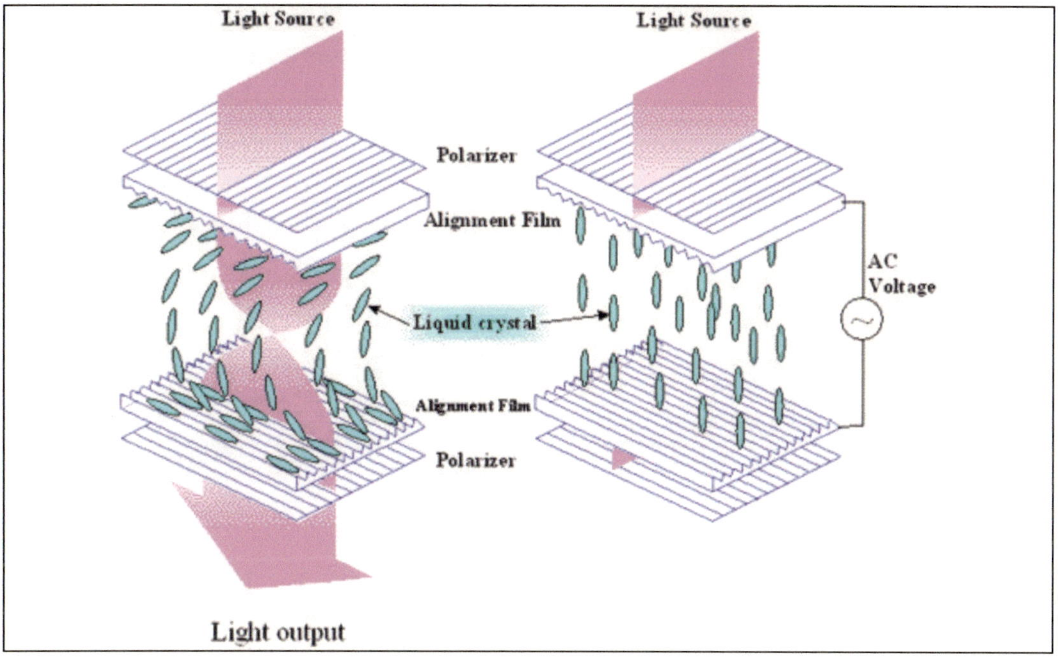

Fig. 10. Principle of operation of a TN pixels cell in a normally white mode.

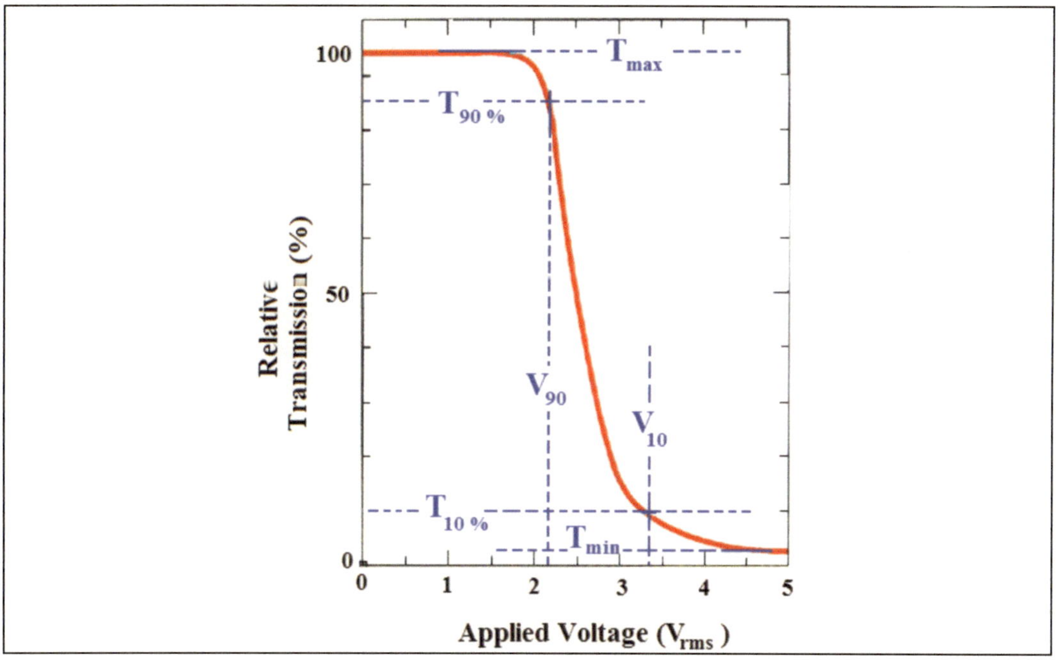

Fig. 11. Transmission versus applied voltage characteristic for normal viewing angle.

The equivalent circuit with the parasitic elements of a pixel cell and a typical TFT-LCD pixel layout are shown in fig. 12. The pixel consists of a switch TFT device, with the gate electrode connected to the row driver lines and the source electrode connected to the column driver lines. Furthermore, a storage capacitor is connected in parallel to the LC pixel capacitance.

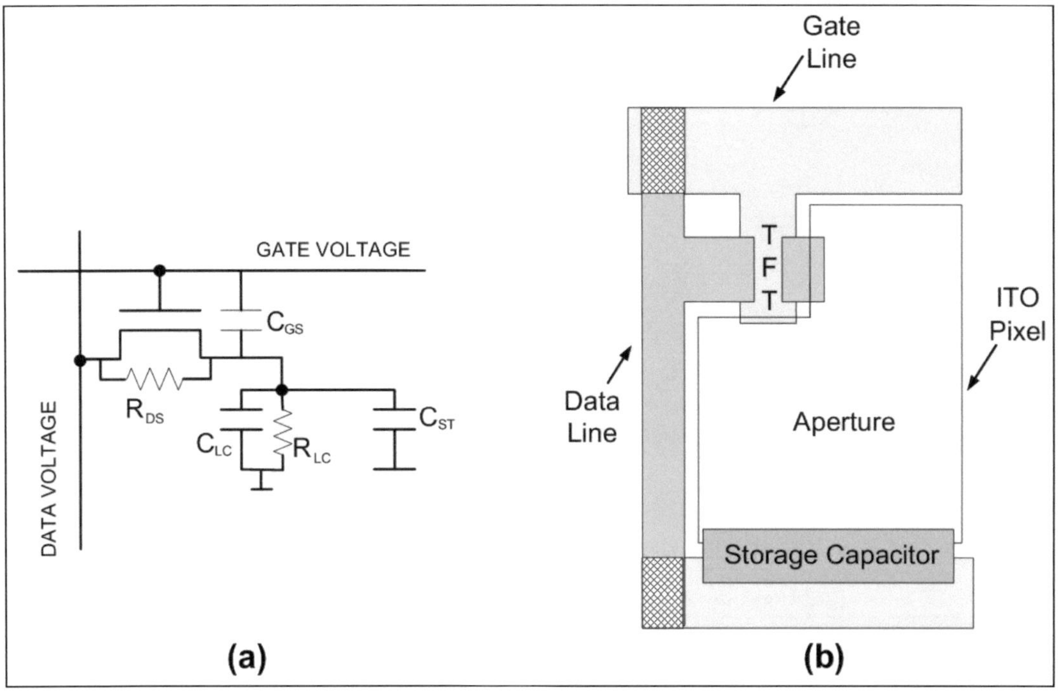

Fig. 12. (a) Equivalent circuit with the parasitic elements of a pixel and (b) typical pixel layout design.

The aperture part is the light transparent part and it is designated for the placement of the liquid crystal while the TFT, voltage lines and storage capacitor areas are non-light transparent. The ratio between the transparent portion of a pixel and its surrounding electronics is called aperture ratio or fill factor. Furthermore, in the shown layout design, the storage capacitor is connected to an adjacent row line resulting in the maximization of the aperture ration but the load capacitance of the row lines is, also, increased. The counter electrode of the LC pixel capacitor is the common ITO electrode on the opposite substrate (Den Boer, 2005). For large displays, this configuration is difficult to be used due to the large RC delay time of the row lines. In order to overcome this problem, a common storage bus can be placed in the aperture area which reduces the load capacitance of the row lines, but also reduces the aperture ration of the pixel.

The crosstalk effect is caused due to the column-line video-signal coupling during one frame and a DC component is being added to the AC data voltage. The DC component can not be entirely eliminated for all gray across the entire pixels matrix, resulting to slight difference in the pixel transmittance between the odd and even frames. A solution to this problem is the polarity inversion method. Apart from elimination of the DC component, the influence of the flicker on the display image quality is also eliminated with the use of a polarity inversion method. Four different polarity inversion methods have been widely used. Figure

13 shows the configuration of the four polarity inversion methods. The type of the polarity inversion method has an impact on the power consumption of the display. In the frame inversion method, all the pixels are driven to + V_P polarity in one frame period and then all of them are driven to – V_P polarity during the next frame period. This method is the most power-efficient method. However, this method is sensitive to the flicker and to vertical and horizontal crosstalk, meaning that this method can not be used in high image quality displays.

In the line and column inversion methods, the polarity of the pixels is alternated in the adjacent rows and columns, respectively. The line inversion method is compatible to the V_{com} modulation (Den Boer, 2005) while column inversion method is not compatible. Furthermore, the line inversion method consumes more power than the column inversion method because the capacitance of all row busses is charged and discharged every row time. Finally, the column inversion method has better results to the compensation of the flicker.

In the pixel inversion method, the polarity of each pixel is inverted from the polarity of its neighbouring pixels, by a combination of simultaneous row and column inversions. This produces the highest quality images by total elimination of the flicker and crosstalk effect. This method is not compatible with the V_{com} modulation and thus it requires high voltage column drivers leading to high power consumption.

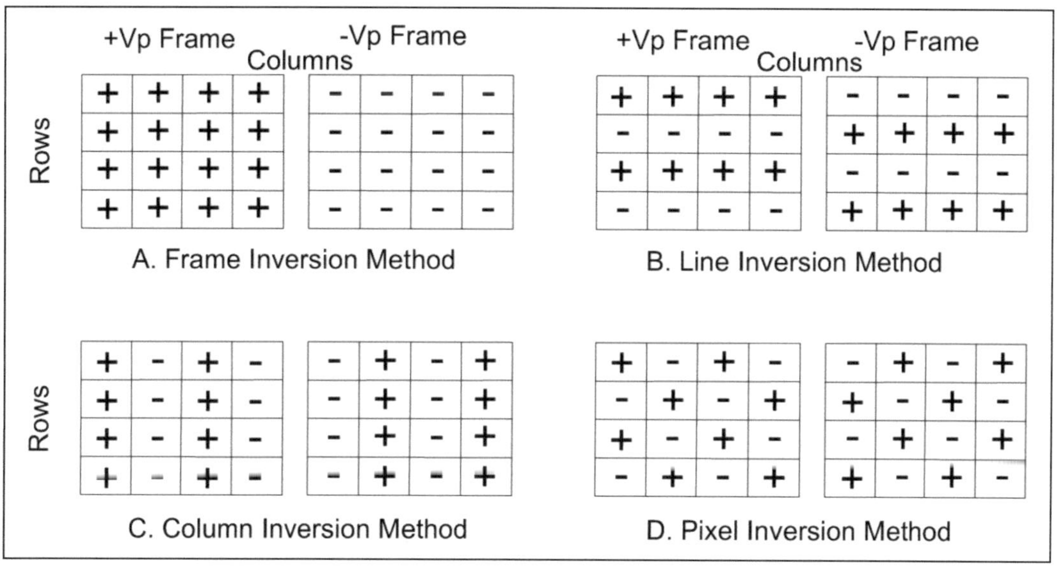

Fig. 13. Typical AMLCD polarity inversion methods.

A full color LCD display can be generated by incorporating red, green and blue color filters at the pixels. In order to produce the desirable color tone, the pixel is divided into three sub-pixels each one having red, green and blue color filter, respectively. The three sub-pixels have the same dimensions and the proper combination of each color tone; by applying the right voltages to the liquid crystals, the desired pixel emissive colour will be produced. The width of each sub-pixel is three times smaller than the sub-pixel length and when the three sub-pixels are very closely placed in parallel, a square full color pixel is produced. Figure 14 shows a full colour square pixel.

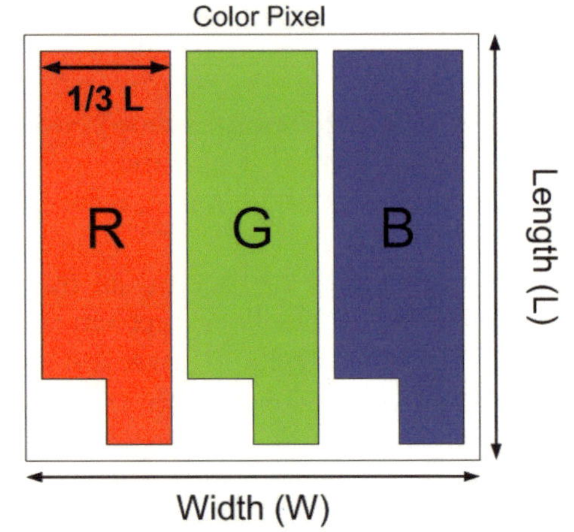

Fig. 14. Typical full color square pixel.

4.3 Row (gate) and column (data) drivers

The pixels matrix on glass substrates is directly connected to the row and column drivers, implemented for integrated circuits. Figure 15 shows the basic architecture of a row driver. The operation of the row driver is based on the generation of the gate pulse, controlling the line of pixels to be programmed by turning "ON" or "OFF" the switch TFT device of each pixel. This is the reason for often being referred as gate driver. The generated, from the row driver, pulse has to satisfy the following specification: the selected gate pulse must be higher than the most positive column voltage by at least one TFT threshold voltage. The non-selected gate pulse must be lower than the lowest column voltage by at least one TFT threshold voltage. The above specifications ensure that the selected pixel stays "turned-on" during the programming period and "turned-off" during the emission period to hold the pixel charge. The duration of the gate pulses is about 10 - 50 μs, depending on the display resolution as described in the systems clock frequency.

A row driver consists of a bi-directional shift register, level shifter, an enable system and an output buffer block. At the beginning, the input start (IS) bit selects a row at a time. If multiple cascaded chips architecture is available, the IS bit can be transferred to the second row driver and play the role of the start bit. This is the reason why IS's bit pin acts as both input and output pin. The DIR input controls the direction of the shift register, enabling the row driver to be mounted either at the left or at the right side of the display area. The enable system allows the output pulse to be turned to the non-selected edge before the end of the line time. The operation of this system is very important especially for large displays with high RC delay time of the lines, in order to avoid cross-talk effects by turning "OFF" the switch TFT devices for a few microseconds just before the data voltage is delivered to the pixels. Level shifters raise the voltages to the desirable output logical levels and the output buffers reduce the output impedance of the row driver so that the row lines can be driven by the output signals.

Column driver is more complex than the row driver. The operation of the column driver is to convert the input video signal into an output analog data signals for one row. The levels of the data signals represent the gray (or color) tone and all data signals are applied to the

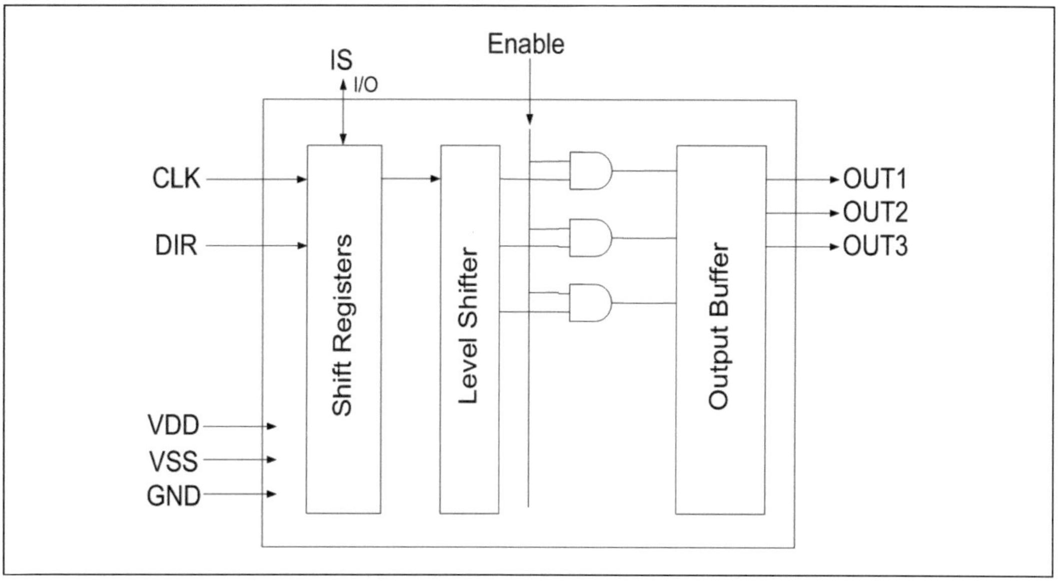

Fig. 15. Basic architecture of a row driver.

source of the switch TFT of all row's pixels simultaneously. The transmission - applied voltage curve (fig. 11) of the liquid crystal indicates that the liquid crystal requires about 5 V for its proper illumination. Considering the need of an AC drive signal, the required voltage swing across the LC material is about 10 V. In order to achieve a 10 V swing, the columns driver uses a 12 V power supply. The basic architecture of a column driver is shown in fig. 16.

A column driver consists of a bi-directional shift register which can be implemented with multiple chip architecture, like the row driver, in order to ensure either top or bottom mounting of the viewing area. The digital data signals for the red, green and blue channels are the input signals of the latch and the following, level shifter raises the signals to the proper voltage levels. The D/A converter is used for transforming the digital signal into analog gray scale levels. The reference input voltage $V_{g\text{-}cor}$ is used for the gamma correction, while the POL signal controls the polarity inversion method that will be used. Finally, the output buffer has the same functionality as the output buffers of the row driver.

5. Analog circuits design

The main disadvantage of the TFTs technologies is the instability of the device electrical characteristics, like threshold voltage and carriers mobility, which have been described in a previous paragraph. These instabilities provoke the repeated implementation of the circuit blocks with constant specifications and identical performance, even if the blocks are placed on the same wafer. In digital blocks, the impact of the parameters variations is negligible due to the fact that TFTs are switching elements. On the other hand, in analog blocks, the effects of the parameters variations are a major problem that the designers have to overcome since the existing, in the literature, analog blocks can not be used. Furthermore, measurements have shown that the threshold voltage variation has stronger impact on the

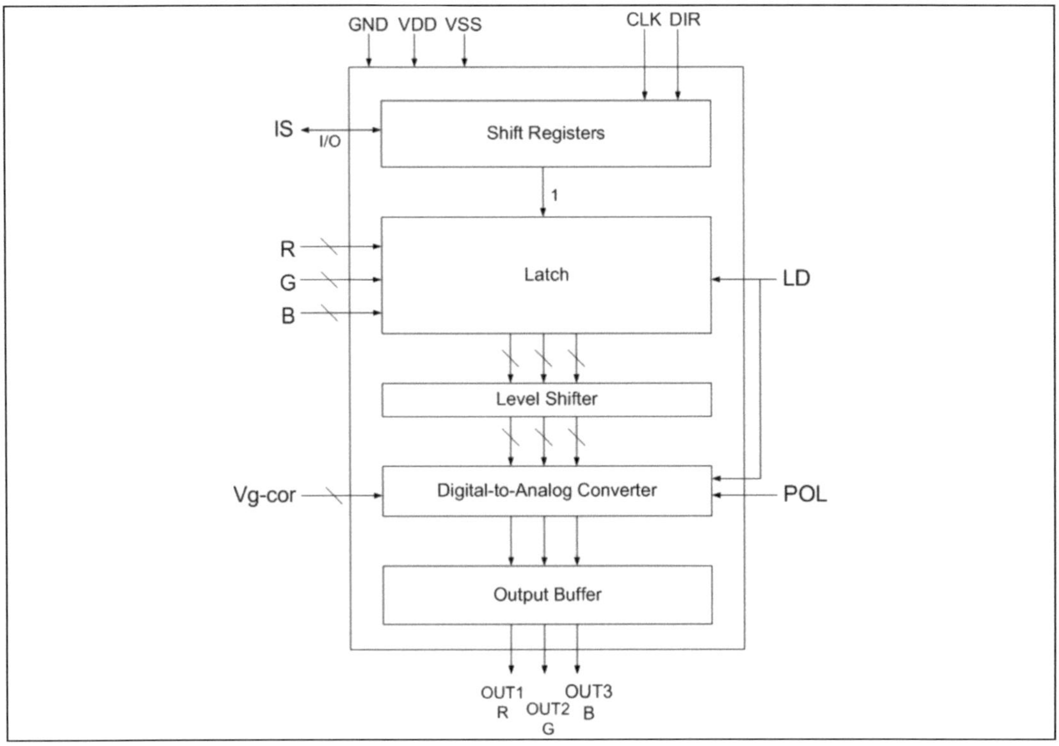

Fig. 16. Basic architecture of a column driver.

analog circuit performance than the mobility variation (Vaidya et. al., 2008), indicating that the new analog topologies have to be designed in such way so that threshold voltage compensation is achieved. In this paragraph, following the example of two compensation methods applied on an analog buffer design, complete description of the analog circuits design procedure with threshold voltage compensation is presented.

5.1 Analog buffer design

As it has already been mentioned, the analog buffer is the output stage of both row and column drivers and provides the necessary voltage to the row and column lines. The most commonly used topology for the analog buffer implementation is the common drain or source follower amplifier. The simple source follower type analog buffer is shown in fig. 17 (a). Although theoretically, the dc level of the output voltage is not the same as the dc level of the input signal, ideally the small-signal voltage gain is close to unity. In practical applications, the source follower exhibits an offset voltage from the input value, in which the offset voltage is mainly determined from the transistor threshold voltage since the final output voltage is equal to V_{GS} - V_{th}. For two source followers implemented with poly-Si TFTs, the output voltages will be different for the same applied input voltage due to the poly-Si TFTs threshold voltage variation. Furthermore, if these two source followers are used in the column driver architecture, a non-well controllable data voltage will be applied to the pixels resulting in the non-uniformity of the pixels brightness and gray scale. Therefore, source follower topologies with immunity in the threshold voltage variations have to be designed, in order to produce high quality image displays.

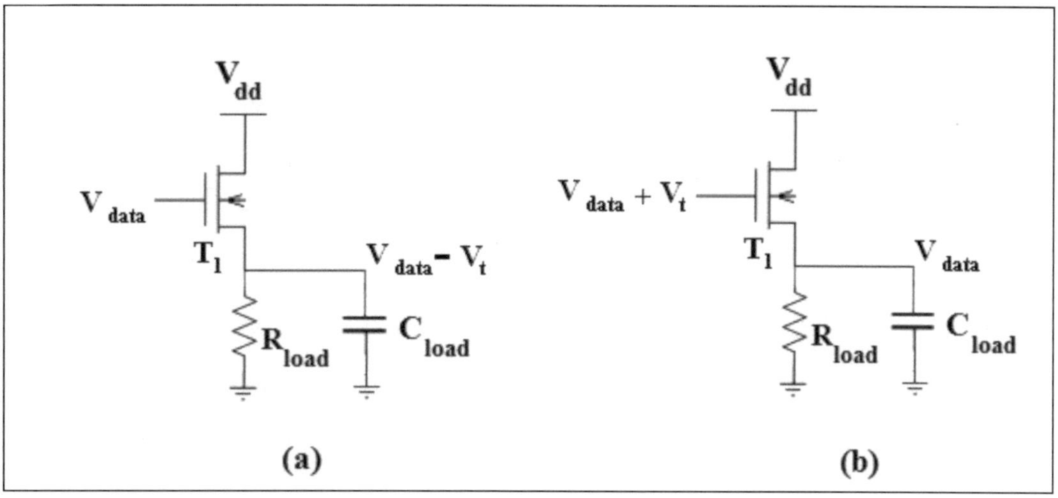

Fig. 17. (a) Typical source follower topology, (b) Source Follower topology with threshold voltage compensation.

Many compensation methods have been proposed including voltage mode or current mode threshold voltage compensation. All of them are based on increasing the input voltage by one threshold voltage by using an additional bias circuit topology, connected at the gate of the driving transistor. Figure 17 (b) shows the source follower topology with threshold voltage variations cancellation technique used in the column / row driver block.
For generation of an output voltage independent of the threshold voltage, a number of circuits have been proposed (Chun & Mok, 2004), (Tai et. al., 2005), (Chung et. al., 2001). The main similarity of the proposed circuits is the use of switches and additional control signals for sensing the threshold voltage of the driving transistor and the storage in a capacitor connected to the gate of the driving transistor. Therefore, the buffer operation has to be divided into compensation phase and data input phase. The disadvantage of these methods is that additional control signals are required in the configuration of the switches, which have to be generated from the row and column driver. Thus, changes in the architecture have to be made and circuit blocks have to be introduced, avoiding the storage capacitor which results in reduction of the buffer maximum frequency of operation. Recently, novel topologies of bias circuits have been proposed without requiring the use of either switches or storage capacitor, since the gate voltage is generated by either a static or a dynamic circuit. Each case is examined in the following paragraphs.

5.2 Static source follower

Figure 18 shows the topology of a static source follower with their timing diagram (Pappas et. al., 2007.a). The static source follower consists of five n-type poly-Si TFTs and only one bias voltage, which is proportional to the input data voltage, is required. For the theoretical analysis of the circuit, all TFTs have to operate in saturation mode. Furthermore, for poly-Si TFT devices which are very closely located on the same wafer and fabricated under the same conditions, their threshold voltage variation can be considered to be negligible compared to the threshold voltage variation over the whole panel. This assumption is generally accepted and it has been verified through measurements on fabricated circuits

(Lin and Tsai, 2007), (Jung et. al., 2007).The gate voltage is produced from the combination of M1 and M2 and M4 is the driving TFT. The operation of the static buffer can be divided onto two phases.

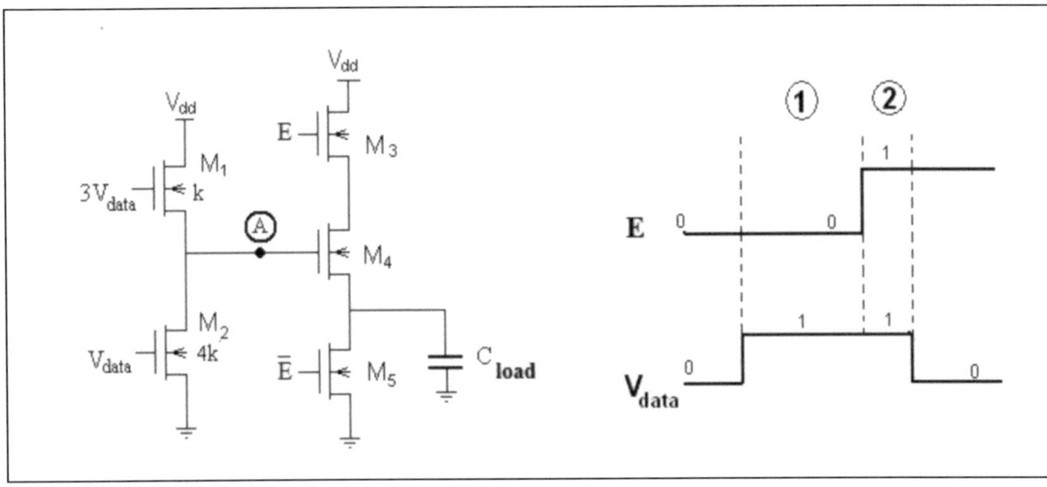

Fig. 18. Static source follower topology with its timing diagram.

During the first phase, the enable signal is at low voltage level causing the load capacitance to be discharged. At the same time, the data and bias voltages are applied to the buffer and node A is charged. The same current flows through transistors M1 and M2, while M2 has four time larger gate width than M1 ($W_2 = 4W_1 \rightarrow k_1 = 4k_2$). Under these conditions, the node A voltage is calculated by the expression

$$I_{DS1} = I_{DS2} \rightarrow \frac{1}{2}k_1\left(V_{GS1} - V_{TH1}\right)^2 = \frac{1}{2}k_2\left(V_{GS2} - V_{TH2}\right)^2$$

$$\rightarrow V_{GS1} - V_{TH} = 2\left(V_{GS2} - V_{TH}\right) \qquad (5)$$

The bias voltage is selected to be equal to $3V_{data}$. Therefore, $V_{GS1} = V_{bias} - V_A = 3V_{data} - V_A$ and $V_{GS2} = V_{data}$. Applying these to eq. 5, we obtain

$$V_A = V_{data} + V_{TH} \qquad (6)$$

Therefore, node A is charged to the proper voltage.

During the second phase of operation, the enable signal (E) is turned to high voltage level. In this situation, the driving transistor is connected as a source follower amplifier causing the load capacitor to be charged up to V_{data}. The theoretical analysis demonstrates that the static source follower exhibits high immunity to threshold voltage variations, since the output voltage is independent on the threshold voltage. The only change that has to be made in the column or row driver architecture is the addition of a circuit block, which will produce the bias voltage and the additional control signal (Enable). This circuit block can be a simple multiplier, because the bias voltage is proportional to the data voltage. The necessary change in the row/column driver architecture is shown in fig. 19.

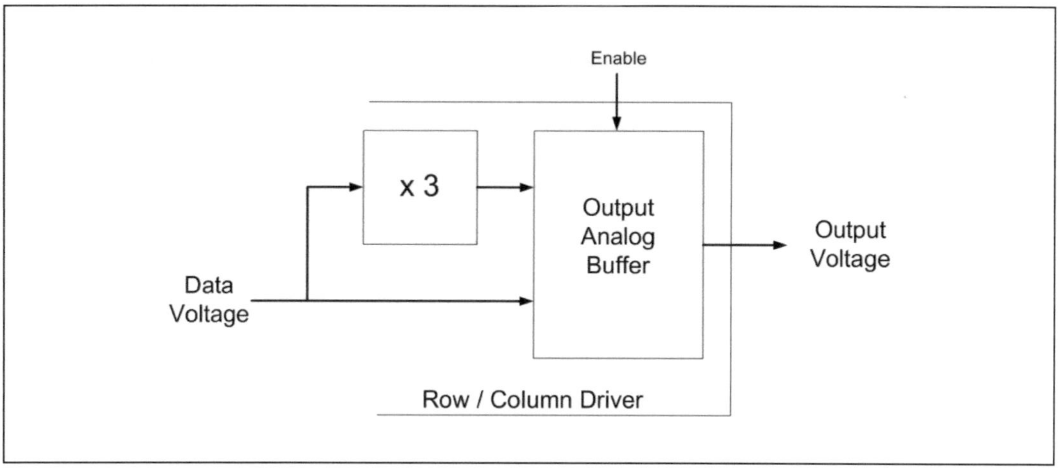

Fig. 19. Row/Column driver architecture with the use of the static source follower.

The advantage of the static source follower is the absence of switches and storage capacitor, leading to less real silicon area and higher maximum frequency of operation. However, the static source follower requires higher supply voltage, since all poly-Si TFTs have to operate in saturation mode and a high bias voltage is used. High power supply results in higher power consumption.

5.3 Dynamic source follower

Figure 20 shows the dynamic source follower topology and its timing diagram (Pappas et. al., 2008). The dynamic source follower consists of five n-type poly-Si TFTs, a 2-to-1 multiplexer and two additional control signals. For the theoretical analysis of the buffer, the same assumption for the threshold voltage variation of closely placed devices is considered and all transistors have to operate in saturation mode. The control signal "Reset" can be used as the multiplexer select signal and as a reset signal for the buffer at the same time.

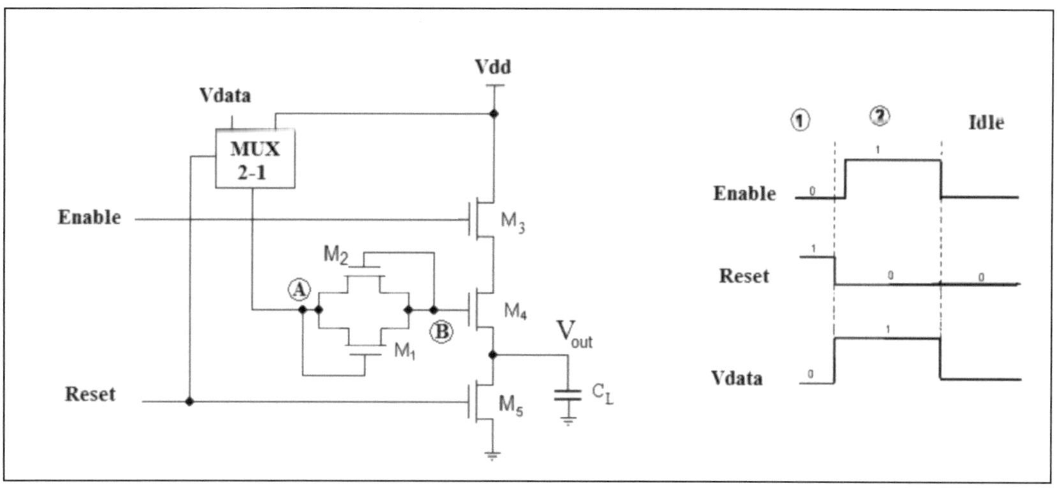

Fig. 20. Dynamic source follower topology with its timing diagram.

The operation of the dynamic source can be divided into three phases. During the first phase, the reset signal is at high voltage level causing the load capacitor to be discharged. At the same time, with the reset signal at high voltage level, the high voltage Vdd is selected from the multiplexer causing node B to be charged up to Vdd – V_{TH} due to the diode-connected transistor M1. During the second phase, the reset signal is turned to low voltage level and the data voltage is selected from the multiplexer causing node A to be charged up to the data voltage. Transistor M1 will turn "OFF" and node B will be discharged through M2 until the voltage at node B is equal to: $V_B = V_{data} + V_{TH}$, due to the diode-connected transistor M2 and the value of the node A voltage. When the voltage at node B obtain its proper value, the enable signal is turned to high voltage level and the driving transistor M4 is connected as a source follower amplifier causing the load capacitor to be charged up to V_{data}. When the load capacitor is completely charged, all signal will turn into their low voltage levels and the buffer will operate in idle mode.

The theoretical analysis demonstrates that the dynamic threshold voltage variation has no impact on the dynamic source follower functionality, without using switches and storage capacitor for sensing the threshold voltage. Furthermore, the dynamic source follower requires lower power supply compared to the static one, leading to lower power consumption. However, the dynamic source follower requires two additional control signals and a 2-to-1 multiplexer, indicating that changes have to be made in the row/column architecture. The row/column architecture with the use of the dynamic source follower is shown in fig. 21.

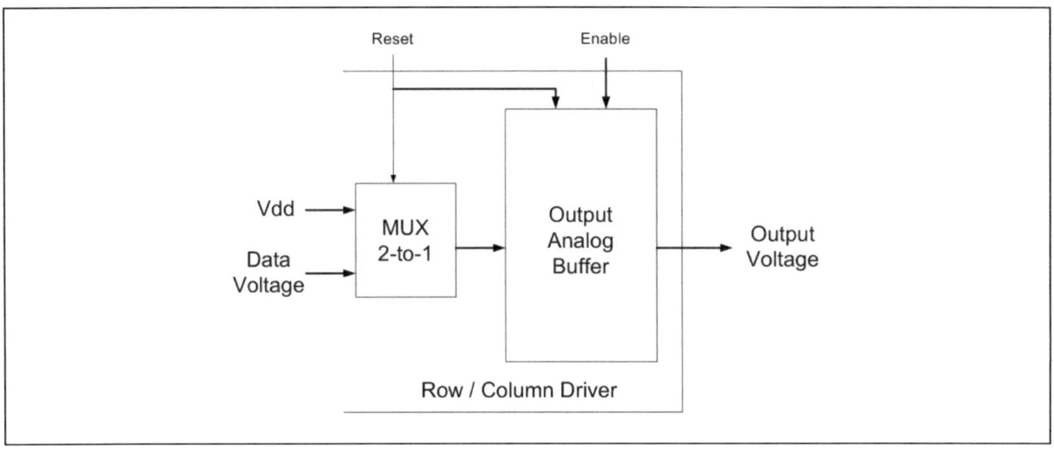

Fig. 21. Row / Column driver architecture with the use of the dynamic source follower.

5.4 Verification of the analog circuits

The verification of the analog circuit functionality can be made through circuit simulations by using simulation programs, like Spice or CADENCE. A poly-Si TFT model is needed for the simulations and the most commonly used is the RPI poly-Si model developed by Renseelaer Polytechnic Institute (Shur et. al., 1997). In order to obtain realistic simulations, the TFT parameters must be extracted from fabricated devices (Pappas et. al., 2007). The parameters extraction can be realized by using, special designed for TFT technologies, CAD tools, like Silvaco ATLAS. For the design of circuits, one of the most significant parameter is the supply voltage. The supply voltage has to be high enough so all transistors will operate in saturation mode and low enough in order to decrease the power consumption. Another trade-off is the dimensions of the transistors. The transistor dimensions have to be large enough to produce drain current able to charge the load capacitors and small enough to reduce the real silicon area of the circuits. Finally, statistical analysis for the threshold voltage variation in fabricated devices has to be made, enabling the modelling of the threshold voltage variations in Monte Carlo analysis (Pappas et. al., 2006). The results of the Monte Carlo analysis can demonstrate the impact of the threshold voltage variation on the functionality and performance of the circuits.

6. Conclusion

Active Matrix Liquid Crystal Displays, implemented with amorphous or polycrystalline silicon thin-film transistors, is the most attractive solution for the display industry. However, as the specifications become more and more demanding, the AMLCDs design and the TFTs device characteristics have to be improved. The AMLCDs electronics modules define the image quality of the display and the power consumption and, therefore, their architecture and design have a great impact on the AMLCD performance. The definition of these aspects and the detailed description of their functionality presented in this chapter can provide a helpful handbook for every designer.

7. References

Arokia, N.; Chaji, G. R. & Ashtiani, S.J. (2005). Driving Schemes for a-Si and LTPS AMOLED Displays. *IEEE Jour. Of Display Technology*, vol. 1, No. 2, pp. 267-277, December.

Bardeen, J. & Brattain, W. H. (1948). The transistor – A Semiconductor Triode. Physics Revision 74, pp. 230-242.

Borkan, H. & Weimer, P.K. (1962). An Analysis of the Characteristics of Insulated-Gate Thin-Film Transistors. *RCA Rev.*, vol. 24, pp. 153-165.

Broody, T.P.; Asars, J. A. & Dixon, G. D. (1973). A 6 x 6 Inch 20 Lines-per-Inch Liquid Crystal Display Panel. *IEEE Tran. Of Electron Devices.* ED-20, 11, pp. 995-1001, 0018-9383.

Chun, L.Y. & Mok, T.K. (2004). Process-indepedent analog data driver for polysilicon TFT AMLCD. *Int. Journal of Electronics*, vol. 41, pp. 199-210.

Chung, H.J.; Lee, S.W. & Han, M.K. (2001). Poly-Si TFT push-pull analog buffer for integrated data drivers of poly-Si TFT-LCDs. *Electronics Lett.*, vol. 37, No. 17, pp. 1093-1095, August.

Den Boer (2005). *Active Matrix Liquid Crystal Displays : Fundamental and Applications.* Newnes. 0750678135. Netherlands.

Depp, S.W.; Juliana, A & Huth, B. G. (1981). Polysilicon FET Devices for Large Area Input/Output Applications. *Proc. 1981 Int. Electron Device Mtg. (IEEE)*, pp. 703-707, New York.

Jagar, S.; Chang, C. F.; Zhang, S.D.; Wang, H. & Poon, M.C. (2003). A SPICE model for thin-film transistors fabricated on grain-enhanced polysilicon film. *IEEE Trans. Electron Devices*, vol. 50, No. 4, pp. 1103-1108, April.

Jung, S.H.; Nam, W.J.; Lee, J.H. and Han, M.K. (2007). A New Analog Buffer Using P-type Poly-Si TFTs for Active Matrix Displays. *IEEE Electron Device Letters*, vol. 27, No. 1, pp. 40-42, January.

Lechner, B.J.; Marlowe, F. J.; Nester E. O. & Yults J. (1971). Liquid Crystal Matrix Displays, *Proc. IEEE 59*, pp. 1566 – 1579.

LeComber, P.G.; Spear, W.E. & Ghaith, A. (1979). Amorphous Silicon Field-Effect Device and Possible Application. *Electronics Letters*, vol. 15, pp. 179-181, 0013-5194.

Lee, S.W.; Han, H. J. & Lee, J.W. (1999). High Performance low-power integrated 8-bit digital data for poly-Si TFTLCDs. *Proc. Soc. For Information Displays Dig.*, pp. 76-79, San Jose, CA.

Lin, C.L. and Tsai, T.T. (2007). A Novel Voltage Driving Method Using 3-TFT Pixel Circuit for AMOLED. *IEEE Electron Device Letters*, vol. 28, No. 6, pp. 489-491, June.

Lin, Y.Y.; Gundlach, D. J.; Nelson, S.F. & Jackson T.N. (1997). Stacked pentacene layer organic thin-film transistors with improved characteristics. *IEEE Electron Devices Lett*, vol. 18, Issue. 12, pp. 606-608, December.

Lueder, E. (1980). Processing of Thin-Film Transistors with Photolithography and Application for Displays, *Digest 1980 SID Intl. Symp. (Soc. For Information Displays)*, pp. 118 – 122, San Jose.

Marlone, F.J. & Nester, E. O. (1972). Alternating Voltage Excitation of Liquid Crystal Display Matrix. U.S. Patent No. 3,654,606.

Pappas, I.; Hatzopoulos, A.T.; Tassis, D.H.; Arpatzanis, N.; Siskos, S. & Dimitriadis, C.A. (2006). *Journal of Applied Physics*, vol. 100, 064506, September.

Pappas, I.; Siskos, S. & Dimitriadis, C.A. (2007.a). A New Analog Buffer Using Low-Temperature Thin-Film Transistors for Active Matrix Displays. *IEEE Trans. On Electron Devices*, vol. 54, No. 2, pp. 219-224, February.

Pappas, I.; Dimitriadis, C.A.; Templier, F.; Oudwan, M. & Kamarinos, G. (2007.b). Above-threshold voltage current model including band tail states in nanocrystalline silicon thin-film transistors for circuit implementation. *Journal of Applied Physics*, vol. 101, 084506, April.

Pappas, I.; Siskos, S. & Dimitriadis, C.A. (2008). A Fast and Compact Buffer Design for Activve MAtrix Liquid Crystal Displays Using Polysilicon Thin-Film Transistors. *IEEE Trans. Circuits and System II : Express Briefs*, vol. 55, No. 6, pp. 537-540, June.

Powell, M. J.; Berkel, C. & Hughes, J. R. (1989). Time and Temperature depedence of instability mechnanisms in amorphous silicon thin-film transistors. *Applied Physics Lett.*, vol. 24, No. 7, pp. 1323-1325, January.

Shur, S.M.; Jacunski, H.C.; Slade, H.C.; Owusu, A.A. & Ytterdal, T (1997). SPICE models for amorphous silicon and polysilicon Thin Film Transistors. *Journal of Electrochemical Society*, vol. 144, issue 8, pp. 2833-2839, August.

Tai, Y.H.; Pai, C.C.; Chen, B.T. & Cheng, H.C. (2005). A source-follower type analog buffer using poly-Si TFTs with large design window. *IEEE Electron Devices Lett.*, vol. 26, No. 11, pp. 811-813, November.

Vaidya, V.; Soggs, S.; Jungbae, K.; Haldi, A.; Haddock, J.N.; Kippelen, B. & Wilson, D.M. (2008). Comparison of pentecene and amorphous silicon AMOLED display driver circuits. *IEEE Trans. On Circuits and Systems-I : Regular Papers*, vol. 55, No. 5, pp. 1177-1184, June.

Valdinoci, M.; Colalongo, L.; Baccarani, G.; Fortunato, G.; Pecore, A. & Policicchio, I. (1997). Floating body effects in polysilicon thin-film transistors. *IEEE Trans. Electron Devices*, vol. 44, No. 12, pp. 2234-2241, December.

Weimer, P.K. (1962). The TFT – A New Thin-Film Transistor. *Proc. IEEE 50*, pp. 1462 – 1471.

Zhang, S.D.; Sin, K.O.; Li, J.N. & Mok, P.T.K. (2000). Ultra-thin elevated channel poly-Si TFT technology for fully-integrated AMLCD system on glass. *IEEE Trans. On Electron Devices*, vol. 47, No. 3, pp. 569-575, 0018-9383.

9

TFT-LCD Driver IC Design

Chih-Wen Lu
National Chi Nan University
Taiwan, R.O.C.

1. Introduction

With the rapid evolution of liquid crystal display (LCD) television (TV), there is a large demand for developing high resolution, high color depth driver integrated circuits (ICs) [1-3]. The panel of an LCD-TV is larger, and has higher definition, than that of a computer monitor. As a result, its color quality requires more accuracy. For example, computer monitors have $2^{6\times3}$ (262,144) or $2^{8\times3}$ (16,777,216) colors. However, a typical LCD-TV has $2^{10\times3}$ (1,073,741,824) colors [2-3]. To develop a high-quality display module, LCD-TV driver systems require higher color depth and resolution. An LCD driver system generally includes column drivers, row drivers, a timing controller, and a reference source. The column drivers are especially critical for achieving a high-quality display [1, 4-5]. For LCDTV applications, drivers must process 10-bit digital input codes and then convert the input codes to analog levels [6-7].

A column driver generally includes shift registers, input registers, data latches, level shifters, DACs, and output buffers [1, 8-9]. Among these components, the DACs occupy the largest area. Due to the hundreds of channels built into a single chip, it is desirable to reduce the area of the DAC, especially for high color depth displays.

To improve the lifetime of the liquid crystal material, the liquid crystals of active matrix liquid crystal displays (AMLCDs) should be driven by the so-called inversion method, which alternates the positive and negative polarities between the liquid-crystal cells with respect to a common backside electrode. Designers use four inversion methods for AMLCD driving: frame, line, column, and dot inversions. High-quality displays prefer the dot inversion method [1, 8]. Figure 1 schematically shows the operation of the dot inversion method. In this method, the backside electrode is at a fixed voltage and a negative-to-positive or positive-to-negative voltage with respect to the fixed voltage of the backside electrode must be driven from the LCD column drivers with alternating polarities between data lines and line times [1, 10]. Hence, the LCD driver IC should supply both positive and negative polarity voltages for a digital sub-pixel code. This increases the resolution of the DAC by one bit, and hence increases the die area.

Figure 2(a) shows the characteristic transmittance-voltage curve of a liquid crystal (LC), which exhibits a nonlinear response to the applied voltage. To obtain a linear luminance output with the digital input code for an LCD, the DAC response is usually set as the inverse of the LC characteristic, as Figure 2(b) indicates [10]. The DAC output should cover both positive and negative polarity voltages. The LCD driver IC usually utilizes R-DACs. To compensate for the nonlinear LC characteristic, gamma correction voltages are applied to

Source: New Developments in Liquid Crystals, Book edited by: Georgiy V. Tkachenko,
ISBN 978-953-307-015-5, pp. 234, November 2009, I-Tech, Vienna, Austria

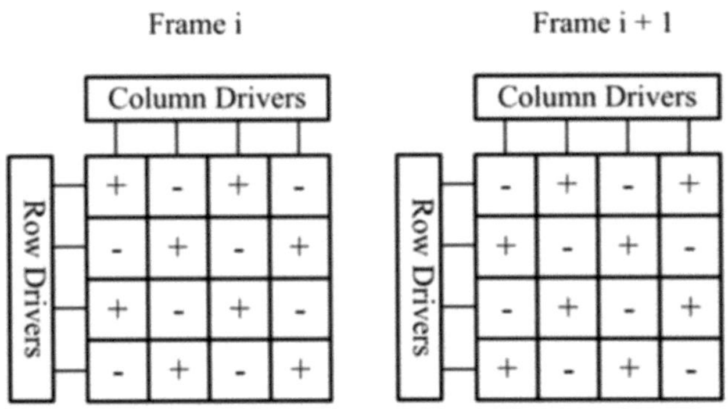

Fig. 1. The operation for the dot inversion method.

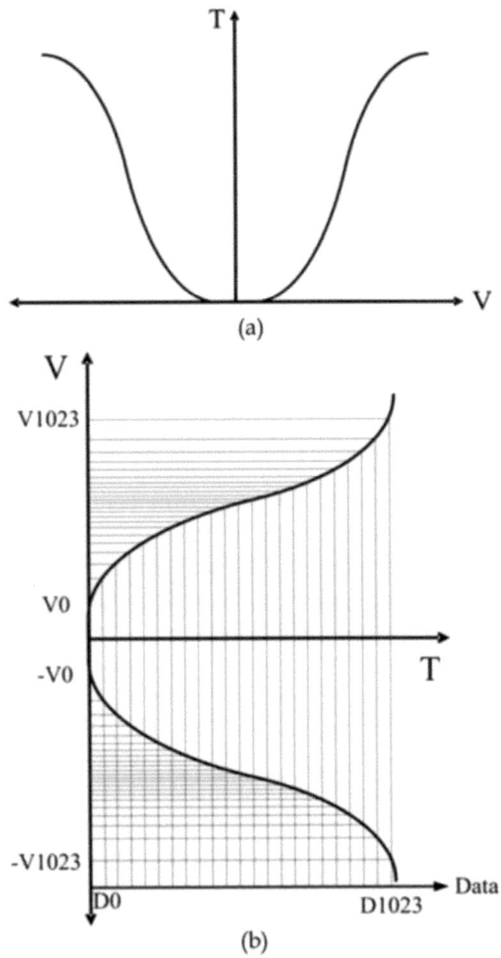

Fig. 2. (a) The characteristic of the liquid crystal. (b) The response of DAC.

the R-DAC resistor string, which makes the resistor values unequal [2-3]. However, the area of the R-DAC and its metal routing will be prohibitively large for a high resolution data converter. This makes the R-DAC impractical for use in column driver ICs for high color depth displays. As an alternative, previous research has proposed an LCD column driver using a linear switched capacitor DAC [2-3]. In this case, the timing controller compensates for the nonlinear LC characteristic, which greatly reduces the die area. To reduce the die area, the current study proposes a 10-bit LCD column driver, the use of an R-DAC, and a charge sharing DAC (C-DAC) for each channel. This approach applies gamma voltages to the R-DAC to fit the inverse of the liquid crystal characteristic. The gamma correction can also be digitally fine-tuned in the timing controller or column drivers.

2. Conventional column drivers

Figure 3 shows the conventional column driver architecture [1, 8]. The column driver supplies high analog voltages to the LCD panel. To reduce power consumption, the digital circuit uses a low voltage power supply. Digital display data is fed to the RGB inputs and sampled into the input registers. A wide data latch presents one row of serial input pixel data to the level shifters inputs. The level shifters then boost the digital signals to higher levels. The DAC of each channel outputs a voltage level corresponding to a digital sub-pixel code. The output buffers drive the highly capacitive data lines of the LCD panel [11-12]. For a dot inversion operation, the DACs offer voltages with positive and negative polarities for the same digital input code. Hence, 11-bit DACs are needed for a 10-bit column driver.

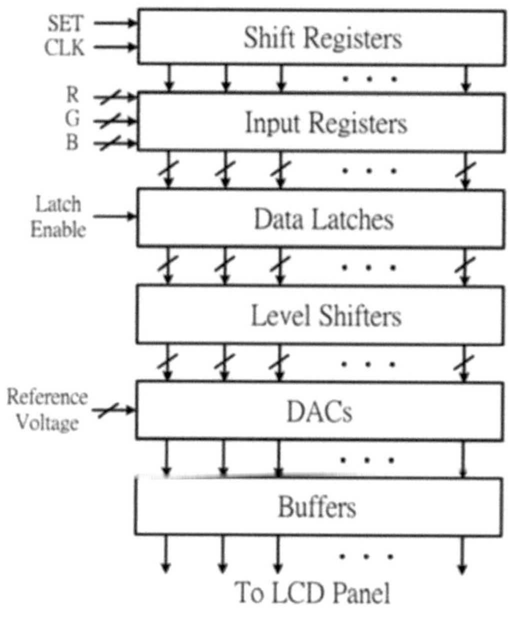

Fig. 3. The conventional column driver architecture.

Since the transmittance response of the liquid crystal to the applied voltage is nonlinear, a nonlinear DAC is needed to obtain a linear transmittance with the digital code. Applying certain gamma voltages to the resistor string of the nonlinear DAC, and the resistor string is made up of unequal resistors to fit the nonlinear curve.

Figure 4 depicts the layout of a conventional column driver. One resistor string is put in the middle of the chip to supply the reference voltages to all channels. Each channel needs a decoder to route the reference voltage, corresponding to the digital input code, to the corresponding output buffer. Since several hundreds of channels are built into a single chip, the die area of the routing lines connecting the resistor string and the decoders is very large. For example, a 10-bit column driver IC requires 2048 metal lines. Hence, these metal lines and the decoders occupy a very large percentage of the column driver IC's area, especially for high color depth displays.

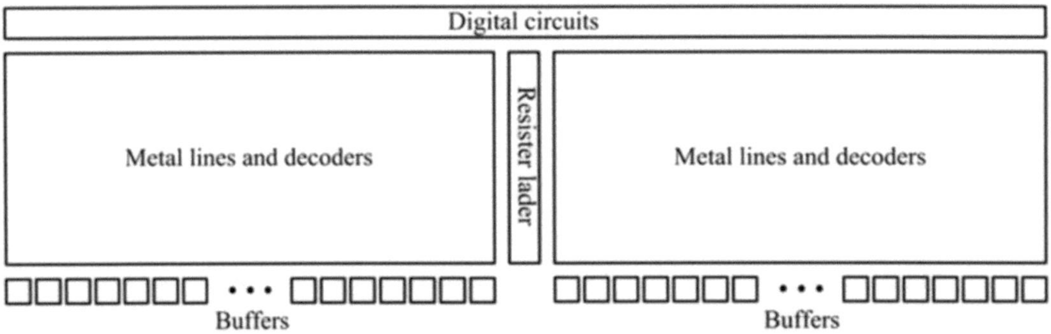

Fig. 4. The layout of the conventional column driver.

To reduce the die area of a column driver for a higher color depth display, Bell employed a linear switched capacitor DAC in his column driver [2-3]. In this arrangement, the timing controller digitally compensates for the nonlinear liquid crystal (LC) characteristic. Since the DAC is linear, additional bits are needed to compensate for the nonlinear LC characteristic. This approach reduces effective resolution.

3. Proposed column driver

This section proposes a 10-bit LCD column driver consisting of piecewise linear DACs.

3.1 Data converter scheme

This study utilizes a piecewise linear compensation mechanism in the proposed column driver to reduce the die area and increase the resolution for a higher color depth display. This design applies gamma voltages to the resistor string of the R-DAC, and uses unequal resistor values to compensate for the nonlinear LC characteristic. Figure 5 shows the characteristic of the piecewise linear DAC and the reverse response of the LC, where V_{G1}, V_{G2}, …, V_{G16} are the external gamma reference voltages. The voltage curve of the piecewise linear DAC is very close to that of the reverse LC response, so these two curves look like identical. Since the characteristic curve of the piecewise linear DAC is much closer to the inverse response of an LC than a fully linear DAC, fewer additional bits are needed to compensate for the nonlinear LC response. Therefore, the effective color depth is much greater than that of a fully linear data conversion. External reference voltages make coarse gamma correction, and a simple digital circuit makes fine compensation adjustments. This digital circuit can be built in the timing controller or the column driver.

Since the proposed column driver IC drives the LCD with positive and negative polarities, the DACs and output buffers are classified into positive and negative components. Figure 6 shows the data conversion scheme. Each channel contains one R-DAC decoder, one C-DAC, and one buffer. Two neighboring channels are grouped together, and take turns driving a pair of adjacent data lines of the LCD panel. One channel is responsible for driving positive

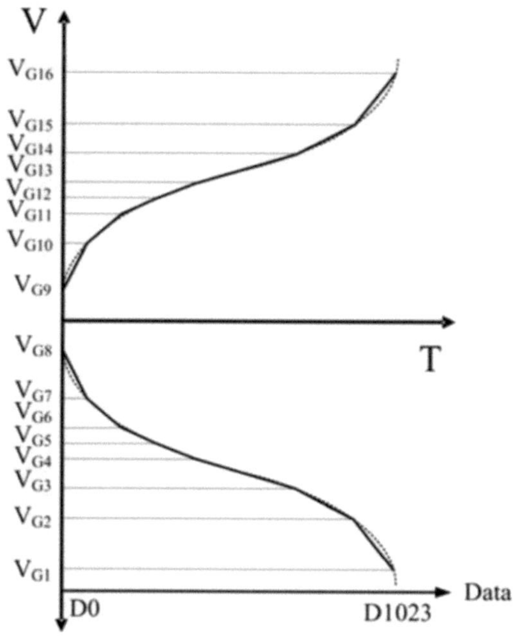

Fig. 5. The characteristic of the piecewise linear DAC and the reverse response of LC.

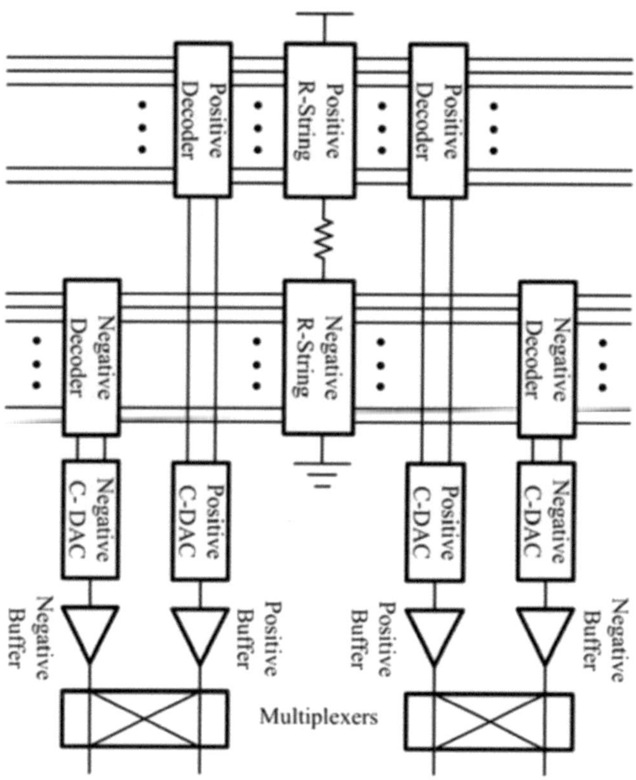

Fig. 6. The data conversion scheme.

polarity and the other for driving negative polarity. The odd DACs and buffers are designed for the negative polarity operation, while the even DACs and buffers drive the positive polarity operation. When the odd column lines are under negative polarity and the even column lines are under positive polarity, the input codes and the output buffers are in a normal order. However, when the polarities of the column lines are exchanged, i.e., the odd and even column lines are alternated to positive and negative polarities, respectively, the orders of the input codes, DACs, and output buffers are exchanged. The negative buffers and the odd DACs still take responsibility for negative polarity operation, and vice versa for the positive buffers and the even DACs. This arrangement reduces the number of decoder bits by one. In other words, only 10-bit decoders are needed for a 10-bit column driver.

3.2 Data converter

The DAC includes a coarse section and a fine section to reduce both the die area and the data conversion time. The 10-bit DAC in this study contains a 7-bit coarse section and a 3-bit fine section implemented by R-DACs and C-DACs, respectively. One resistor string generates the voltage references for all R-DACs in a column driver. Since the DACs cover the positive and negative polarities, the column driver requires an 8-bit resistor string. Each channel contains one 7-bit nonlinear R-DAC and one 3-bit linear C-DAC. Reducing the bit number from 11 to 7 greatly reduces the area of the R-DACs.

Data conversion is serially implemented by the R-DAC and C-DAC. The R-DAC decoder selects two neighboring voltages according to the 7 MSBs and sends them to the CDAC. The C-DAC then uses the two neighboring voltages to perform voltage division and passes the final voltage to the buffer. Figure 7 shows a schematic of the R-DAC, in which the resistor string divides the voltage and generates 256 voltage segments. The upper half of the voltage segments and decoder are used for the positive polarities; the lower ones for negative polarities. The decoders are controlled by the 7 MSBs (b9 ~ b3) in Figure 7. An offset switch array obtains two neighboring voltages (v_i and v_{i+1}) for the C-DAC.

The voltage division in the C-DAC is based on precharging and charge redistribution. Figure 8 shows the schematic of the C-DAC, which consists of 3 binary weighted capacitors, an additional unit capacitor, and a set of switches that can connect the capacitors to the input voltages. Two phases are needed to accomplish the voltage division in this circuit. In the pre-charge phase ($\phi = 0$), the weighted capacitors are connected to v_{i+1} or v_i depending on the 3-bit code (b2 ~ b0). In the evaluation phase ($\phi = 1$), all capacitors are disconnected from the inputs and connected to the output. A charge-redistribution then occurs, and the reconstructed analog value finally appears at the output. The output voltage can be expressed as

$$
\begin{aligned}
v_{out} &= \frac{2^2 \cdot \left(v_{i+1}b_2 + v_i\bar{b}_2\right) + 2 \cdot \left(v_{i+1}b_1 + v_i\bar{b}_1\right) + \left(v_{i+1}b_0 + v_i\bar{b}_0\right) + v_i}{2^3} \\
&= \frac{\left(4b_2 + 2b_1 + b_0\right)}{8}\left(v_{i+1} - v_i\right) + v_i
\end{aligned}
\qquad (1)
$$

Equation (1) shows that the C-DAC divides the voltage for each segment voltage of the R-string and exhibits a 3-bit DAC behavior.

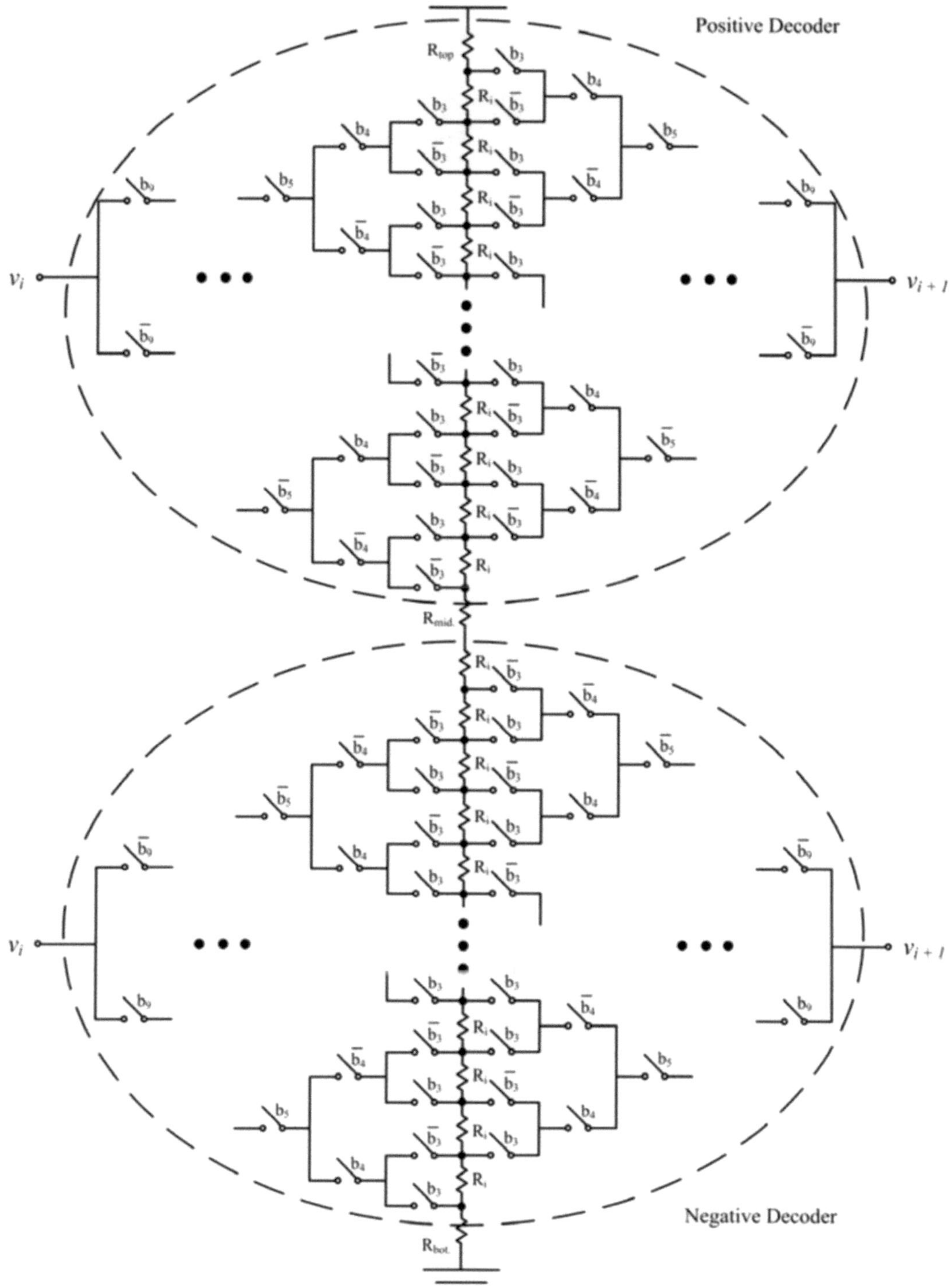

Fig. 7. Schematic of the R-DAC for the proposed column driver.

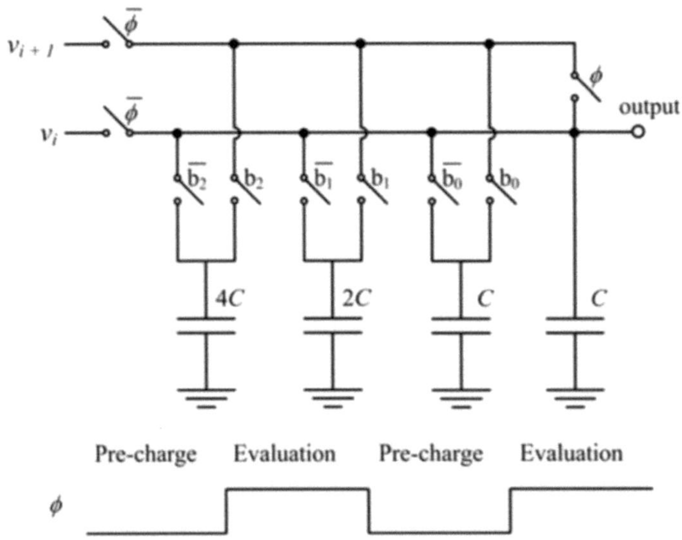

Fig. 8. Schematic of the C-DAC for the proposed column driver.

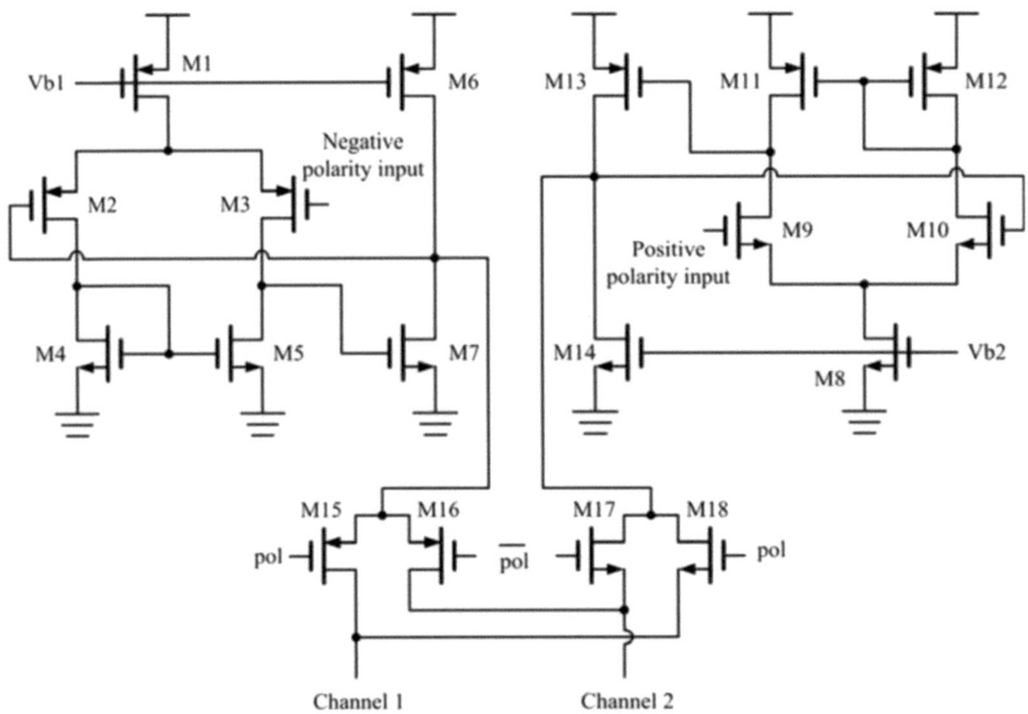

Fig. 9. Schematic of a pair of complementary differential buffers with the switches.

3.3 Output buffer

To drive the data lines of the LCD panel, each channel needs an output buffer. The output buffers, which are usually made of operational amplifiers, drive highly capacitive data lines.

Since the approach in this study classifies the buffers into positive and negative components, rail-to-rail amplifiers are not needed. The PMOS input buffer has a large discharge capability and its common mode input voltage can reach a very low level. Hence, it is used to drive positive-to-negative polarity operation. Similarly, an NMOS input buffer is suitable for the negative-to-positive polarity transition. Figure 9 shows a schematic of a pair of complementary differential buffers with switches [1]. The switches M15-M18, which are controlled by the polarity control signal "pol," are used for the inversion operation. The transistors M1-M7 and M8-M14 are constructed as a PMOS input differential amplifier and an NMOS input differential amplifier, respectively. The compensation scheme has been described previously [1, 4].

3.4 Column driver architecture

Figure 10 shows a block diagram of the LCD column driver based on the data conversion scheme above. In this arrangement, data exchange circuits are attached between the latches and the level shifters to implement the proposed data conversion scheme. The digital input codes are serially read into the input register, which is controlled by the shift registers. After all the data of one row is read and latched on the data latches, it is sent to DACs through the data exchange circuits in parallel. The"pol" signal controls the polarity inversion. The data conversion is implemented by R-DACs and C-DACs in series. The buffers described in the prior section drive the capacitive column lines.

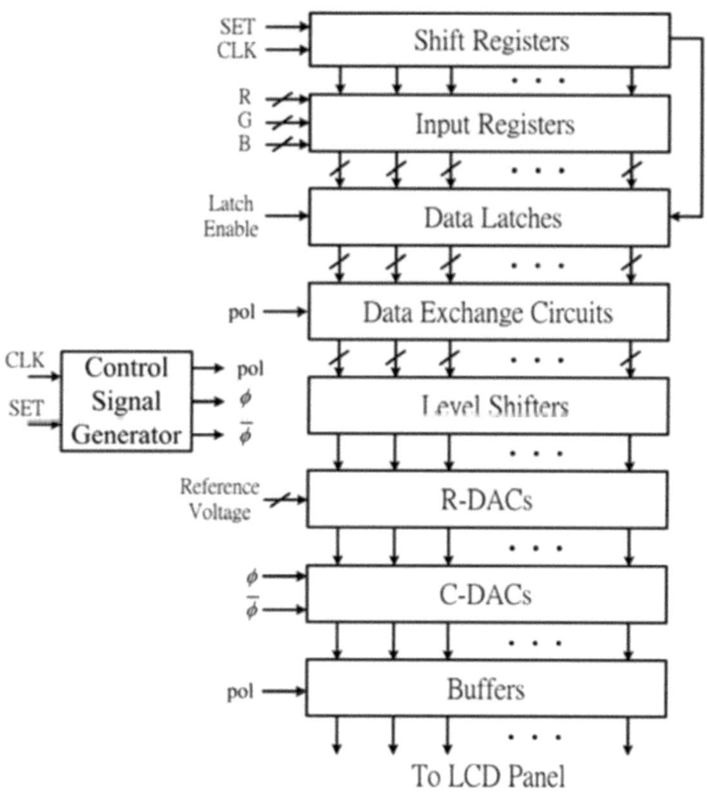

Fig. 10. The implemented block diagram of the LCD column driver architecture.

4. Experimental results

Since the reference voltages are connected to all channels, many DACs may use the same reference voltage. The more DACs there are connected to a single reference voltage, the larger the required C-DAC settling time. This study simulates the settling time for different numbers of connected DACs using a 0.35-µm 5-V CMOS model. Figure 11 shows the simulated results where the settling time is measured at 99.9% of its final voltage for a full swing (0.266 V ~ 4.75 V). The settling time is 5.2 µs when 200 DACs are connected to a single reference voltage. Although a column driver IC contains several hundreds or even up to a thousand DACs, these DACs are distributed to 256 (2^8) reference voltages. This means that not all the DACs are connected to a single reference voltage. A typical UXGA (1600×1200) display has a pixel clock frequency of 162 MHz and a horizontal scanning time of 9.877 µs [4]. Hence, the proposed column driver is suitable for UXGA displays.

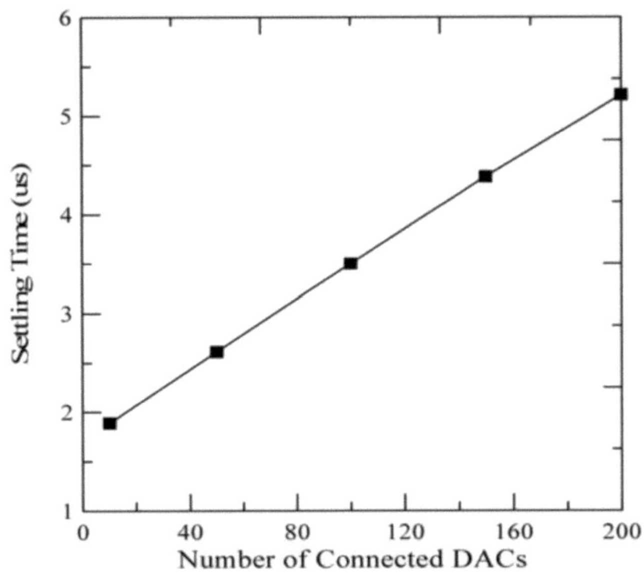

Fig. 11. The simulated maximum settling time of the DAC for different numbers of connected DACs.

Due to the limited silicon area, the proposed LCD column driver has only four channels. The 10-bit LCD column driver with R-DAC and C-DAC was fabricated using a 0.35-µm 5-V CMOS technology. Table I shows the device sizes used in the proposed column driver, where R_{top}, R_{mid}, R_{bot}, and R_i are designated in Figure 7. Figure 12 is a photograph of the die. Except for the resistor string of the R-DAC, the die area is 0.2×1.26 mm^2 for four channels. Each RGB digital input code is 10-bits wide.

The Differential Nonlinearity (DNL) and Integral Nonlinearity (INL) are typically measured for a DAC. However, it is difficult to determine these two specifications for a nonlinear DAC. To demonstrate the performance of the proposed circuit, the nonlinear gamma voltages are not applied to the R-string and the resistor values of the resistor string are made equal. Since an LCD panel needs several column drivers, the uniformity of different drivers is very important. Figure 13 shows the measured transfer curves of a DAC for eight off-chip column drivers. To show the deviation between different chips, Figure 14 provides an

Buffer	M1 $\frac{3\mu\times 4}{3\mu}$	M6 $\frac{7\mu\times 3}{0.5\mu}$	M11 $\frac{4.3\mu\times 3}{2.8\mu}$
	M2 $\frac{8\mu\times 5}{3\mu}$	M7 $\frac{11\mu\times 2}{0.5\mu}$	M12 $\frac{4.3\mu\times 3}{2.8\mu}$
	M3 $\frac{8\mu\times 5}{3\mu}$	M8 $\frac{4\mu\times 4}{3\mu}$	M13 $\frac{10\mu\times 3}{0.5\mu}$
	M4 $\frac{4\mu\times 3}{2.5\mu}$	M9 $\frac{6\mu\times 5}{3\mu}$	M14 $\frac{8\mu\times 2}{0.5\mu}$
	M5 $\frac{4\mu\times 3}{2.5\mu}$	M10 $\frac{6\mu\times 5}{3\mu}$	
Resistance of R-DAC	R_{top}		800Ω
	$R_{mid.}$		550 Ω
	$R_{bot.}$		800 Ω
	R_i		50 Ω
Switch size	PMOS		$\frac{3\mu}{0.5\mu}$
	NMOS		$\frac{1\mu}{0.5\mu}$
Capacitance of C-DAC	C		1pF

Table I The device sizes used in the proposed column driver.

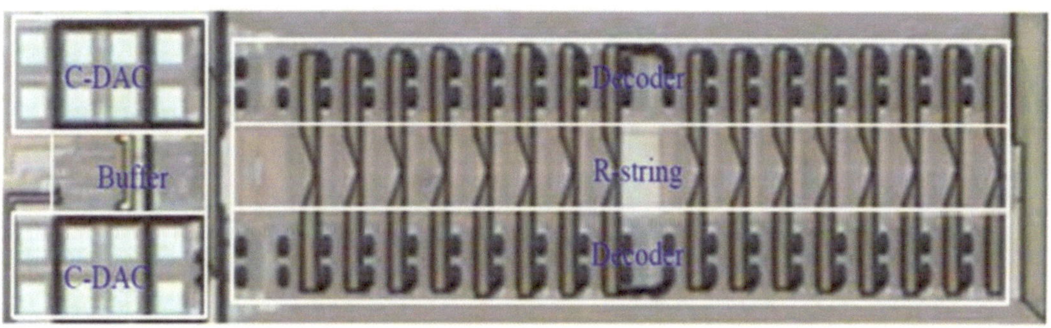

Fig. 12. Photograph of the proposed column driver.

enlarged view of the transfer curves, where the maximum deviation is 3.5 mV from the mean. This deviation is mainly due to process variations. The approach in this study uses no error correction. Hence, the deviation can be reduced by applying an offset canceling technique to the buffer amplifier. Figures 15(a) and (b) show the DNL values for positive and negative polarities, respectively. Figures 16(a) and (b) show the INL values for positive and negative polarities, respectively. The combination of R-DACs and C-DACs creates two groups of DNL values. The maximum DNL and INL values are 3.83 and 3.84 LSB, respectively. This study uses a 1-LSB voltage of 2.44mV to calculate the INL and DNL values. The linearity, however, is less important than the deviations between off-chip drivers for LCD drivers [2].

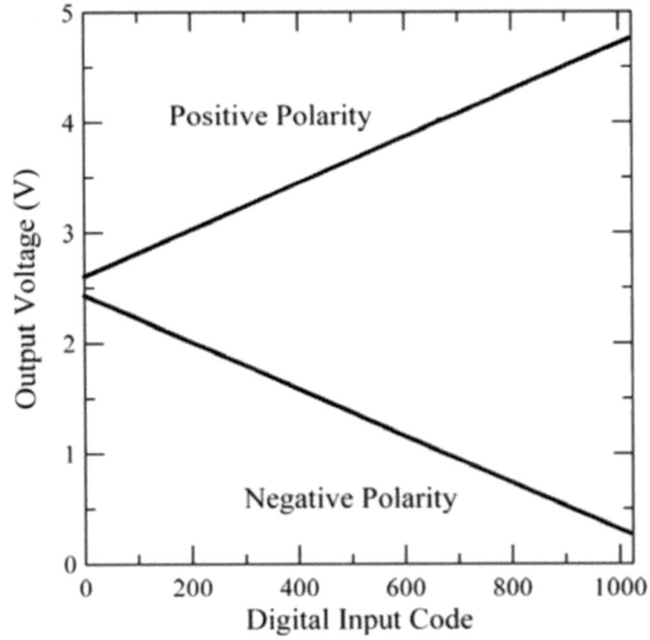

Fig. 13. The measured output responses of column drivers for 8 chips.

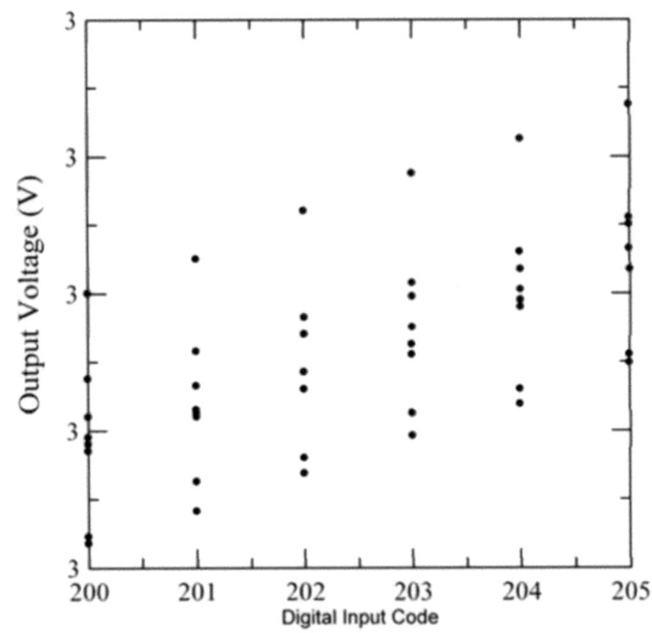

Fig. 14. The enlargement of the output response of the column driver for the digital input code of 200 ~ 205.

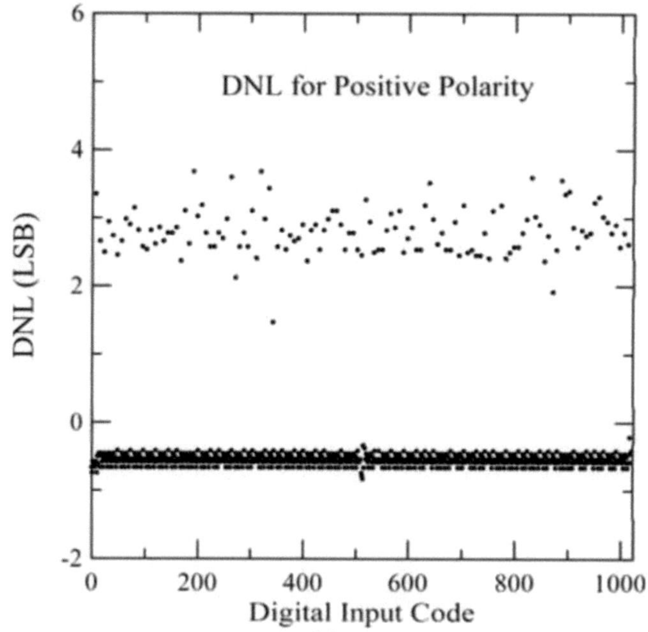

(a)

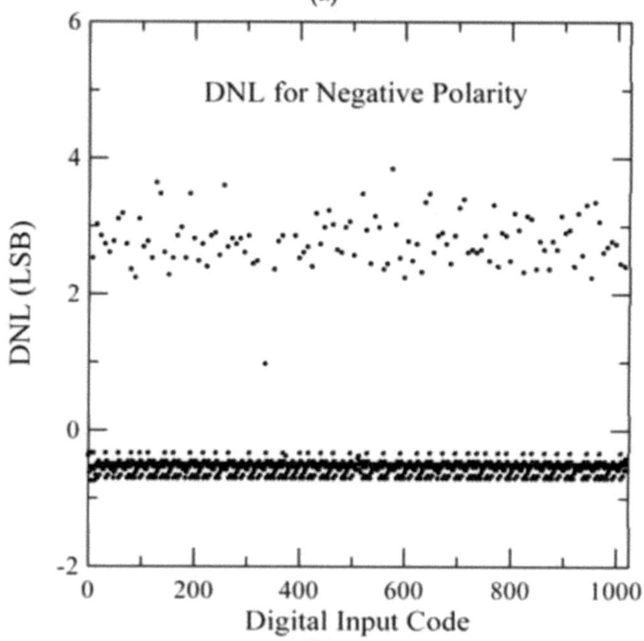

(b)

Fig. 15. The measured DNL for (a) positive polarity (b) negative polarity of the proposed column driver.

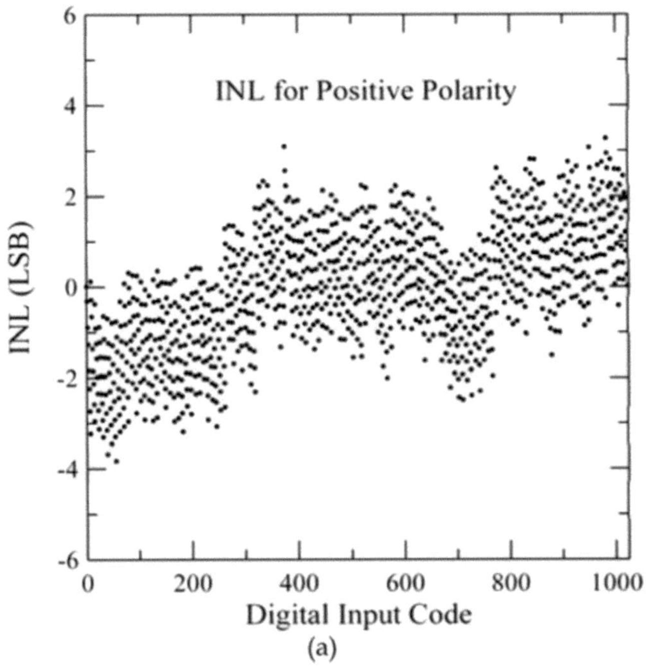

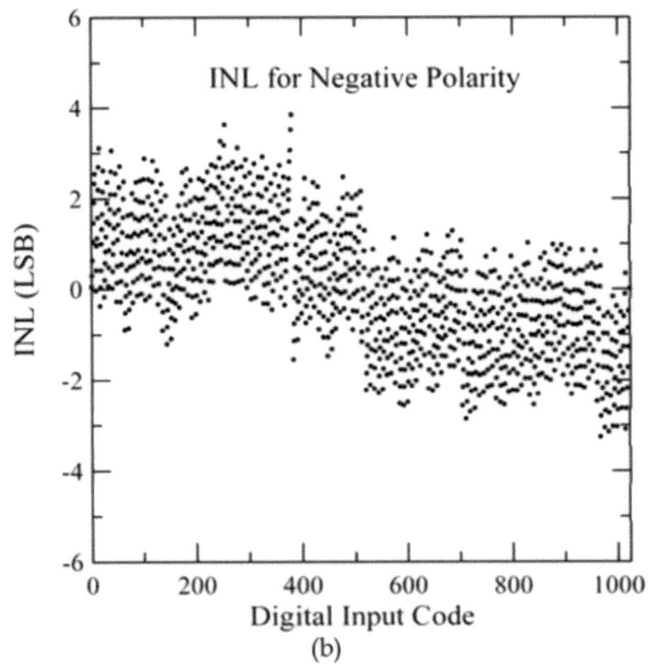

Fig. 16. The measured INL for (a) positive polarity (b) negative polarity of the proposed column driver.

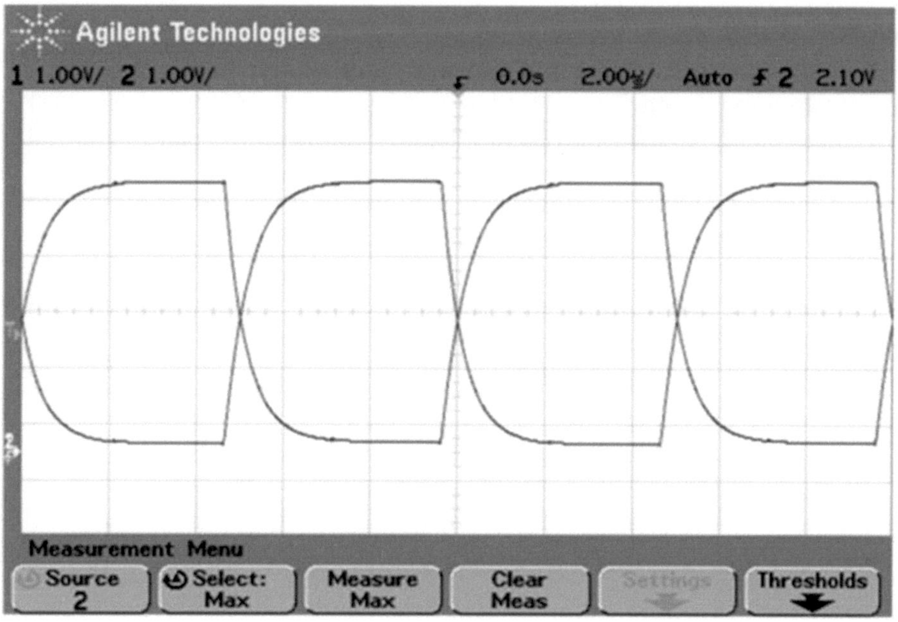

Fig. 17. The measured output waveform of two neighboring channels under dot inversion for the RGB digital inputs of '1111111111', where the voltage levels for negative and positive polarities are 0.266 V and 4.75 V, respectively.

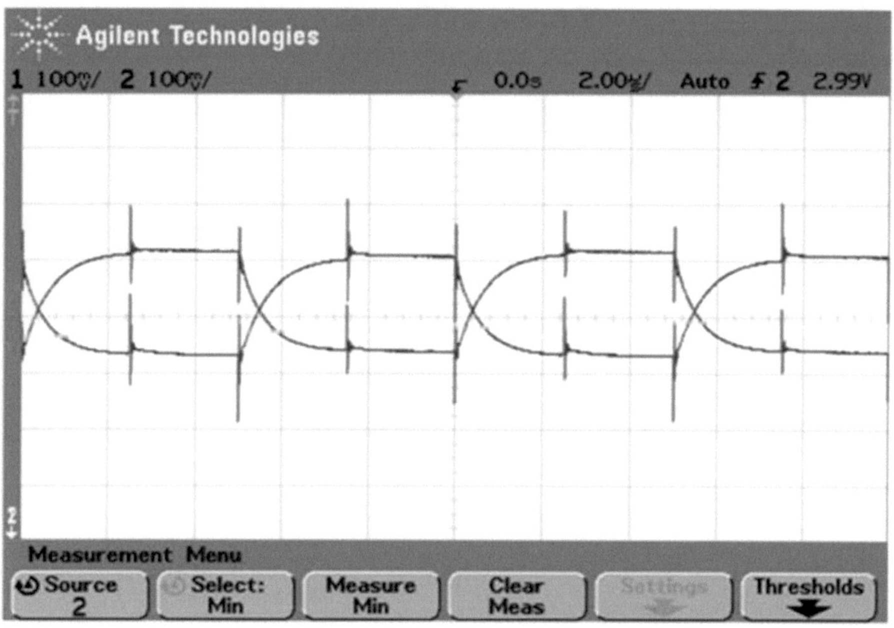

Fig. 18. The measured output waveform of two neighboring channels under dot inversion for the RGB digital inputs of '0000000000', where the voltage levels for negative and positive polarities are 2.425 V and 2.598 V, respectively.

Figure 17 shows the measured output waveforms of two neighboring channels under dot inversion for the RGB digital inputs of '1111111111.' Here, the voltage levels for negative and positive polarities are 0.266 V and 4.75 V, respectively. A load resistor of 5 kΩ and a capacitor of 90 pF were used. Figure 18 shows a similar waveform for '0000000000' inputs, where the corresponding voltage levels for negative and positive polarities are 2.425 V and 2.598 V, respectively. These two figures show that the settling time is within 3 μs, which is smaller than that of previously published work [2] and standard UXGA displays [5]. Table II summarizes the performance of the proposed column driver IC. The average area per channel is 0.063 mm^2, which is smaller than the reported areas of fully R-DAC-based column drivers [5, 8]. These experimental results show that the proposed column driver is suitable for UXGA LCD-TV applications.

	This work	[2]	[5]	[8]
Process technology	0.35 μm CMOS	N/A	0.6 μm CMOS	0.8 μm CMOS
Power supply	5 V	N/A	5 V	10 V
Number of bits	10	8 ~ 12	6	6
Maximum Deviation	1.43 LSB	±1 LSB@12 bits	N/A	N/A
Maximum DNL of DAC	3.83 LSB	N/A	N/A	N/A
Maximum INL of DAC	3.84 LSB	N/A	N/A	N/A
Settling time	3 μs for R_L = 5 kΩ and C_L = 90 pF	16 μs	8.3 μs for C_L = 30 pF 28.3 μs for C_L = 30×402 pF	2 μs for C_L = 100 pF
Buffer Power	3 μA/buffer	N/A	8.2 μA/buffer	N/A
Silicon area	0. 2x1.26 mm^2 for 4 channels*	16.5×0.9 mm^2 for 420 channels	18.1x2.35 mm^2 for 402 channels	100x2 mm^2 for 18 channels

*except for resistor string

Table II Performance summary of the proposed column driver.

5. Conclusion

This study presents a 10-bit LCD column driver consisting of piecewise linear DACs. This design uses external reference voltages and unequal resistor values in the resistor string to make coarse gamma correction. A simple digital circuit built into the timing controller or the column driver makes fine compensation adjustments. These features decrease the die area and increase effective resolution. The experimental results show that the settling time is within 3 μs. The average die area per channel is 0.063 mm^2, which is smaller than those of full R-DAC-based column drivers. The maximum deviation from the channel mean is 3.5 mV for 8 off-chip drivers. This deviation can be further reduced by employing an offset canceling technique in the buffer amplifier. The measured maximum DNL and INL values are 3.83 and 3.84 LSB, respectively. Therefore, the proposed column driver is suitable for UXGA LCD-TV applications.

6. References

[1] C-W. Lu and K.J. Hsu, (2004). A high-speed low-power rail-to-rail column driver for AMLCD application. *IEEE Journal of Solid-State Circuits*, Vol. 39, No. 8, Aug. 2004, pp. 1313- 1320, ISSN 0018-9200

[2] M.J. Bell, (2005). An LCD column driver using a switch capacitor DAC. *IEEE Journal of Solid-State Circuits*, Vol. 40, No. 12, Dec. 2005, pp. 2756-2765, ISSN 0018-9200

[3] M.J. Bell, (2005). An LCD column driver using a switch capacitor DAC, *IEEE Int. Solid-Sate Circuits Conf. Dig. Tech*, pp. 556-557, San Francisco, Feb. 2005

[4] C-W Lu, (2004) High-speed driving scheme and compact high-speed low-power rail-to-rail class-B buffer amplifier for LCD applications. *IEEE Journal of Solid-State Circuits*, Vol. 39, No. 11, November, 2004, pp. 1938-1947, ISSN 0018-9200

[5] T. Itaku, H. Minamizaki, T. Satio, and T. Kuroda, (2003) A 402-output TFT-LCD driver IC with power control based on the number of colors selected. *IEEE Journal of Solid-State Circuits*, Vol. 38, No.3, March, 2003, pp. 503-510, ISSN 0018-9200

[6] A. Gordon F. Dingwall, "Matrix addressed LCD display having LCD age indication, and autocalibrated amplification driver, and a cascaded column driver with capacitor-DAC operating on split groups of data bits," US Patent 5,739,805, Apr. 14, 1998.

[7] J-S Kang, J-H Kim, S-Y Kim, J-Y Song, O-K Kwon, Y-J Lee, B-H Kim, C-W Park, K-S Kwon, W-T Choi, S-K Yun, I-J Yeo, K-B Han, T-S Kim, S-i Park, (2007). A 10b driver IC for a spatial optical modulator for full HDTV applications. *Digest of Technical Papers. IEEE International Solid-State Circuits Conference*, pp. 138-139, San Francisco, Feb. 2007

[8] J-S Kim, D-K Jeong, and Gyudong Kim, (2000) A multi-level multi-phase charge-recycling method for low-power AMLCD column drivers. *IEEE Journal of Solid-State Circuits*, Vol. 35, No. 1, January, 2000, pp. 74-84, ISSN 0018-9200

[9] Z. Wei, "Simplified multi-output digital to analog converter (DAC) for a flat panel display," US Patent, 6,781,532 B2, Aug. 24, 2004.

[10] D. McCartney, "Designing with TFT LCD column drivers", [online seminar]. Available: http://www.national.com/AU/design/0,4706,11_0_,00.html. Published date: July 2002.

[11] P-C Yu and J-C Wu, (1999) A class-B output buffer for flat-panel-display column driver. *IEEE Journal of Solid-State Circuits*, Vol. 34, No.1, Jan. 1999, pp. 116-119, ISSN 0018-9200

[12] T. Itakura and H. Minamizaki, (2002) A two-gain-stage amplifier without an on-chip Miller capacitor in an LCD driver IC. *IEICE Trans. Fundamentals*, Vol. E85-A, No. 8 August, 2002, pp. 1913-1920.

[13] TFT-LCD source drivers NT39360, NT3982, and NT3994. Novatek. [Online]. Available: http://www.novatek.com.tw/

10

ACF Curing Process Optimization for Chip-on-Glass (COG) Considering Mechanical and Electrical Properties of Joints

Bo Tao, Han Ding, Zhouping Yin and Youlun Xiong
State Key Laboratory of Digital Manufacturing Equipment and Technology
Huazhong University of Science and Technology
P.R.China

1. Introduction

In the field of flat panel displays (FPD), packaging technology has significant influence on display performance. The electrical and mechanical interconnect between the liquid crystal displays (LCD) and its driver integrated circuit (IC) is a key issue that needs improvement to achieve finer pitch, easier assembly and greater connection reliability. With the decrease of the pixel size and the increase of pixel count for high-density LCD, the overall trend of the driver IC is packaged closer and closer to the LCD itself, even is onto the backside glass of the LCD. Bonding the driver IC chips directly to the glass substrate of the LCD panel might be a better choice when the pitch becomes less than 70–100 μm (Helge & Liu, 1998). Since the announcement by Citizen back in 1983 of a chip-on-glass (COG) driver assembly process for their LC pocket TV, many different types of COG assembly processes have been developed (Helge & Liu, 1998). In COG technology, the driver ICs are bonded directly to the indium-tin-oxide (ITO) traces on the glass without increasing the size of the panel, except for the finer bump pitch and smaller contact resistance, which can bring a significant reduction in the size of the FPD module. Because LCD is particularly heat sensitive and cannot withstand normal soldering temperatures, conductive adhesives are widely used to connect the driver IC to LCD. Usually, there are two different mechanisms used to cure the conductive adhesives, heat curing for thermosetting adhesives and UV curing for thermoplastic adhesives. Among them, heat curing appears to be most common. Nowadays interconnection using anisotropic conductive film (ACF) is the most major packaging method for production of FPD modules, to provide electrical conduction and mechanical adhesion between the driver ICs and the glass substrate with a high resolution, light weight, thin profile, and low power consumption (Myung & Kyung, 2006). Figure 1 presents a schematic illustration of a typical COG connection process using ACF.

ACF is a thermosetting epoxy impregnated with small amount of electrically conductive particles, which can be pure metals such as gold, silver, or nickel, or metal-coated ones with plastic or glass cores. During ACF curing, when the heat and force are applied, the conductive particles are trapped between the mating bumps of IC and substrate to provide electrical conductivity, and the adhesive matrix is used to provide the necessary electrical insulation, to protect the metallic contacts from mechanical damage, and to provide stable

Source: New Developments in Liquid Crystals, Book edited by: Georgiy V. Tkachenko,
ISBN 978-953-307-015-5, pp. 234, November 2009, I-Tech, Vienna, Austria

adhesion. This arrangement allows ACF to conduct in z-direction, i.e. normal to the plane of adhesive film, while remaining insulation in the x-y plane due to the particles concentrated is far below the critical value to achieve percolation conduction. ACF has many distinct advantages over it counterparts. First, it is environment friendly, avoiding the toxicity and concerning from the lead and chlorofluorocarbon-based flux cleaners. Second, lower curing temperature is required that reduces joint fatigue and stress cracking problems. Third, it has higher flexibility and closer match in coefficient of thermal expansion (CTE) that enables a more compliant connection and minimizes failures. Furthermore, the smaller filling particle size facilitates finer line resolution, and the placement of adhesives is not critical. More recently, jointing technique based on ACF is playing an increasingly important role in the design and production of electronic packaging applications, such as the COG technique for LCD, flip-chip bonding of radio frequency chips, and so on (Yim, et al, 2005).

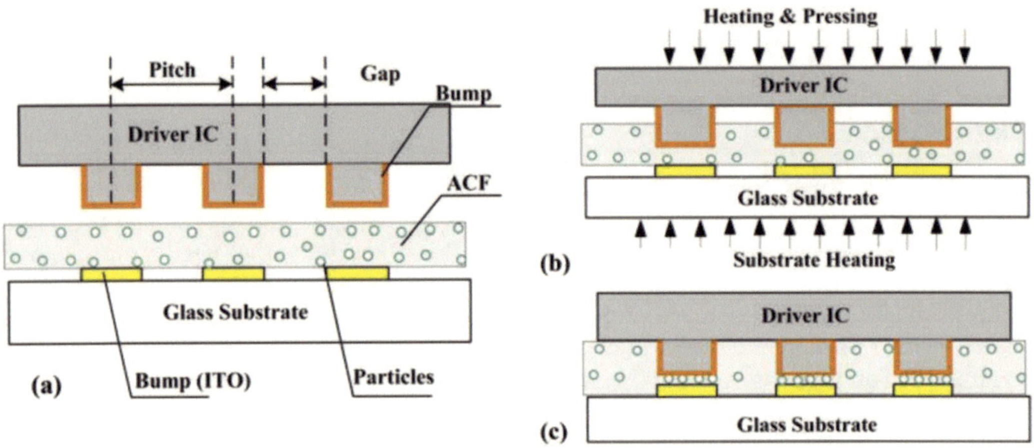

Fig. 1. A typical COG packaging process using ACF interconnection

The reliability of COG packaging is an important aspect in the electronics industry. It is found that the reliability is much dependant on the properties of the ACF (Lin & Zhong, 2008). Missing strong metal connection, ACF jointing usually has poor adhesion strength and unstable contact resistance, which are the two most critical reliability issues of ACF applications (Kim & Kim, 2008). High adhesion strength is a critical parameter of fine pitch interconnection that fragile to shocks encountered during assembly, handling and lifetime. During the curing process, voids are generated between the chip, adhesion and glass panel. These voids may affect the adhesion strength of the ACF, resulting in low reliability (Uddin, et al, 2004). Mechanisms to affect the stability of contact resistance include water absorption, electrochemical corrosion and metal oxidation, resulting in ever-increasingly unstable resistance through time, particularly under high temperature and high humidity conditions (Wu & Chau, 2002). These degradation mechanisms interfere with the contact resistance of ACF joints and hence limiting the ultimate electrical and mechanical performance of FPD module, which depends to a large extent on the curing conditions, including curing time, temperature, pressure, etc. Successful bonding involves the selection of proper bonding parameter, during which chemical reactions proceed to completion, in order to develop its strong adhesion strength and stable contact resistance.

During ACF bonding process, heat and pressure are applied concurrently to the component, and the conductive particles between an IC bump and a glass substrate pad change their shape from spheres to ovoid to form the z-direction conduction paths. On the one hand, when heat is applied during bonding, the epoxy matrix of ACF is cured and becomes soft first and then rubbery. This transformation allows the ACF to flow, which in turn allows the conductive particles within to move and distribute themselves evenly throughout the ACF joints. When the curing process is completed, the ACF becomes hardened and the mobility of the conductive particles loses. A reliable electrical interconnect should have sufficient amount of conductive particles captured between the bump and pad, and they do not flow away after cured. The fluidity of the conductive particles during ACF bonding is strongly dependent on the curing temperature and time. Higher curing temperature and shorter time will limit the fluidity of particles and chemical reaction of epoxy, resulting in large and uneven contact resistance due to the less particles captured and their uneven distribution. Furthermore, inadequate chemical reaction will decrease the capability to endure the high hydrothermal impact during operations, which will speed the contact resistance shift (Hwang & Yim, 2008). On the other hand, the deformation amount of the conductive particles, determined by the amount of the pressure applied during the bonding process, has also a great influence on the contact resistance and adhesion strength of the ACF joints. Too much pressure will make the particle a larger deformation degree, meaning a larger recovery rate. When the external pressure is cancelled after cured, a larger residual stress will be present, due to the difference of the expanding fore of the deformed particles and the compressive force of the polymer matrix resin, which will speed the interface breakdown between ACF and IC or ACF and substrate. In fact, too much spread of the particles between adjacent bumps or pads, caused by too large pressure applied, will also increase possibility of short-circuiting. Whereas if the bonding force is too low, the particles may not be able to make contact between the connecting bumps and pads (Masahiro & Katsuaki, 2006). Moreover, in some cases an excessive pressure on bumps can cause glass breakage, so the bonding pressure imposed on the backside of the IC must be controlled precisely. Hence, to have a reliable ACF interconnection in a fine-pitched COG module, the thermo-compression bonding conditions need to be optimized except for the material properties of ACF.

Extensive studies have been done on the ACF bonding in past decades. However, most of them focus on certain bonding process parameter optimization, such as bonding pressure, temperature, time, and its correlations with the reliability of ACF joints subjected to various thermal, mechanical or environmental stresses. In fact, these bonding parameters not only influence the contact resistance of ACF joints but also determine their adhesive strengths greatly. Therefore, the electrical performance and adhesive strengths must both be considered to determine the optimum bonding parameters for reliability of interconnection from a systematic viewpoint of ACF curing reaction mechanism. Usually, the curing reaction of ACF is characterized by the curing degree of epoxy resin, defined as the fraction or extent to which the maximum possible cross links has been produced in a reaction. However, little work has been done to reveal the correlations of adhesion strength and contact resistance of flip chip with the curing degree of ACF matrix (Chung, et al, 2008). In the present work, the effect of different curing degrees on the electrical and mechanical properties of a typical ACF is studied through a systematic joints reliability evaluation method, and the optimum curing degree as well as its corresponding curing conditions for the given ACF is suggested to achieve highly reliable ACF joints, where the performance variations of the adhesion strength and contact resistance are considered simultaneously. The later sections are organized as follows. In section two, the degradation data of the contact resistance of some ACF assemblies, bonded with several curing degrees, is collected during a standard high hydrothermal fatigue test. The resistance distribution of the ACF assemblies for each curing degree is verified and the distribution parameters are estimated respectively. In section three, a reliability analysis method based on the degradation data of contact resistance is adopted, and the reliability index as well as the mean-time-todegradation (MTTD) of ACF joints, as a function of the curing degree, is deduced, through which, the optimization curing degree is suggested. In section four, combining with the mechanism analysis and actual experiments, the curing kinetics model of the given ACF is built. Based on which, the optimum bonding parameters are suggested and are verified by way of actual ACF curing experiments. Finally, section five organizes this work and the value of this work is evaluated.

2. Degradation test and probability distribution analysis of ACF joints contact resistance

Investigators found that the worst environment for ACF joints was the thermal cycling and high hygrothermal (Wu & Chau, 2002). The contact resistance of ACF joints will become unstable through time, particularly under high temperature and high humidity conditions, where the mechanisms that affect the stability of contact resistance include water absorption, metal oxidation and electrochemical corrosion. The residual stress due to thermal compression during bonding, the oxidation of metal conductive bump and particles, hygrothermal expansion of adhesive and the CTE mismatch between components, were the main factors to result in the contact resistance increase of the ACF joints in the high hygrothermal environment. In this work, some high hygrothermal fatigue tests of ACF assemblies, bonded with several different curing degrees, are conducted under the high temperature (85°C) and high humidity (85%RH) conditions (so called 85/85 conditions), which are well known as the qualification standards throughout the electronic industry. The corresponding contact resistance is examined and is recorded. After that, the distributions to model the data collected are checked, and the distribution parameters are estimated respectively, which can be fitted as a function of the test time.

2.1 Experimental procedure

Four groups of ACF-based joints specimens were prepared with various curing degrees, which were achieved by controlling the curing time accurately and keeping the curing pressure 3N and temperature 170°C unchanged for all specimens. In each group, there were four specimens. In the test, a thermosetting epoxy-based ACF was adopted, which contains Ag particles with an average diameter 3.5 μm and occupies 5% volume fraction or so. After that, all specimens were put into a chamber with the temperature 85°C and humidity 85%RH to reveal the contact resistance degradation, which will be used to evaluate the influence of curing degree on the ACF joints reliability. The contact resistance of all the specimens were measured and recorded every three days averagely, as listed in table 1.

2.2 Probability distribution of contact resistance degradation data

The contact resistance degradation data of each group specimens during the hygrothermal fatigue tests is appraised with the aid of the well-known two-parameter weibull distribution method. Figure 1 shows the weibull probability plots of each group specimens at various

Curing Degree	Test No.	Time of Contact Resistance Degradation Testing (Days)				
		0	3	6	9	12
80%	1	298	573	1337	NA*	1687
	2	218	958	1053	NA	1518
	3	273	855	1728	NA	1428
	4	168	NA	938	NA	1033
85%	1	278	550	735	NA	883
	2	225	705	740	NA	755
	3	215	752	1070	NA	1128
	4	195	763	885	NA	1143
90%	1	168	NA	758	NA	1100
	2	250	NA	878	NA	1263
	3	315	650	1037	NA	1363
	4	148	815	948	NA	1493
95%	1	373	958	NA	1185	1580
	2	205	488	NA	903	1278
	3	355	1270	NA	2288	2210
	4	138	290	NA	715	1350

Table 1. Contact resistances of specimens during the 85°C/85%RH hygrothermal test for various curing degrees (unit: mΩ)

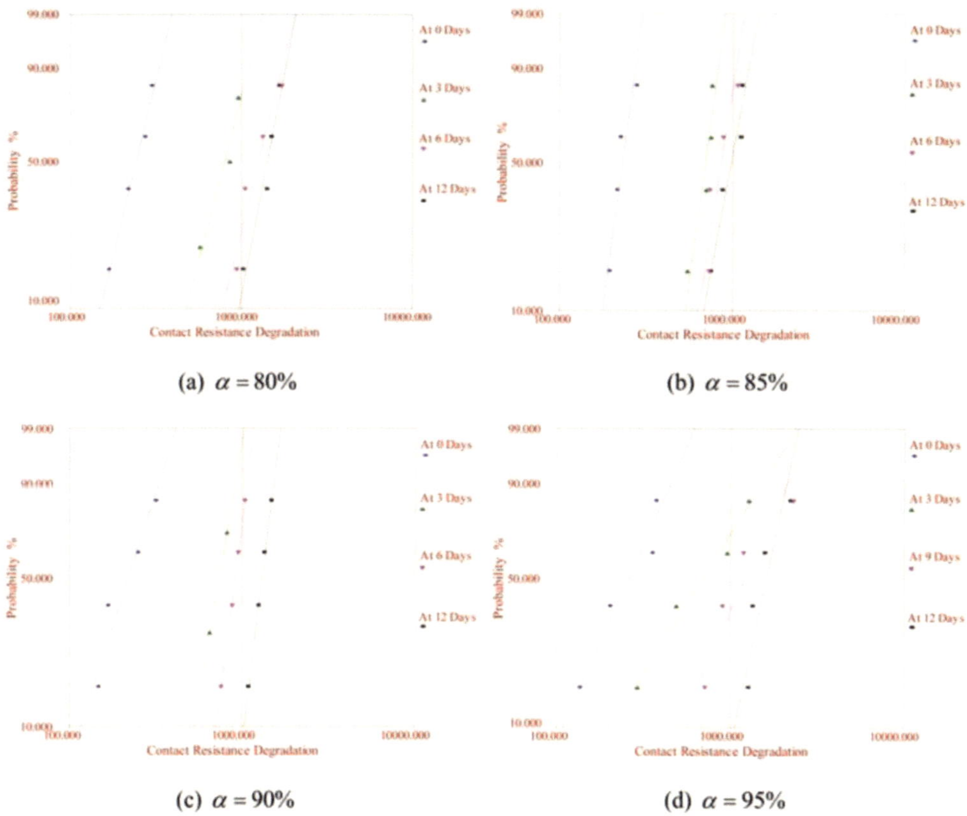

Fig. 2. Weibull distribution of contact resistance degradation data for each group of specimens characterized by different curing degree α

observation time. From figure 2, it can be seen that for each group specimens, most of the data fall on the straight-line plots except for several occasional outliers. This suggests that the two-parameter weibull distribution is a reasonable candidate to model the contact resistance degradation data of ACF joints, so the probability density function (PDF) of the contact resistance of specimens can be given by:

$$f(x)=\frac{\beta}{\eta}\cdot\left(\frac{x}{\eta}\right)^{\beta-1}\cdot\exp\left(-\left(\frac{x}{\eta}\right)^{\beta}\right) \tag{1}$$

Herein, t is the hygrothermal testing time, β and η are the shape parameter and scale parameter of the weibull distribution respectively. Usually, both β and η are timedependent and can be expressed as a certain function of the hygrothermal testing time t. The shape and scale parameters of each weibull distribution plot, corresponding to different curing degree, are recorded, as listed in table 2.

Curing Degree	Distribution Parameters	Time of Contact Resistance Degradation Testing (Days)				
		0	3	6	9	12
80%	β	3.97	3.67	3.87	NA	4.92
	η (mΩ)	263.54	880.63	1390.69	NA	1536.46
85%	β	7.20	7.09	6.27	NA	5.21
	η (mΩ)	241.82	734.69	914.23	NA	1055.37
90%	β	3.00	5.59	7.60	NA	7.80
	η (mΩ)	246.64	785.62	957.74	NA	1378.58
95%	β	2.19	1.53	NA	2.14	4.53
	η (mΩ)	307.65	880.29	NA	1438.02	1740.97

Table 2. Weibull distribution parameters of each specimen group corresponding to figure 1

2.3 Estimation of the time-dependent distribution parameters

From table 2, it is found that the shape parameter keeps unchanged approximatively except for certain occasional outlier for each group, while the scale parameter all vary obviously with an incremental trend for each specimen group. Least squares fitting is used to model the data and the resultant time-dependent functions for each specimen group characterized by four different curing degree, namely 80%, 85%, 90% and 95%, are expressed by the equations (2) to (5) respectively, which are graphically shown in figure 3 correspondingly.

$$\begin{cases}\beta=4.11\\ \eta=102.2\times x+481.2\end{cases} \tag{2}$$

$$\begin{cases}\beta=6.44\\ \eta=896.3\times\exp(0.014\times x)-654.5\times\exp(-0.395\times x)\end{cases} \tag{3}$$

$$\begin{cases}\beta=6.0\\ \eta=666.0\times\exp(0.061\times x)-419.3\times\exp(-1.153\times x)\end{cases} \tag{4}$$

$$\begin{cases}\beta=2.6\\ \eta=819.9\times\exp(0.063\times x)-512.3\times\exp(-0.514\times x)\end{cases} \tag{5}$$

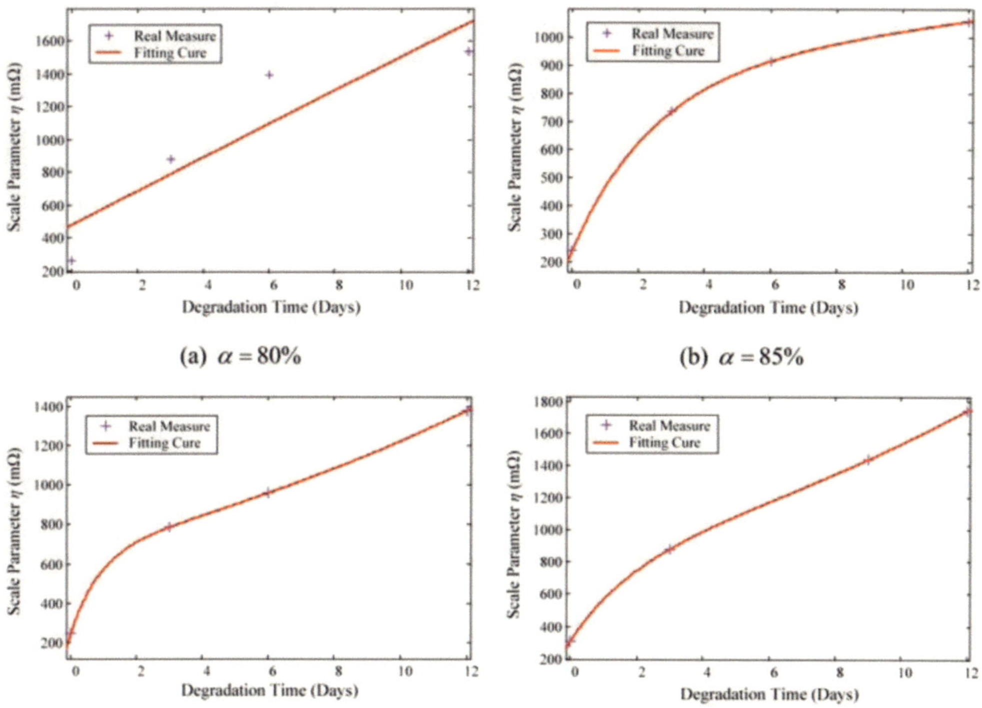

(a) $\alpha = 80\%$ (b) $\alpha = 85\%$

(c) $\alpha = 90\%$ (d) $\alpha = 95\%$

Fig. 3. Plots of weibull distribution shape parameters versus test time for each group specimens

As shown in figure 3, for each group, the scale parameter η increases with the increment test time t. That means the contact resistance of the ACF joints degrades in an exponential way except for the case where the curing degree is 80%. In the next analysis, all the shape parameters are characterized by the mean value of the test results.

3. Reliability analysis and curing degree optimization

Although the equations (2) to (5) are drawn from the given test data, it is still reasonable to conclude that the weibull distribution parameter η is the function of test time t while β is time-independent constant value for each specimen group. So submitting β and parametric $\eta(t)$ into the equation (1) yields the conditional probability density function of the contact resistance for each group at a given test time, written as:

$$f(x \mid t) = \frac{\beta}{\eta(t)} \cdot \left(\frac{x}{\eta(t)}\right)^{\beta-1} \cdot \exp\left(-\left(\frac{x}{\eta(t)}\right)^{\beta}\right) \tag{6}$$

Generally, the interfacial delamination of ACF bonding emerges during its application that can result in the failure of whole COG module. The failure criterion is usually defined as the resistance increase to certain threshold value, denoted by a constant d. Then the reliability function of the ACF joints at a specific time t for each group specimen is defined as:

$$\Re(d,t) = P(x \le d) = \int_0^d f(x|t)dx$$
$$= \int_0^x \left\{ \frac{\beta}{\eta(t)} \cdot \left(\frac{d}{\eta(t)} \right)^{\beta-1} \cdot \exp\left(-\left(\frac{d}{\eta(t)} \right)^{\beta} \right) \right\} dx = 1 - \exp\left(-\left(\frac{d}{\eta(t)} \right)^{\beta} \right). \tag{7}$$

From equation (7), it is found that the joints reliability is the function of time t and the failure threshold value *d*. Similarly, for each specimen group, the mean value function of the contact resistance at a specific time *t* is defined as:

$$\bar{x}(t) = \int_0^{+\infty} \{x \cdot f(x|t)\} dx$$
$$= \int_0^{+\infty} \left\{ \beta \cdot \left(\frac{x}{\eta(t)} \right)^{\beta} \cdot \exp\left(-\left(\frac{x}{\eta(t)} \right)^{\beta} \right) \right\} dx = \eta(t) \cdot \Gamma\left(\frac{1}{\beta} + 1 \right) \tag{8}$$

Herein, Γ (•) is the Gamma function. Obviously, if the resultant mean value according to equation (8) equals to the failure threshold value *d*, the corresponding time *t* is the meantime- to-degradation of the specimen, denoted by MTTD, i.e.

$$d = \eta(MTTD) \cdot \Gamma\left(\frac{1}{\beta} + 1 \right) \tag{9}$$

Solving equation (9) will obtain MTTD value. Typically, for the ACF joints formed under the given curing degrees, namely 80%, 85%, 90% and 95%, β and $\eta(t)$ are given by the equation (2) to (5) respectively.

3.1 Time-dependent analysis of joints reliability

Substituting equation (2) to (5) into equation (7) respectively, the ACF joints' reliability functions, as a function of the hygrothermal test time for the four given curing degrees, are given respectively, by which the joints reliability at certain specific time *t* can be estimated and calculated if the resistance failure threshold value *d* is given. Two group curves of the reliability against the time for the given four group specimens according to two different failure criterions, i.e. 1000 mΩ and 1400 mΩ, are comparatively plotted together, as shown in the figure 4.

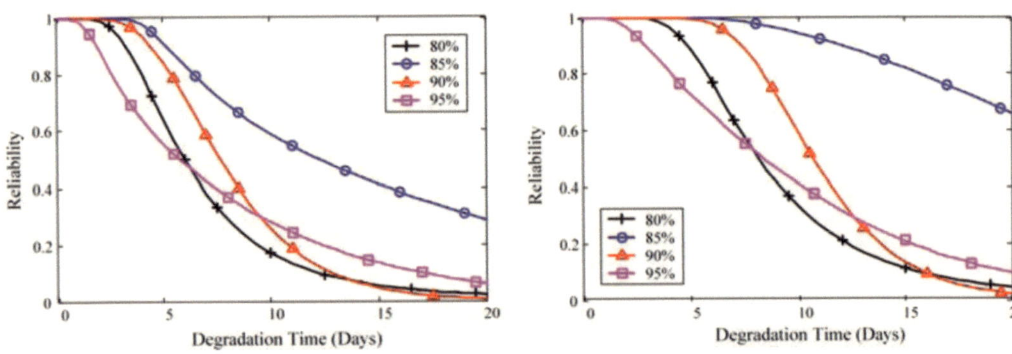

(a) d =1000mΩ (b) d = 1400mΩ

Fig. 4. Reliability curves of joints versus time for different failure criterion *d*

From figure 4, it is found that whichever the threshold value is used, the reliability of ACF joints reliability $\Re(d,t)$ versus hygrothermal test time t for each curing degree decreases monotonously while in different ways. This means that all the ACF joints, bonded under different curing degrees conditions, degrade by a single same or similar damage mechanism when suffering from same hygrothermal fatigue test. For the ACF joints tested, those, cured with a curing degree 85%, have a highest reliability than else obviously. That implies that the optimum curing degree is a certain middle value, near the 85%, in the range of 80% to 90%. For the ACF joints cured with other curing degrees, the reliability curves are interlaced to each other. For those with curing degree 95%, the reliability is lowest in the early time but the curve is flater than other two cases. That means that the ACF joints with high curing degree have a better endurance under the high hygrothermal environment.

3.2 Time-dependent analysis of joints resistance

Similarly, the mean resistance of ACF joints' can be quantitatively calculated by substituting equation (2) to (5) into equation (8) respectively. Clearly, from equation (8) it can be seen that the contact resistance of ACF joints is only correlated to the test time t. Numerical calculations are achieved for each group specimens, and the resultant resistance against hygrothermal test time t is graphically shown in figure 5. From figure 5, it is found that the resistance of all the ACF joints' increases monotonously with the increment of time t. The similar sigmoid shape except for the case 80% also implies that the ACF joints enough cured will degrade by a single same or similar mechanism under high hygrothermal environment. From figure 5, it is also found that the joints with the curing degree 85% have a lower contact resistance and a slower degradation rate than other specimens. That also implies that the optimal curing degree does exist near the 85% in the range of 80% to 95%. The optimum curing degree needs to be investigated further according to the mean time to degradation of joints in the following section.

3.3 MTTD calculation for given failure criterion

As mentioned above, the MTTD value of the joints can be estimated using equation (9) for a given failure threshold value of contact resistance. Usually, equation (9) is a highly nonlinear equation, and directly solving the optimal solution of MTTD for certain given threshold value is very difficult. Fortunately, figure 5 shows that there is a one-to-one relationship between contact resistance and fatigue time for each group specimens. That means that there will be only one optimal MTTD value for any given failure criterion. Herein, a numerical calculation method based an improved Golden Section Search arithmetic is used to calculate the desirable MTTD, described as follows:

1. Pick two large enough time values t_L and t_U that bracket the optimal *MTTD* range, and construct the goal function denoted by equation (9).
2. Calculate two interior values from $t_1 = 0.382 \times (t_U - t_L) + t_L$ and $t_2 = 0.618 \times (t_U - t_L) + t_L$, then calculate the corresponding $d(t_1)$ and $d(t_2)$.
3. Check if both the condition $abs(1 - t_1/t_2) \leq \varepsilon$ and $d(t_1) < d < d(t_2)$ are met, where ε is the convergence criterion and d is the resistance failure threshold value defined. If met, stop calculation and set MTTD = $0.5 \times (t_1 + t_2)$, otherwise, turn to the step (4).
4. If $d > d(t_2)$, set $t_L = t_2$ with t_U unchanged. If $d < d(t_1)$, set $t_U = t_1$ with t_L unchanged. If $d(t_1) < d < d(t_2)$, set $_U = t_2$ and $t_L = t_1$. Repeat steps (2), (3) and (4) till the optimal estimation of MTTD or other calculation restriction is met.

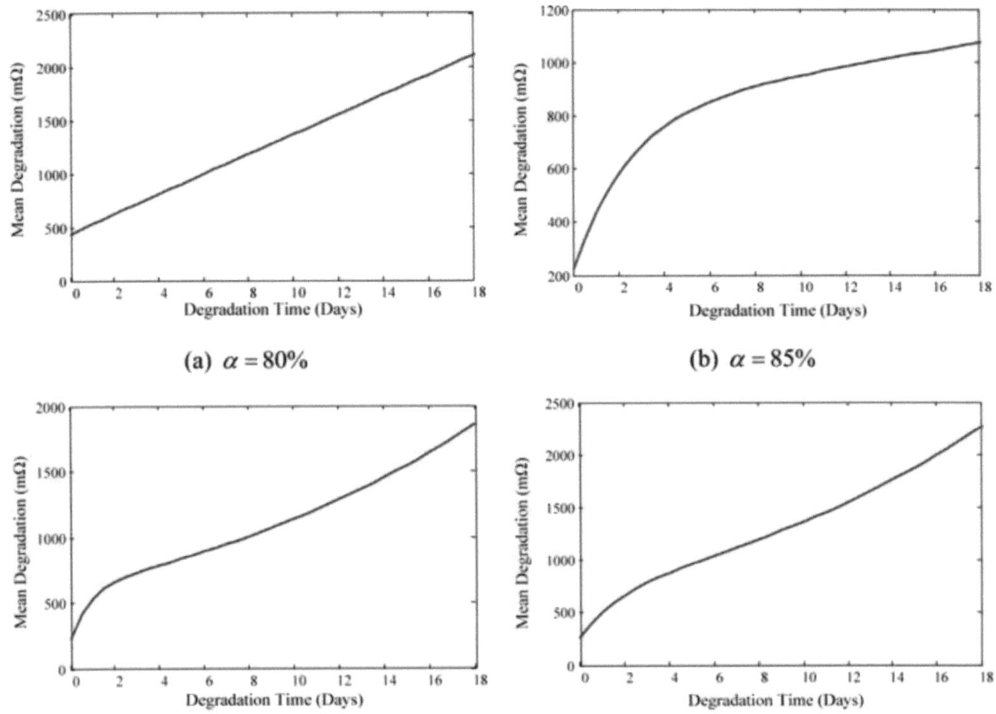

Fig. 5. Mean degradation value of contact resistance versus time for joints with varous cure degree

A C++ program agreeing with the above procedure is developed to compute the optimal MTTD value for certain given failure criterions, namely 1000 mΩ, 1100 mΩ, 1200 mΩ and 1400 mΩ, as listed in table 3.

Failure Criterion	Curing Degree			
	80%	85%	90%	95%
$D = 1000\text{m}\Omega$	6.07	14.85	8.25	5.19
$D = 1100\text{m}\Omega$	7.15	21.54	9.81	6.44
$D = 1200\text{m}\Omega$	8.23	27.75	11.24	7.70
$D = 1400\text{m}\Omega$	10.38	38.76	13.77	10.06

Table 3. MTTD value of ACF joints tested for different resistance failure criterions (unit: days)

3.4 Curing degree optimization analysis

To find the optimum value of curing degree, the influence of curing degree on the MTTD of the joints is analyzed. For a more reliable conclusion, least square fitting is used herein to model the data listed in table 3 for the four different failure criterions respectively, which are comparatively shown in fatigue 6.

From figure 6, it is found that for each failure criterion, the resultant MTTD value firstly increases and then decreases with the increment of the curing degree, and the maximum MTTD value of the ACF joints occurs at the curing degree 83% or so for each failure criterion. Although this conclusion is drawn from the given test data, it is still reasonable to conclude that the optimum curing degree for the ACF tested is 83% or so, and the desirable range of the curing degree is 82% to 85% considering 95% confident interval. In fact, more failure criterions else have also been done, and same conclusions have also been drawn.

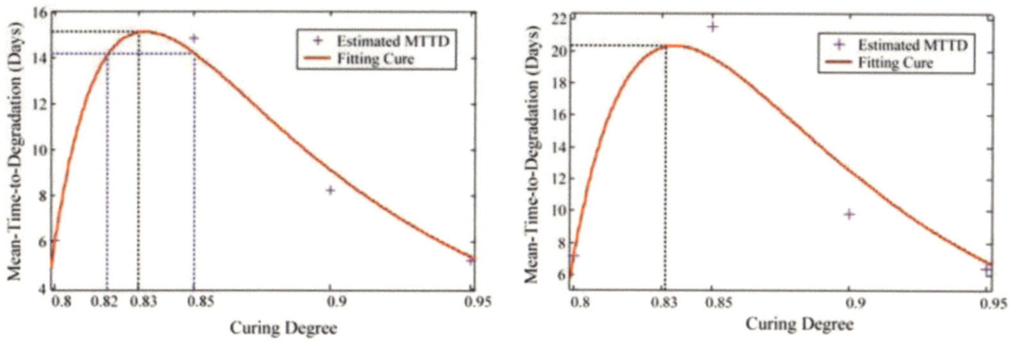

(a) $D = 1000\text{m}\Omega$ (b) $D = 1100\text{m}\Omega$

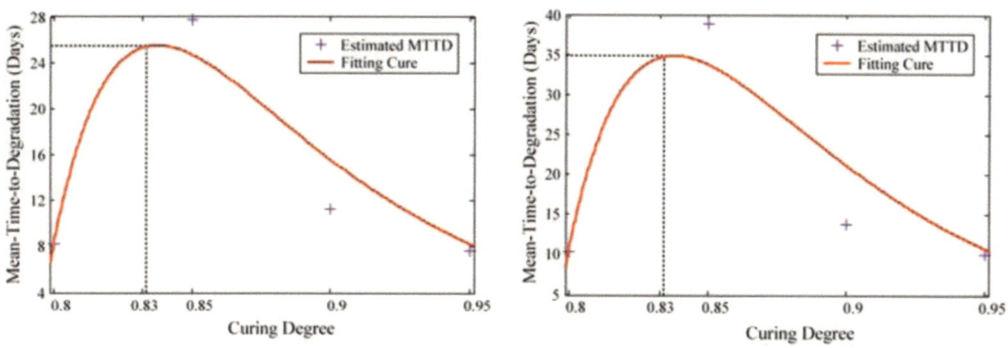

Fig. 6. MTTD value of joints versus curing degree for different resistance failure criterions

4. Curing parameters choice and optimization

Usually, the curing process of the ACF joints is achieved through controlling some key curing parameters, such as curing time, temperature and so on accurately, instead of controlling the curing degree directly. Therefore, we need to correlate the curing degree to those key parameters through curing kinetics modeling, by which the optimum curing parameters can be chosen for a given curing degree necessary.

4.1 Modeling for curing kinetics of ACF

To study cure kinetics of epoxy resin, several different methods have been proposed over the past decades, such as Fourier transform IR spectroscopy (FTIR), high pressure liquid chromatography (HPLC), nuclear magnetic resonance (NMR), differential scanning calorimeter (DSC), chemical titrations, and so on. Among them, DSC analysis is one of the best-known methods, which is mainly classified into two categories. One is isothermal test and the other is dynamic test. Both of them are based on the assumption that the exothermic heat evolved during the curing reaction is proportional to the extent of monomer conversion. That means that, for an ACF curing process, the measured heat flow dH/dt is proportional to the curing reaction rate $d\alpha/dt$. This assumption is valid if there are no other enthalpic events except for chemical reactions occurring, such as evaporation, enthalpy relaxation, or significant changes in heat capacity conversion. Usually, the instantaneous change of the conversion rate is defined as:

$$\dot{\alpha}=\frac{d\alpha}{dt}=\frac{dQ/dt}{\Delta Q}. \tag{10}$$

Herein, ΔQ is the exothermic heat, expressed as heat per mol of reacting groups ($KJ\cdot mol^{-1}$) or per mass of materials ($J\cdot g^{-1}$). Usually, the curing kinetics equations of thermosetting materials are classified into two general categories: nth order and autocatalytic, which represents the overall process if more than one chemical reaction occurs simultaneously during curing (Chan, et al, 2003).
For thermosetting materials that follow nth order kinetics, the rate of conversion is usually proportional to the concentration of unreacted sections (Chan, et al, 2003), i.e.

$$\dot{\alpha}=k(1-\alpha)^{n}. \tag{11}$$

Herein, n is the reaction order, and k is the temperature-dependent rate constant given by the Arrhenius equation:

$$k=A\cdot\exp\left(-\frac{E}{R\cdot T}\right). \tag{12}$$

Herein, E is the activation energy, R is the gas constant, T is the absolute temperature, and A is the frequency factor. Equation (11) assumes that the reaction rate $\dot{\alpha}$ is dependent only on the amount of unreacted materials and the reacted sections do not participate in the remaining reactions. As such, a logarithmic plot of the equation (11) would result in a linear relationship, from which the reaction order can be estimated to the linear slope.
Autocatalyzed curing reactions, on the contrary, assume that at least one of the reacted sections will participate in the remaining reactions, and usually are characterized by an accelerating isothermal-conversion rate. The kinetics of autocatalyzed curing reactions is generally expressed by (Lee, et al, 1997):

$$\dot{\alpha}=\left(k_{0}+k\cdot\alpha^{m}\right)\cdot(1-\alpha)^{n}. \tag{13}$$

Herein, m and n are the reaction orders. k_0 is the initial rate constant and is zero if no reactions occur at initial time. k is the temperature-dependent rate constant given by the equation (12). For autocatalytic reactions, at least two reaction orders, i.e. m and n , are needed to be determined. Boey and Qiang (2000) propose a numerical method to estimate the parameters that depend on the extent of reaction at the exothermic peak as well as the rate of the reaction at the peak. Usually, to simplify the calculation, the total reaction orders are assumed to two, i.e. let $m+n=2$. Thus, the coefficients of the kinetics equation modeled by equation (13) can be estimated from the following equation group:

$$\begin{cases}\alpha_{p}=\dfrac{m}{m+n}\\ m+n=2\\ \dot{\alpha}_{p}=\dfrac{k\cdot m^{n}\cdot n^{m}}{(m+n)^{(m+n)}}\end{cases}. \tag{14}$$

Herein, α_p and $\dot{\alpha}_p$ are the curing degree and curing rate correspondingly at the exothermic peak, which can be easily obtained from the DCS thermogram.
It should be mentioned that in order to model the curing kinetics, it need to check the curing reaction of ACF given, nth order or autocatalytic. For the former, equation (11) indicates the maximum curing rate occurs at the time zero, while the maximum curing rate occurs at a certain middle time during the cure for the latter, which typically reaches its maximum between 20% and 40% conversion. Usually, the criterion mentioned here are used to check the curing kinetics of the undergoing reactions is nth order or autocatalytic. Whichever the kinetic model is adopted, the activation energy *E* and frequency factor *A* for the cure process must to be calculated. There are two different methods to estimate them according to the DSC test method adopted. For the isothermal DSC test, they can be estimated from the linear logarithm plot of the Arrhenius equation based on the isothermal curing test data. For the dynamic DSC test, the estimation of the activation energy and frequency factor can be achieved through the well-known Kissinger equation, as:

$$\ln\left(\frac{\beta_i}{T_{pi}^2}\right) = \ln\frac{A\cdot R}{E} - \frac{E}{R}\cdot\frac{1}{T_{pi}} \tag{15}$$

Herein, the subscript *i* is the specimen number, β is the heating rate and T_p is the peak temperature of reaction curve. Equation (15) indicates that there is a linear relationship between $\ln(\beta/T_p^2)$ and $1/T_p$. If the least square linear fitting is met, then *E* and *A* can be estimated from the slope and intercept of the linear plot fitted.

4.2 Coefficients estimation of curing kinetics

The estimation of the coefficients for the curing kinetics aforementioned is achieved through a group of dynamic DSC experiments, where a thermosetting epoxy-based ACF was tested using a DSC with a computerized data acquisition system in this study. The ACF contains Ag particles with an average diameter 3.5 μm and occupies 5% volume fraction or so. Some dynamic DSC tests were performed from 80°C to 180°C with four different ramp rate, namely 20°C/min, 15°C/min, 10°C/min, 5°C/min, during which, the rates of heat generation as a function of the temperature and time were recorded correspondingly. The plots of the dynamic DSC scans are shown in figure 7 and the resultant data were listed in table 4.

β (K•min^{-1})	T_p (K)	$\ln(\beta/T_p^2)$	$\ln\beta$ (K/sec)	$1/T_p$ ($\times10^{-3}$K^{-1})
5	379.32	-10.267	-2.488	2.636
10	389.85	-9.629	-1.795	2.565
15	396.14	-9.255	-1.386	2.524
20	400.43	-8.989	-1.099	2.497

Table 4. Dynamic DSC data of ACF tested

From figure 7, it is found obviously that the larger the heating rate is, the sharper the curve does be. That means the curing process of the ACF is quickened with the increase of the heating rate. The calorimetric curve of figure 7(a) was integrated in order to obtain the integral curing curves, indicating the time dependence of the curing degree using a digital integral method. Herein, the curing degree was estimated by the division of the cumulative heat at a certain time ΔQ_t over the total exothermic heat ΔQ of the curing process, as shown in figure 8. The zero-initial sigmoid shape of the curves means the maximum curing rate occurs at certain middle time of the overall cure process, which reveals that the undergone cure process follows an autocatalytic mechanism, and equation (13), keeping k_0 zero, will be adopted to model the cure kinetics of the ACF.

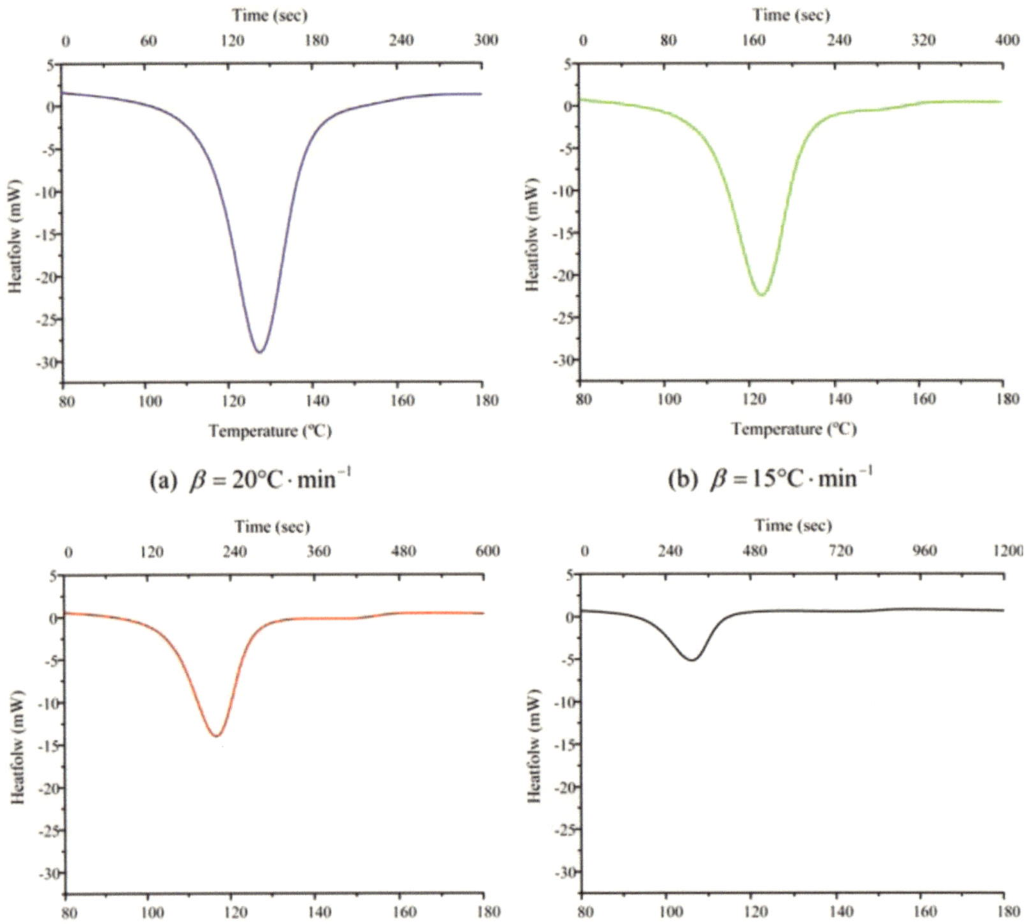

(a) $\beta = 20°C \cdot min^{-1}$ (b) $\beta = 15°C \cdot min^{-1}$

(c) $\beta = 10°C \cdot min^{-1}$ (d) $\beta = 5°C \cdot min^{-1}$

Fig. 7. Dynamic DSC plot of ACF for different heating rate

From the DSC thermograms, as shown in figure 7(a) and figure 8, the total exothermic heat ΔQ of the curing process is 1758.9 mJ and the cumulative heat ΔQ_t at the exothermic peak time is 935.4 mJ. Therefore, the cure degree at the exothermic peak, namely α_p, are 0.53. Taking it into equation (14), and the resultant coefficients m and n are 1.06 and 0.94 respectively.

Next, the activation energy E and frequency factor A will be estimated through Kissinger equation. Test data listed in table 4 is used to model the relationship of $-\ln(\beta/T_p^2)$ and $1/T_p$, and it is found that they follow a linear relationship with an ultra-high regression index 1.0,

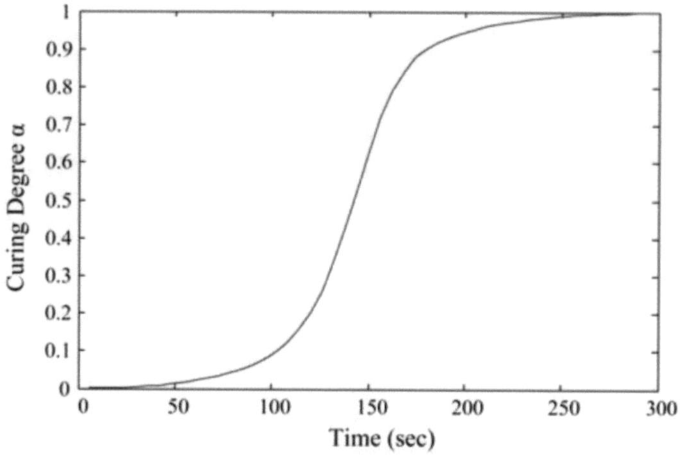

Fig. 8. ACF curing degree versus time under constant heating rate ($20°\,C \cdot min^{-1}$)

or so, as shown in figure 9. From the linear plot, the slope and intercept are found to be 9156.12 and -13.86. From equation (15), we can get the following equation group:

$$\begin{cases} \dfrac{E}{R} = 9156.12 \\ \ln \dfrac{A \cdot R}{E} = 13.86 \end{cases} \tag{16}$$

Let the gas contast R is 8.314 $J \cdot mol^{-1} \cdot K^{-1}$, solving the equation group (16) can get that the activation energy E is 76.12kJ$\cdot$mol^{-1} and frequency factor A is 1.60×10^8 sec^{-1}.

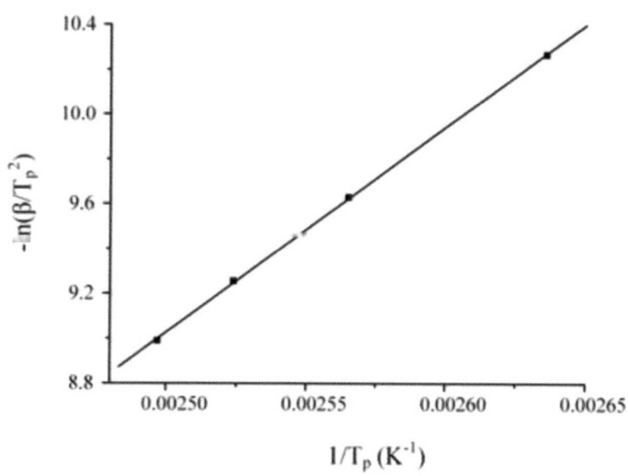

Fig. 9. Curve of $-\ln(\beta/T_P{}^2)$ versus $1/T_P$ for the ACF tested

4.3 Choice of curing parameters

Usually, there are mainly two curing patterns. One is the isothermal curing and the other is the non-isothermal curing. For the isothermal curing process, the function of the curing degree versus curing time, for a certain curing temperature, can be obtained by directly integrating equation (13), which can be rewritten as the following form:

$$\frac{d\alpha}{\alpha^m \cdot (1-\alpha)^n} = A \cdot \exp\left(-\frac{E}{R \cdot T}\right) dt \,. \tag{17}$$

By taking the integral firstly and then the natural logarithm, the equation (17) can be expressed as follow:

$$\ln(F(\alpha)) = \ln t + \ln A - \frac{E}{R \cdot T}, \tag{18}$$

where

$$\begin{cases} f(\alpha) = \alpha^m \cdot (1-\alpha)^n \\ F(\alpha) = \int_0^\alpha \frac{d\alpha}{f(\alpha)} \end{cases} \,. \tag{19}$$

Usually, the glass substrate is temperature sensitive, so the bonding temperature during the COG packaging should be controlled accurately.
For the non-isothermal curing process, characterized by the heating rate β, the equation (13) can be rearranged as:

$$\frac{d\alpha}{dt} = \frac{d\alpha}{dT} \cdot \frac{dT}{dt} = \beta \cdot \frac{d\alpha}{dt} = A \cdot \exp\left(-\frac{E}{R \cdot T}\right) \cdot f(\alpha) \tag{20}$$

By integrating equation (20), the relationship of curing degree and curing temperature for certain heating rate can be given. According reference (Ozawa, 1970), the integral equation (20) can be expressed by means of polynomial form, i.e.

$$\lg \beta = \lg\left(\frac{A \cdot E}{R}\right) - \lg(F(\alpha)) - 2.32 - 0.457 \times \frac{E}{R \cdot T} \,. \tag{21}$$

For the isothermal curing of the ACF tested, taking activation energy, frequency factor as well as the coefficients m and n into equation (18), the function of cure degree is written as:

$$\ln\left(\int_0^\alpha \frac{d\alpha}{\alpha^{1.06} \cdot (1-\alpha)^{0.94}}\right) = \ln(1.60 \times 10^8 \cdot t) - \frac{9.16 \times 10^3}{T} \tag{22}$$

Herein, t is curing time with the unit second, T is curing temperature with the unit K, and α is curing degree. Through the equation (22), the bonding time necessary to reach certain cure degree can be determined for certain bonding temperature. Figure 10 shows some typical relationship curves of curing temperature and time to reach some certain curing degree. It is found that for each curing degree, the curing time needed is lengthened with the temperature increasing of the isothermal curing process. Herein, some typical valuepairs for the curing time and temperature is chosen for the optimum curing degree range, i.e. 82%~85%, and listed in table 5.

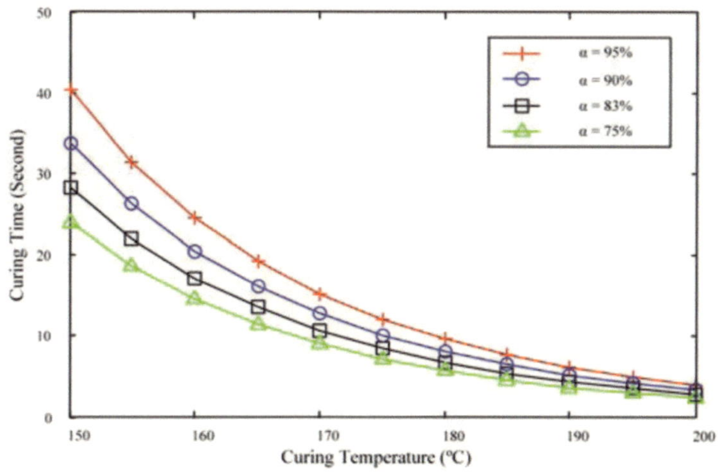

Fig. 10. Relationship of curing temperature versus time to reach some certain curing degrees

	Curing Degrees											
	82%			83%			84%			85%		
T (°C)	160	165	170	160	165	170	160	165	170	160	165	170
t (sec)	16.8	13.2	10.4	17.2	13.5	10.6	17.6	13.8	10.9	18.0	14.1	11.1

Table 5. Some typical value-pairs of curing temperature and time to reach certain curing degree in the recommended optimum range

5. Conclusions

In the work, the ACF curing process is optimized from the viewpoint of curing degree to find out the desirable curing process parameters. First of all, the influence of various curing degrees on the contact resistance of ACF joints is studied, using a systematic joints reliability evaluation method through some typical high hygrothermal fatigue tests. Degradation analysis is achieved, instead of the traditional failure time analysis, and the dependence of ACF joints' mean time to degradation on curing degree is analyzed, by which the optimum curing degree value as well as the recommend range is suggested. Results show that the optimum value for curing degree is 83% and the recommend range is from 82% to 85% for the ACF tested considering 95% confident interval. After that, the recommened curing parameters to reach certain desirable curing degree are also investigated, which is achieved by building the curing kinetics model of the ACF tested. The study of this work will provide an important support to optimize the curing process for various ACF-based packaging applications, such as the COG packaging for LCD, flip-chip bonding of radio frequency chips, and so on COG.

6. Acknowledgement

This work is supported by the National Science Foundation of China under grant 50805060, 50625516, the National Fundamental Research Program of China under Grant 2009CB724204, and the China Postdoctoral Fund under grant 20070420173.

7. References

Boey, F.Y.C.; Qiang, W. (2000). Experimental modeling of the cure kinetics of an epoxyhexaanhydro- 4-methylphthalicanhydride (MHHPA) system. Polymer, Vol.41, 2081- 2094, 0032-3861

Chan, Y.C.; Uddin, M.O.; Chan, H.P. (2003). Curing Kinetics of Anisotropic Conductive Adhesive Film, *Journal of Electronic Materials*, Vol. 32, No. 3, 2003, 131-136

Chung, C.K,; Kwon, Y.M.; et al. (2008). Theoretical Prediction and Experimental Measurement of the Degree of Cure of Anisotropic Conductive Films (ACFs) for Chip-On-Flex (COF) Applications. *Journal of Electronic Materials*, Vol. 37, No. 10

Helge, K.; Liu, Johan (1998). Overview of Conductive Adhesive Interconnection Technologies for LCD's. IEEE Trans. on Components, Packaging, and Manufacturing Technology – Part A, Vol.21, No.2, 208-214, 1070–9886

Hwang, J.S.; Yim, M.J.; Paik, K.W. (2008). Effects of bonding temperature on the properties and reliabilities of anisotropic conductive films (ACFs) for flip chip on organic substrate application. *Microelectronics Reliability*, Vol. 48, 293-299, 0026-2714

Jarmo, M. (2003). Contact resistance of metal-coated polymer particles used in anisotropical conductive adhesives. Soldering & Surface Mount Technology,12–15, 0954-0911

Kim, J.W.; Kim, D.G; et al. (2008), Analysis of Failure Mechanism in Anisotropic Conductive and Non-Conductive Film Interconnections. *IEEE Trans. on Components and Packaging Technologies*, Vol. 31, No. 1, 65-73, 1521-3331

Lee, J.Y.; Shim, M.J.; Kim, S.W. (1997). Autocatalytic cure kinetics of natural zeolite filled epoxy composites. *Materials Chemistry and Physics*, Vol.48, 36-40, 0254-0584

Lin, Y. C.; Zhong Jue (2008). A review of the influencing factors on anisotropic conductive adhesives joining technology in electrical applications. *J. Material Science*, Vol. 43, 3072–3093

Masahiro I.; Katsuaki S. (2006). Effect of curing conditions on the electrical properties of isotropic conductive adhesives composed of an epoxy-based binder. *Soldering & Surface Mount Technology*, Vol.18, No.2, 40–45, 0954-0911

Myung J.Y.; Kyung W.P. (2006). Recent advances on anisotropic conductive adhesives (ACAs) for flat panel displays and semiconductor packaging applications. *Int. Journal of Adhesion & Adhesives*, Vol.26, 304-313, 0143-7496

Ozawa, T. (1970). Kinetic analysis of derivative curves in thermal analysis. *J. Thermal Analysis*, Vol.2, 301-324

Uddin, M.A.; Alam, M.O. et al. (2004). Adhesion strength and contact resistance of flip chip on flex packages -- effect of curing degree of anisotropic conductive film. *Microelectronics Reliability*, Vol. 44, 505–514, 0026-2714

Wu, C.M.; Chau, M.L. (2002). Degradation of flip-chip-on-glass interconnection with ACF under high humidity and thermal aging. *Soldering and Surface Mount Technology*, Vol. 14, No.2, 51–58, 0954-0911

Yim, M.J.; Jeong, I.H.; et al. (2005). Flip Chip Interconnection with Anisotropic Conductive Adhesives for RF and High-Frequency Applications. *IEEE Trans. on Components and Packaging Technologies*, Vol. 28, No. 4, 789-796, 1521-3331

11

Introduction to LED Backlight Driving Techniques for Liquid Crystal Display Panels

Huang-Jen Chiu[1], Yu-Kang Lo[1], Kai-Jun Pai[1], Shih-Jen Cheng[1], Shann-Chyi Mou[2] and Shih-Tao Lai[3]
[1]*National Taiwan University of Science and Technology,*
[2]*Ching-Yun University,*
[3]*Chung-Yuan Christian University,*
Taiwan

1. Introduction

Liquid crystal display (LCD) is widely used in various display applications such as cellular phones, PC monitors, televisions (TVs), multimedia products, among others. An LCD backlight module usually includes backlight sources, a light-diffusion plate, a reflector, a brightness-enhancement film (BEF) and a light-guide plate (LGP). Conventionally, cold cathode fluorescent lamps (CCFLs) are required to provide sufficient backlighting for LCD panels [1-5]. Growing concerns about environmental issues will inhibit the use of CCFLs that contain poisonous mercury. Owing to improvement in long operative life, wider operation temperature range, and the simplicity of driver circuit work with low and safe voltages, light emitting diode (LED) has gradually substituted the CCFL as backlight [6-9]. This chapter will introduce some LED backlight driving techniques for LCD panels. Some dimming control methods will also be introduced and compared for regulating the LED current and brightness of the LED backlight system. The principal goal of this chapter is to ensure that readers become familiar with LED backlight driving techniques for LCD panels. We begin this chapter with a look at three LCD backlight structures: edge-light type, bottom-light type and hollow type. Figure 1(a) shows the bottom-light structure. Because of its high-luminance feature, the bottom-light structure is commonly used for PC monitors and TVs. With regard to the luminance uniformity on an LCD panel, a light-mixed zone is necessary between the diffusion plate and the light sources. This zone causes undesirable thickness for large-sized TV applications. Figure 1(b) shows the hollow type structure in which an LGP is used to reduce the thickness of the light-mixed zone. The good qualities of this type pertain to its compact shape, high luminance and good thermal dissipation. Figure 1(c) shows the edge-light structure commonly used in a small-scale LCD panel. This type is of compact shape and low power consumption, so it is suitable for notebook PCs and personal digital assistant (PDA) products. There are two types of LEDs for backlight sources, the white-light LEDs and RGB LEDs. The white-light LED is composed of a blue LED coated with yellow phosphor. Simple driving feature make it as a popular choice for new generation of LCD backlight sources in portable display products. Its color filter divides the emitted white light into RGB sub-pixels to present color pictures. Thicknesses of

Source: New Developments in Liquid Crystals, Book edited by: Georgiy V. Tkachenko,
ISBN 978-953-307-015-5, pp. 234, November 2009, I-Tech, Vienna, Austria

RGB sub-pixels must be adjusted according to the corresponding wavelengths to correct the white balance on LCD panel. This results in the difficulty of manufacturing process. The white-color point may vary after a long working time. Thus, the RGB LEDs mixing three-color lights to white light are more suitable for medium-scale, or even large-size screens [10-16]. White balance of the LCD panel with RGB LED backlight can be easily corrected by regulating the emission luminance of the RGB LEDs individually.

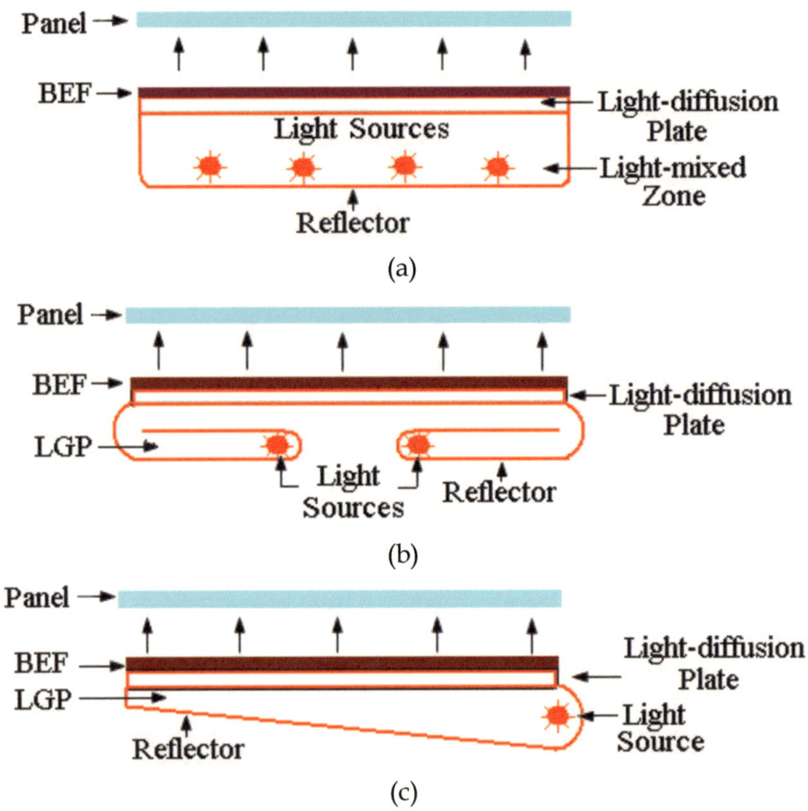

Fig. 1. (a) Bottom-light Type, (b) Hollow Type and (c) Edge-light Type Backlight Structures.

2. RGB LED backlight circuit

Figure 2 shows the block diagram of an LCD TV power supply with RGB LED backlight design. The LCD TV power provides a 12V output for the signal-process board, a 24V output for the backlight driving circuit and an additional 5V standby-power output. As shown in Figure 3, the backlight driving circuit consists of three power converters. Backlight LEDs are connected in series and parallel in the RGB LED backlight modules. LED current/ voltage characteristic variations cause brightness difference. Therefore, dimming control is an important design consideration for LED backlight applications. We studied three dimming methods for current regulation of the parallel connected LED arrays: the transconductance-amplifier (TA) dimming, the current-mirror (CM) dimming and the burst-mode (BM) dimming.

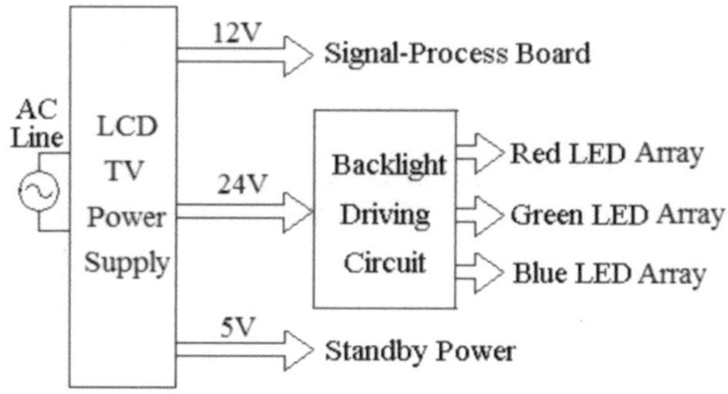

Fig. 2. Block Diagram of an LCD TV Power Supply.

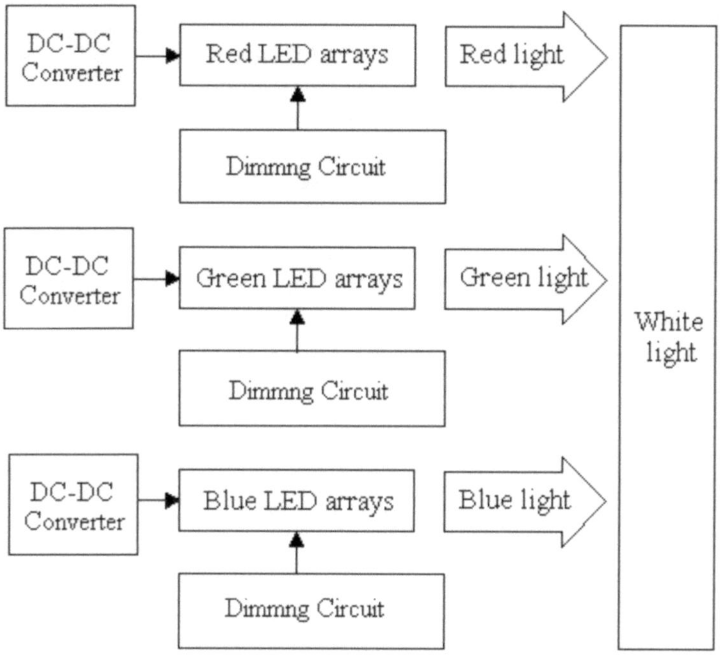

Fig. 3. RGB LED Backlight Driving Circuit

Figure 4(a) shows the TA dimming circuit. The LED current can be expressed as Equation (1).

$$I_{LED} = \frac{V_d}{R}, \tag{1}$$

Figure 4(b) shows the CM dimming circuit. The LED current can be expressed as Equation (2).

$$I_{LED} \approx I_r = K_n(V_{GS} - V_{TN})^2 = \frac{V_d - V_{GS}}{R}, \tag{2}$$

where K_n and V_{TN} are the conduction parameter and threshold voltage of the dimming transistors Q_r and Q_d, respectively. By using the TA dimming and CM dimming circuits, the current regulation of paralleled LED arrays can be achieved. However, the conduction losses of the dimming transistors will be difficult to solve [17]. An adaptive voltage output for the DC-DC converter is usually designed to retain the minimum drain-source voltage on the dimming transistors. As shown in Figure 4(c), the backlight LED current can be also controlled with a BM dimming circuit. Considering the switching loss for the dimming transistors, the burst-mode frequency f_b is designed at 400Hz that are unperceivable to the human eye. The duty ratios of the dimming transistors are varied to regulate the LED average current that can be represented as Equation (3).

$$I_{LED(av)} = I_m \delta, \tag{3}$$

where I_m denotes the peak value of the LED array current. The dimming transistors are operated as low-frequency switches, the thermal problem on the dimming transistors can be improved significantly. The current variations can be minimized by using the TA or BM dimming methods while the CM dimming has the simplest circuit configuration. Anyway, the TA and CM dimming methods are unsuitable to be used in the high-power LED backlight design of LCD panels due to the significant conduction losses of the dimming transistors under dimming operations. The emission luminance of the RGB LEDs is able to be regulated individually for achieving the white balance of the LCD panel. The luminance of the red light is always highest and the luminance of the blue light is lowest among three color lights. In practical applications, the blue light is most sensitive to human eyes such that lower luminance of blue LED is enough to compose white light with the red and green LEDs.

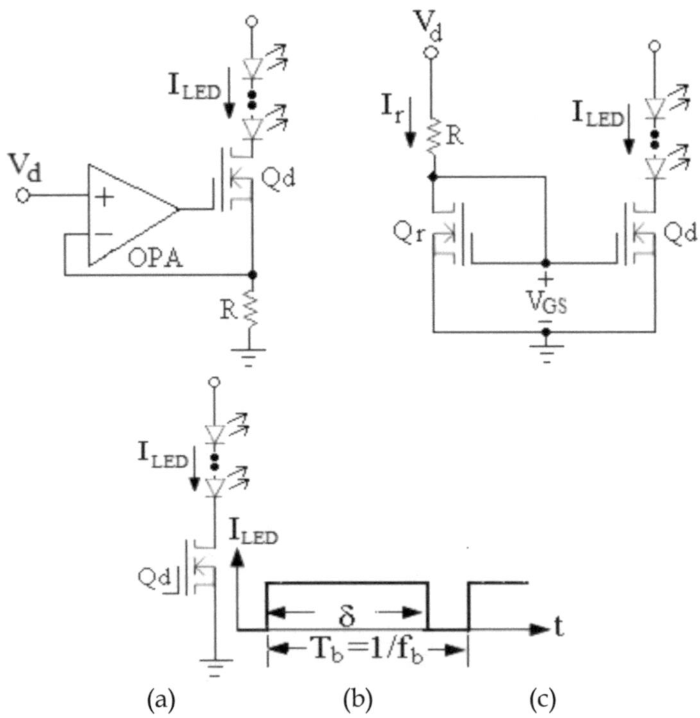

Fig. 4. (a) TA, (b) CM and (c) BM Dimming Methods

3. Soft-switched LED backlight circuit

Figure 5(a) shows a half-bridge DC-DC Series-Resonant Converter (SRC) topology for driving the RGB LEDs. The soft-switched DC-DC resonant converter includes power switches Q_1 and Q_2, resonant inductor L_r, resonant capacitor C_r, transformer T_1, rectifier diodes D_{f1} and D_{f2}, filter capacitor C_f, and the LED arrays represented by an equivalent resistance R_o. The characteristic impedance and the resonant frequency are respectively [18-21].

$$Z_c = \sqrt{\frac{L_r}{C_r}}, \tag{4}$$

$$f_r = \frac{1}{2\times\pi\times\sqrt{L_r C_r}}, \tag{5}$$

From (4) and (5), L_r and C_r can be expressed as

$$L_r = \frac{Z_c}{2\times\pi\times f_r}, \tag{6}$$

$$C_r = \frac{1}{2\times\pi\times f_r \times Z_c}, \tag{7}$$

The turn number of the primary winding is

$$N_p = \frac{(V_{Np} - 1)\times 10^8}{4\times f_s \times B_r \times A_e}, \tag{8}$$

where V_{Np} is the peak-to-peak amplitude of the transformer primary voltage v_{Np}, f_s is the switching frequency, B_r is the magnetic flux density, and A_e is the effective core area. To simplify the analysis, the first-order harmonic approximation has been applied, and the circuit elements of the primary side in Figure 5(a) are reflected to the secondary side, as depicted in Figure 5(b). The equivalent impedances Z_1 and Z_2 are respectively

$$Z_1 = \frac{n^2}{j\omega C_r} + j\omega n^2 L_r, \tag{9}$$

$$Z_2 = j\omega n^2 L_m // R_o, \tag{10}$$

where $n=N_s/N_p$.

The voltage divider rule can be used to obtain the output voltage phasor V_L as

$$V_L = V_{in}\left(\frac{Z_1}{Z_1 + Z_2}\right), \tag{11}$$

where the amplitude of V_L is assumed equal to V_o. V_{in} is the phasor of the fundamental component of v_a in Figure 5(a).

$$\left|V_{in}\right| = \frac{n}{\pi}V_s, \tag{12}$$

Substituting (9), (10) and (12) into (11), the voltage gain transfer function can be expressed as

$$\left|\frac{V_o}{V_s}\right| = \frac{n\omega^2 L_m C_r R_o}{\pi\sqrt{[R_o - \omega^2 C_r R_o (L_r + L_m)]^2 + [n^2 \omega L_m (1-\omega^2 L_r C_r)]^2}}, \tag{13}$$

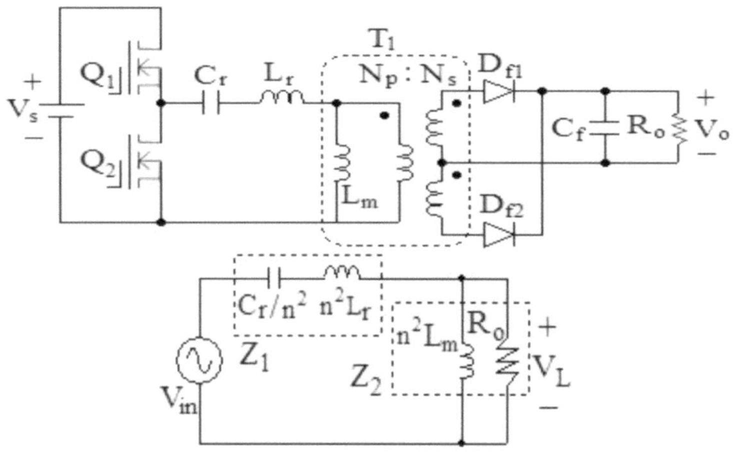

(a) (b)

Fig. 5. (a) Half-bridge DC-DC SRC Topology and (b) Its Equivalent Circuit.

It is clearly seen from (13) that the switching frequency must be varied to regulate the output voltage. The highest switching frequency appears at the highest input voltage and the lightest load. On the other hand, the lowest switching frequency happens at the lowest input voltage and the heaviest load. For the SRC to operate in the zero-voltage-switching (ZVS) region, the lowest switching frequency must be higher than the resonant frequency as expressed in (5). Moreover, due to the switching speed limitations of the power devices, the highest switching frequency is below a specified value. In other words, the variations of the input DC voltage and the load variations must be confined to a small range. Usually a power factor corrector (PFC) is added in front of the DC-DC converter to raise the input power factor and reduce the input current harmonics. A phase-shift pulse width modulation (PSPWM) dimming control can effectively confine the load variation of the DC-DC SRC. Consequently, the output voltage variation of the PFC can be limited to a smaller extent. This results in a better operating condition for the SRC. For the PSPWM dimming strategy, the working durations of the shunt LED arrays are properly phase-shifted to confine the variation of the output current of the SRC. Figure 6 illustrates the circuit arrangement for N shunt single-colored LED arrays with PSPWM dimming method. It is almost the same as the conventional one, except that the dimming signals are applied with a specified phase difference. With the PSPWM dimming, there are always overlaps between the LED driving currents. The maximum duty cycle, or the overlap, is 100 %, corresponding to the highest brightness. To prevent the DC-DC SRC from operating at no load, the minimum duty cycle of the PSPWM dimming signal is 1/N, where N is the number of the shunt LED arrays.

Under this circumstance, the overlap is zero, corresponding to the lowest brightness. Compared with the conventional dimming scheme, it is apparently recognized that the load variation of the SRC is less with the proposed PSPWM dimming function. To further investigate the operating principle of the PSPWM dimming, a more general case with N shunt LED arrays is discussed as follows. Figure 7 shows the waveforms of the N driving currents and the output current of the SRC. As stated earlier, the duty cycle range of the dimming signal is from 1/N to 100 %. In terms of the phase angle, if a complete period is 360°, the duty cycle range is from 360°/N to 360°. Assuming that the dimming signal for the LED array 1 starts at 0°, then the dimming signal for the k-th LED array would start at

$$\phi_k = 360^o \times (k-1)/N, \tag{14}$$

If the duty cycle of each dimming signal is ϕ_d, then the average driving current of one LED array is

$$I_{avg} = \phi_d \times I_p / 360^o, \tag{15}$$

where I_p is the amplitude of the driving current for each LED array. Therefore the average output current of the SRC is

$$I_{o,avg} = \phi_d \times I_p \times N / 360^o, \tag{16}$$

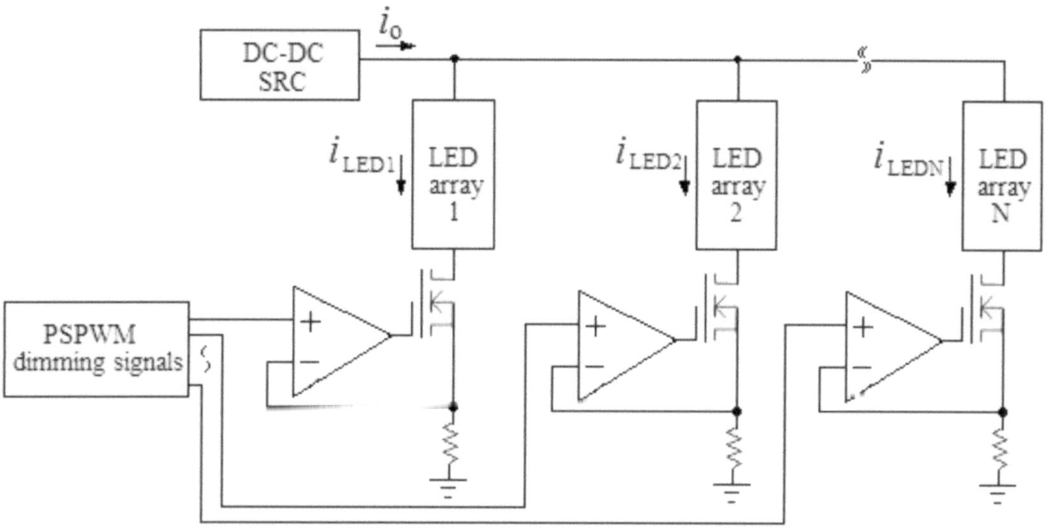

Fig. 6. The PSPWM Dimming Method.

It can be observed from Figure 7 that if the end of the dimming signal for LED array 1 is at ϕ_d, where ϕ_d is between ϕ_k and ϕ_{k+1} and $k \neq 1$, then the output current of the SRC in the range of ϕ_k to ϕ_{k+1} is

$$i_o = \begin{cases} kI_p & for\ \phi_k \le \phi \le \phi_d \\ (k-1)I_p & for\ \phi_d \le \phi \le \phi_{k+1} \end{cases}, \tag{17}$$

This is also the SRCs output current in each duration from ϕ_j to ϕ_{j+1}, where j = 1 to N. Therefore, the average output current of the SRC is now

$$I_{o,avg} = \frac{kI_p \times \lfloor \phi_d - 360^o / N \times (k-1) \rfloor + (k-1)I_p \times \lfloor 360^o / N \times k - \phi_d \rfloor}{360^o / N}, \quad (18)$$
$$= N \times \phi_d \times I_p / 360^o$$

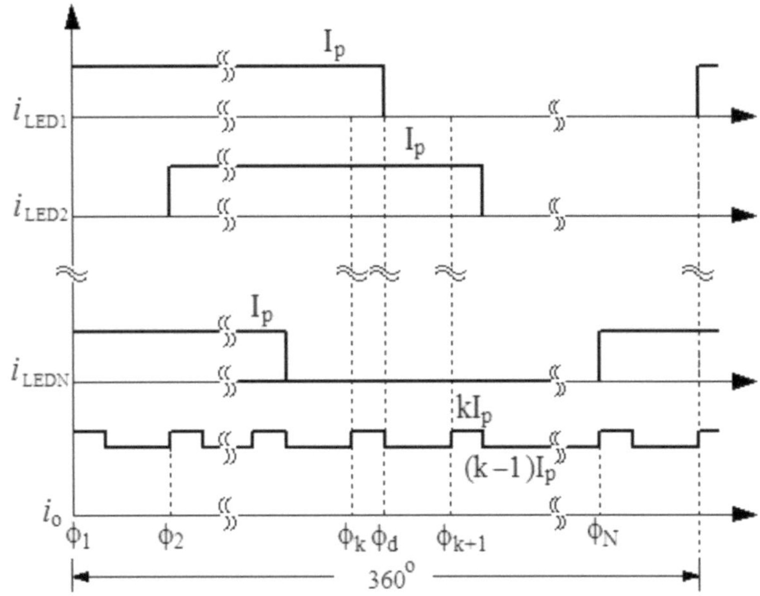

Fig. 7. Current Waveforms of N Shunt LED Arrays for the PSPWM dimming.

A favored feature is that the load variation of the SRC is always within one step change of I_p, no matter what the load level is. Therefore, by carefully designing the duty cycle and the amplitude of the driving current for each LED array, the no load operation of the DC-DC SRC may be precluded. Moreover, the output transient of the SRC is improved due to the confined load change. The number of the LED array for one color, and the peak driving current of each LED array are first determined according to the specifications of the LED and the spectrum of the white color. Then a suitable duty cycle is chosen allowing a reasonable span of variation for dimming control.

4. Single-stage LED backlight circuit

Figure 8 shows a single-stage LED backlight driving system. The backlight driving system consists of an AHB DC/DC cell integrated with a charge-pump PFC cell. The power MOSFETs Q1 and Q2, operate with asymmetrical duty ratios, δ and 1-δ, which require short and well-defined dead time between the conduction intervals. D1, D2 and C_{p1} and C_{p2} are the body diodes and the parasitic capacitors of power MOSFETs, respectively. The charge-pump PFC cell is composed of resonant inductor L_r, charge-pump capacitors C_{r1} and C_{r2}, input diodes D_{i1}, and D_{i2}, clamping diodes D_{c1}, and D_{c2}. The capacitor C_{bus} is used as the DC bus capacitor between the charge-pump PFC cell and the post-stage AHB DC/DC cell. The transformer leakage inductor L_l resonates with the parasitic capacitors C_{p1} and C_{p2} during dead-time intervals to achieve zero-voltage switching for the power MOSFETs. The blocking capacitor C_b is used to assure that the power sent into the transformer primary winding is a pure AC type. A DC voltage is supplied to the LED arrays through the secondary rectifier and filter circuit that are composed of D3, D4, L_o and C_o.

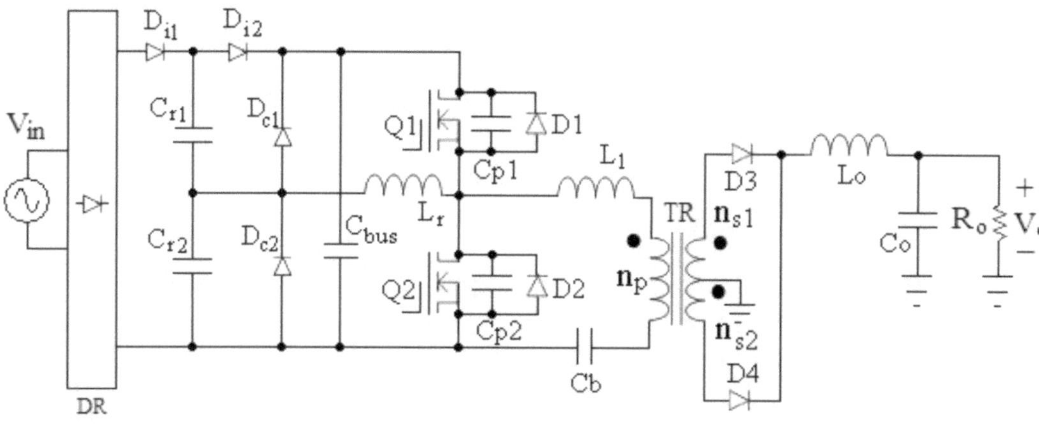

Fig. 8. Single-stage LED Backlight Driving System.

The average rectified input current $|I_{in}|_{,av}$ can be expressed as follows.

$$|I_{in}|_{,av} = \frac{\Delta Q}{T_s} = f_s C_{r1} |V_{in}|, \tag{19}$$

where ΔQ is the charge variation of C_{r1}. From Equation (19), we can see that the average rectified input current is proportional to the rectified input voltage. Thus, high power factor can be achieved. Based on the power balance between the input and output of the AC/DC converter, the following equation has to be satisfied.

$$|I_{in}|_{,av} = \frac{2P_o}{\eta V_{in}^2} |V_{in}|, \tag{20}$$

where η and Po are the overall efficiency and output power of the converter. From Equations (19) and (20), the design equations for the resonant inductor L_r and the charge-pump capacitor C_{r1} can be derived as follows [22-25].

$$L_r = \frac{\eta V_{in}^2}{8\pi^2 f_s P_o}, \tag{21}$$

$$C_{r1} = \frac{2P_o}{\eta f_s V_{in}^2}, \tag{22}$$

The ZVS conditions for power switches depend on the resonant inductance current I_{Lr} related with the input voltage. At the zero-crossing of input voltage, the resonant inductance current I_{Lr} will be ignorable. Considering the ZVS condition during an entire a line period, the transformer leakage inductance L_l could be determined by using Equation (23).

$$L_l \geq (C_{p1} + C_{p2})[\frac{n_p V_{bus}}{min(n_{s1}, n_{s2})I_o}]^2, \tag{23}$$

In practical design, an external inductor L_e is usually needed to be added in series connected with L_l for satisfying ZVS condition [26-28]. The input current has a near sinusoidal waveform and in phase with the input voltage. High efficiency and high power factor can be achieved because of single-stage power conversion with soft-switching features.

5. Conclusion

The advantages of LED backlighting over conventional CCFLs are numerous: fast response, broader color spectrum, longer life span, and no mercury. However, CCFLs still have cost advantages. For a LED backlighting, luminous efficacy and thermal management are the most important issues need to be solved before commercialization. Anyway, rapid advances in material and manufacturing technologies will enable significant developments in high-luminance LEDs for backlighting applications. In this chapter, we introduced some LED backlight driving systems for LCD panels. Dimming control methods are then discussed to regulate the LED current and brightness for the LED backlight system.

6. References

[1] C. H. Lin, "The Design and Implementation of a New Digital Dimming Controller for the Backlight Resonant Inverter," *IEEE Trans. Power Electronics*, vol. 20, no. 6, pp. 1459-1466, Nov. 2005.

[2] C. G. Kim, K. C. Lee, and B. H. Cho, "Modeling of CCFL using Lamp Delay and Stability Analysis of Backlight Inverter for Large Size LCD TV," *IEEE APEC'05*, Vol. 3, pp. 1751-1757.

[3] Y. H. Liu, "Design and Implementation of an FPGA-Based CCFL Driving System With Digital Dimming Capability," *IEEE Transactions on Industrial Electronics*, Vol. 54, Issue 6, pp. 3307-3316, Dec. 2007.

[4] C. H. Lin, "Digital-Dimming Controller with Current Spikes Elimination Technique for LCD Backlight Electronic Ballast," *IEEE Transactions on Industrial Electronics*, Vol. 53, Issue 6, pp. 1881-1888, Dec. 2006.

[5] Y. K. Lo, and K. J. Pai, "Feedback Design of a Piezoelectric Transformer-based Half-bridge Resonant CCFL Inverter," *IEEE Trans. Industrial Electronics*, vol. 54, no. 4, pp. 2716-2723, Oct. 2007.

[6] K. H. Lee, and S. W. R. Lee, "Process Development for Yellow Phosphor Coating on Blue Light Emitting Diodes (LEDs) for White Light Illumination," *in Proc. Electronics Packaging Technology Conference*, 2006, pp. 379-384.

[7] T. Taguchi, Y. Uchida, and K. Kobashi, "Efficient White LED Lighting and Its Application to Medical Fields," *Journal of physica status solidi (a)*, vol. 201, no. 12, pp. 2730-2735, Sept. 2004.

[8] N. Mohan, T. M. Undeland, and W. P. Robbins, "Power Electronics," USA: John Wiley & Sons, 2003, pp. 301-313.

[9] H. van der Broeck, G. Sauerlander, and M. Wendt, "Power Driver Topologies and Control Schemes for LEDs," *in IEEE Proc. APEC'07*, 2007, pp. 1319-1325.
[10] C. C. Chen, C. Y. Wu, Y. M. Chen, and T. F. Wu, "Sequential Color LED Backlight Driving System for LCD Panels," *IEEE Transactions on Power Electronics*, Vol. 22, Issue 3, pp. 919-925, May 2007
[11] H. J. Chiu and S. J. Cheng; "LED Backlight Driving System for Large-Scale LCD Panels," *IEEE Transactions on Industrial Electronics*, Vol. 54, Issue 5, pp.:2751-2760, Oct. 2007.
[12] G. Park; T. S. Aum, J. H. Bae, J. H. Kwon, S. K. Lee; M. H. Lee and H. S. Soh, "Optimization of Direct-type LCD Backlight Unit," *Pacific Rim Conference on Lasers and Electro-Optics*, Aug. 2005, pp. 205-206.
[13] S. Y. Lee, J. W. Kwon, H. S. Kim, M. S. Choi and. S. Byun, "New Design and Application of High Efficiency LED Driving System for RGB-LED Backlight in LCD Display;" *the 37th IEEE Power Electronics Specialists Conference*, June 2006, pp.1-5.
[14] S. Muthu, F. J. Schuurmans, and M. D. Pashley, "Red, Green, and Blue LED based White Light Generation: Issues and Control," *Annual Meeting. Conference Record of the Industry Applications Conference*, Oct. 2002, Vol. 1, pp. 327-333.
[15] F. Bernitz, O. Schallmoser, and W. Sowa, "Advanced Electronic Driver for Power LEDs with Integrated Colour Management," *Annual Meeting. Conference Record of the Industry Applications Conference*, Vol. 5, Oct. 2006, pp. 2604-2607.
[16] C. C. Chen, C. Y. Wu, and T. F. Wu, "Fast Transition Current-Type Burst-Mode Dimming Control for the LED Back-Light Driving System of LCD TV," *IEEE Power Electronics Specialists Conference*, June 2006, pp. 1-7.
[17] Donald A. Neamen, "Electronic Circuit Analysis and Design, 2e," McGraw-Hill, 2001.
[18] C. C. Chen, C. Y. Wu, and T. F. Wu, "LED Back-light Driving System for LCD Panels," *IEEE APEC '06*, pp. 381-385.
[19] S. Y. Lee, J. W. Kwon, H. S. Kim, M. S. Choi, and K. S. Byun, "New Design and Application of High Efficiency LED Driving System for RGB-LED Backlight in LCD Display, " *IEEE PESC '06*, pp.1-5.
[20] M. Rico-Secades, A. J. Calleja, J. Ribas, E. L. Corominas, J. M. Alonso, J. Cardesin, and J. Garcia-Garcia, "Evaluation of a Low-Cost Permanent Emergency Lighting System based on High-Efficiency LEDs," *IEEE Transactions on Industry Applications*, Vol. 41, No. 5, Sept.-Oct. 2005, pp.1386-1390.
[21] H. Sugiura, S. Kagawa, H. Kaneko, M. Ozawa, H. Tanizoe, T. Kimura, and H. Ueno, "Wide Color Gamut Displays using LED Backlight- Signal Processing Circuits, Color Calibration System and Multi-Primaries," *IEEE ICIP'05*, Vol. 2, pp. 9-12.
[22] G. Moschopoulos and P. Jain, "Single-Phase Single-Stage Power-Factor-Corrected Converter Topologies," *IEEE Transactions on Industrial Electronics*, Vol. 52, Issue 1, pp.23–35, Feb. 2005.
[23] F. S. Kang, S. J. Park, and C. U. Kim, "ZVZCS Single-Stage PFC AC-to-DC Half-Bridge Converter," *IEEE Transactions on Industrial Electronics*, Vol. 49, Issue 1, pp.206-216, Feb. 2002.
[24] J. Qian, and F. C. Y. Lee, "A High-Efficiency Single-Stage Single-Switch High-Power-Factor AC/DC Converter with Universal Input," *IEEE Transactions on Power Electronics*, Vol. 13, No. 4, July 1998, pp.699-705.

[25] J. Qian, and F. C. Lee, "Charge Pump Power-Factor-Correction Technologies. II. Ballast Applications," *IEEE Transactions on Power Electronics*, Vol. 15, No.1, pp. 130-139, Jan. 2000.

[26] F. Bernitz, O. Schallmoser, and W. Sowa, "Advanced Electronic Driver for Power LEDs with Integrated Colour Management," *IEEE IAS'06*, Vol. 5, pp. 2604-2607.

[27] S. Muthu and J. Gaines, "Red, Green and Blue LED-based White Light Source: Implementation Challenges and Control Design," *IEEE IAS'03*, Vol. 1, pp. 515-522.

[28] S. Muthu, F. J. Schuurmans, and M. D. Pashley, "Red, Green, and Blue LED based White Light Generation: Issues and Control," *IEEE IAS'02*, Vol. 1, pp. 327-333.

12

Optoelectronic Device using a Liquid Crystal Holographic Memory

Minoru Watanabe
Shizuoka University,
Japan

1. Introduction

Recently, the technologies related to liquid crystal spatial light modulators have progressed dramatically [1]-[4]. Such modulators are classifiable as two types: transmissive and reflective. Both types are used widely for various applications, e.g. liquid crystal television panels, personal computer displays, and projector systems. In particular, the resolution of the latest liquid crystal spatial light modulators in projectors has reached 1,920 pixels × 1,080 pixels, the pixel size of which has also reached 8.5 μm × 8.5 μm [1], [2] as portrayed in Fig. 1 and Table 1. Therefore, their current resolution and pixel size make them available for use as holographic media.

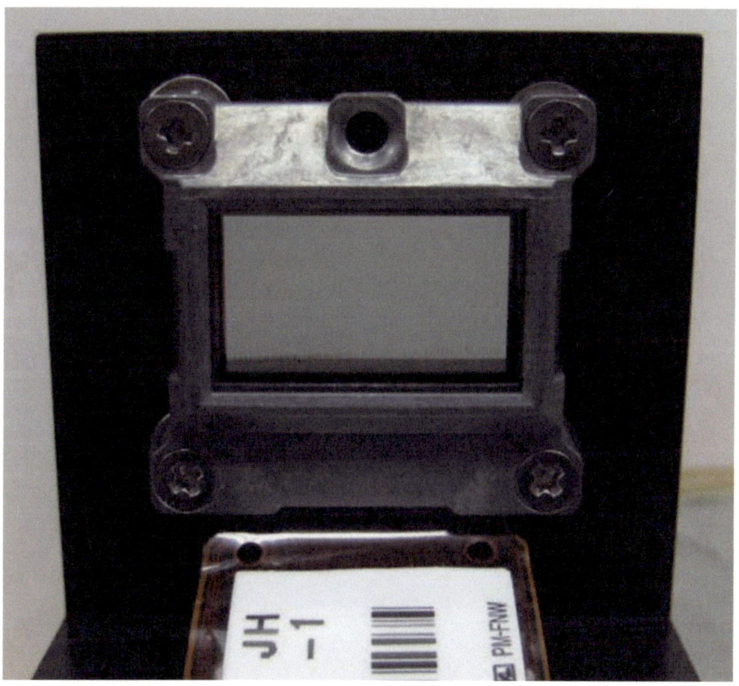

Fig. 1. Photograph of a liquid crystal – spatial light modulator (LC-SLM). The modulator is an LCD panel (L3D07U-81G00 Seiko Epson Corp.)

Source: New Developments in Liquid Crystals, Book edited by: Georgiy V. Tkachenko,
ISBN 978-953-307-015-5, pp. 234, November 2009, I-Tech, Vienna, Austria

LCD type	L3D07U-81G00
Resolution	1,920 x 1,080
Panel size	0.7 inch
Pixel pitch	8.5 μm
Aperture ratio	55 %

Table 1. Specifications of the L3D07U-81G00 LC-SLM Panel.

Moreover, recently, optically reconfigurable gate arrays (ORGAs) with a holographic memory have been developed [5]–[7], [11]–[14], [21]–[23]. The gate array of this optoelectronic device has a fine grain gate array structure similar to those of field programmable gate arrays (FPGAs) [8]–[10]. Computations or circuit operations on the gate array are executed electrically, as they are on FPGAs, whereas configurations and reconfigurations for the gate array are optically executed. The ORGA architecture has features of rapid reconfiguration and numerous reconfiguration contexts. Such an optical reconfiguration architecture often uses liquid crystal spatial light modulators as holographic memory media [11]–[14], [21]–[23].

Therefore, this chapter first presents the characteristics of a liquid crystal holographic memory to generate binary patterns. In addition, as an illustration of one application of liquid crystal devices, this chapter presents discussion of the research of optically reconfigurable gate arrays (ORGAs).

2. Transmissive-type computer-generated hologram

2.1 Calculation of a holographic memory

This section presents a description of a transmissive-type computer-generated hologram that can provide two-dimensional binary patterns. Figure 2 presents coordinates of a hologram plane and an observation plane. Both planes are placed in parallel at a distance of L. The observation plane is given by the coordinate (x, y); the holographic plane is given by the coordinate (x_0,y_0). An incident light for the holographic memory is assumed as a collimated monochromatic laser source. The collimated laser beam is incident from the left side of the holographic memory plane.

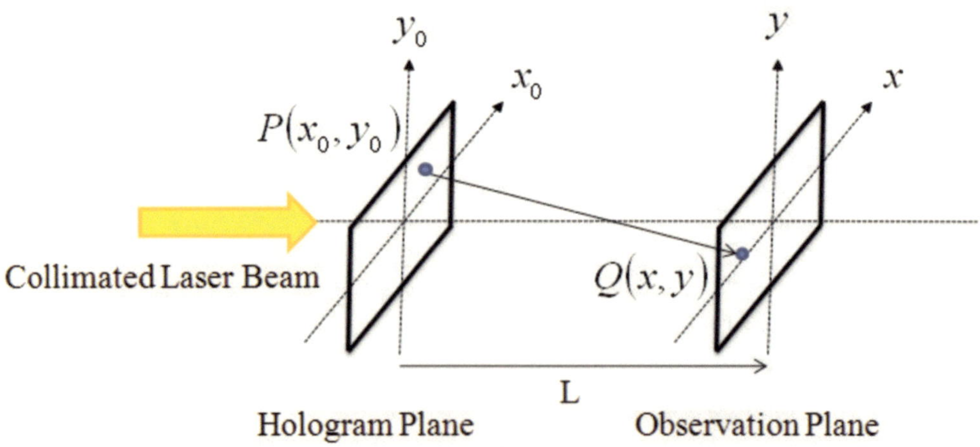

Fig. 2. Coordinates for diffraction from a liquid crystal holographic memory.

Here, a two-dimensional binary pattern on the observation plane is assumed to be given as a function $O(x,y)$, which represents a configuration or reconfiguration context in optically reconfigurable gate arrays (explained later). At that time, the intensity distribution of a holographic medium is calculable using the following equations.

$$H(x_0, y_0) \propto \int_{-\infty}^{\infty}\int_{-\infty}^{\infty} O(x,y)\sin(\frac{2\pi}{\lambda}r)dxdy,$$

$$r = \sqrt{L^2 + (x_0 - x)^2 + (y_0 - y)^2}. \quad (1)$$

In those equations, λ signifies the wavelength, L signifies the distances between the holographic plane and the observation plane, and r stands for the distance between the point source $P(x_0, y_0)$ on the holographic memory plane and the point of observation $Q(x,y)$. The distance L is expected to take $(n+1/4)\lambda$, where n is an arbitrary natural number, to receive the perpendicular incident beam on the observation plane efficiently with the shortest distance from the holographic memory plane. The value $H(x_1, y_1)$ is normalized as 0–1 for the minimum intensity H_{min} and maximum intensity H_{max}, as shown below.

$$H'(x_0, y_0) = \frac{H(x_0, y_0) - H_{min}}{H_{max} - H_{min}}. \quad (2)$$

Finally, the normalized image H' is used for implementing a holographic memory.

2.2 Diffraction from a holographic memory

Next, the diffraction pattern is estimated from the above calculated holographic memory pattern. The complex light distribution at the coordinate (x, y) are calculated using the following equations as

$$u(x,y) \propto \int_{-Y_{min}}^{Y_{max}}\int_{-X_{min}}^{X_{max}} H'(x_0, y_0)\exp(i\frac{2\pi}{\lambda}r)dx_0dy_0,$$

$$r = \sqrt{L^2 + (x_0 - x)^2 + (y_0 - y)^2}, \quad (3)$$

where $H'(x_0, y_0)$ denotes the calculated and normalized holographic memory pattern, λ represents the wavelength, L stands for the distances between the holographic plane and the observation plane, and X_{max}, X_{min}, Y_{max}, and Y_{min} respectively represent the holographic memory sizes. Finally, the diffraction intensity from a holographic memory is calculable as

$$I(x,y) = u(x,y)u^*(x,y), \quad (4)$$

where the superscript asterisk denotes the complex conjugate.

2.3 Single bright bit example in the Fresnel region

In this section, once again, the holographic memory pattern described in section 2.1 is treated, but in the Fresnel region. If distance L between the two coordinate planes can be assumed to be large compared with the sizes of a holographic memory and observation area, when the following condition is satisfied,

$$\frac{1}{4\lambda}\left\{(x_0-x)^2+(y_0-y)^2\right\}^2 << L^3, \tag{5}$$

then r can be approximated to

$$r \simeq L+\frac{(x_0-x)^2+(y_0-y)^2}{2L}, \tag{6}$$

where (x_0,y_0) is the coordinate of the holographic memory plane and (x,y) is the coordinate of the observation plane. Here, assuming that the condition $L=(n+1/4)\lambda$ (n = an arbitrary natural number) is satisfied, then $(n+1/4)\lambda$ can be substituted into the first term L of Eq. 6 shown above. Then, substituting Eq. 6 with the condition into Eq. 1, the following equation is accomplished.

$$H(x_0,y_0) \propto \int_{-\infty}^{\infty}\int_{-\infty}^{\infty} O(x,y)\cos\left(\frac{\pi}{\lambda L}\left\{(x_0-x)^2+(y_0-y)^2\right\}\right)dxdy. \tag{7}$$

Assuming that the single bright bit is located on the coordinate (α,β), the equation $O(x,y)$ can be considered as $\delta(x-\alpha,y-\beta)$. The two-dimensional Dirac delta function $\delta(x,y)$ is defined as shown below.

$$\delta(x,y)=\begin{cases}\infty, & for\ \ x=y=0\\ 0, & otherwise\end{cases} \tag{8}$$

and

$$\int_{-\infty}^{\infty}\int_{-\infty}^{\infty}\delta(x,y)dxdy=1. \tag{9}$$

When $O(x,y)=\delta(x-\alpha,y-\beta)$, Eq. 7 can be simplified to the following equation.

$$H(x_0,y_0) \propto \cos\left(\frac{\pi}{\lambda L}\left\{(x_0-\alpha)^2+(y_0-\beta)^2\right\}\right). \tag{10}$$

The maximum and minimum of the above equation are, respectively, 1 and -1. Therefore, the above equation can be substituted into Eq. 2. Finally, the following equation of a holographic memory pattern including a single bright bit in Fresnel region can be derived.

$$H'(x_0,y_0)=\frac{1}{2}\cos\left(\frac{\pi}{\lambda L}\left\{(x_0-\alpha)^2+(y_0-\beta)^2\right\}\right)+\frac{1}{2}. \tag{11}$$

This equation represents a Fresnel zone lens, the center of which is located at coordinate (α,β). An example of a holographic memory of size of 1.632 mm × 1.632 mm to generate a single bright bit is shown in Fig. 3. In this example, the holographic memory pattern was calculated using the condition that the target laser wavelength is 532 nm, the distance L is 100 mm, and the coordinate (α , β) of a bright bit is (0, 0).

Fig. 3. Holographic memory pattern to generate a single bright bit. The size of the holographic memory pattern is 1.632 mm × 1.632 mm. The target laser wavelength is 532 nm. The distance L is 100 mm. The coordinate (α , β) of a bright bit is (0, 0).

It can be confirmed that the holographic memory pattern represents the Fresnel zone lens. Therefore, a holographic memory pattern to generate a two-dimensional binary pattern with multi-bright bits becomes a superimposition of the Fresnel zone lens. Next, the diffraction pattern from the Fresnel zone lens is estimated. The complex light distribution at the coordinate (x, y) is calculated using the following equation.

$$u(x,y) \propto \int_{-Y_{min}}^{Y_{max}} \int_{-X_{min}}^{X_{max}} \left\{ \cos\left(\frac{\pi}{\lambda L} \left\{ (x_0-\alpha)^2 + (y_0-\beta)^2 \right\} \right) + 1 \right\}$$

$$\times \exp\left(i \frac{\pi}{\lambda L} \left\{ (x_0-x)^2 + (y_0-y)^2 \right\} \right) dx_0 dy_0 .$$

Therein, λ is the wavelength, L signifies the distances between the holographic plane and the observation plane, and X_{max}, X_{min}, Y_{max}, and Y_{min} respectively represent the holographic memory sizes. Finally, the diffraction intensity from a holographic memory is calculable as follows.

$$I(x,y) = u(x,y)u^*(x,y). \tag{12}$$

Therein, the superscript asterisk denotes the complex conjugate. To produce a compact system, the system parameters are not always in the Fresnel region. Therefore, at that time, the Fresnel approximation is inapplicable for calculations and Eqs. 1, 2, 3 and 4 must be used. However, when the system parameters are in the Fresnel region, the approximation described above is useful for holographic memory estimations.

3. Optically Reconfigurable Gate Array (ORGA)

Among applications using liquid crystal devices, studies of Optically Reconfigurable Gate Arrays (ORGAs) exist. Such an ORGA is an optoelectronic device using a liquid crystal device as a holographic memory. This device is being developed as an alternative device of current VLSIs or as a next-generation general-purpose programmable VLSI. The following section presents a description of the ORGA background and architecture.

3.1 Background

In recent years, SRAM-based Field Programmable Gate Arrays (FPGAs) have been used widely for large-item small-volume production because of their flexible programmable capabilities [8]–[10]. Moreover, demand for high-speed reconfigurable devices has been increasing. If circuit information can be downloaded rapidly from a configuration memory, idle circuits on a gate array can be removed. At that time, other necessary circuits can be downloaded from the configuration memory into the gate array, thereby increasing the gate array's activity. In so doing, high-speed dynamic reconfiguration can increase the performance of programmable gate arrays. However, since reconfiguration of FPGAs requires more than several milliseconds, FPGAs are unsuitable for use as dynamically reconfigurable devices [8]–[10].

However, high-speed reconfigurable devices have been developed: DAP/ DNA chips, DRP chips, and multi-context FPGAs [15]–[20]. Those devices package reconfiguration memories and microprocessor arrays or gate arrays onto a chip. The internal reconfiguration memory stores reconfiguration contexts of 4–16 banks, which can be changed from one to another on a clock. Consequently, the arithmetic logic unit or gate array of such devices can be reconfigured on every clock cycle in a few nanoseconds. Nevertheless, an important problem remains: simultaneously increasing the internal reconfiguration memory while maintaining the gate density is extremely difficult.

For that reason, optically reconfigurable gate arrays (ORGAs) [5]–[7], [11]–[14], [21]–[23] have been developed to provide two capabilities: rapid reconfiguration and numerous reconfiguration contexts. Such optical reconfiguration architecture often uses liquid crystal spatial light modulators as a holographic memory [11]–[14], [21]–[23]. This chapter presents a description of the studies of ORGAs with a liquid crystal spatial light modulator.

3.2 Entire construction

An overview of an Optically Reconfigurable Gate Array (ORGA) is shown in Fig. 4. An ORGA consists of a gate-array VLSI (ORGA-VLSI), a holographic memory, and a laser diode array. The holographic memory can store numerous reconfiguration contexts. A laser array mounted on the top of the holographic memory addresses the reconfiguration contexts. The diffraction pattern from the holographic memory can be received as a reconfiguration context on a photodiode-array of a programmable gate array on an ORGA-VLSI. Such ORGA architecture enables microsecond-order reconfiguration and multiple reconfiguration contexts. Therefore, virtually, the architecture can achieve gate counts larger than the physical gate count on a VLSI.

3.3 Gate array structure

The basic functionality of an ORGA-VLSI is fundamentally identical to that of currently available field programmable gate arrays (FPGAs). Figure 5 depicts the gate array structure of a first prototype ORGA-VLSI. The ORGA-VLSI chip was fabricated using a 0.35 μm triple-metal CMOS process [12]. A photograph of the board is portrayed in Fig. 6. The specifications are presented in Table 2. Here, the fundamental function of an ORGA-VLSI is

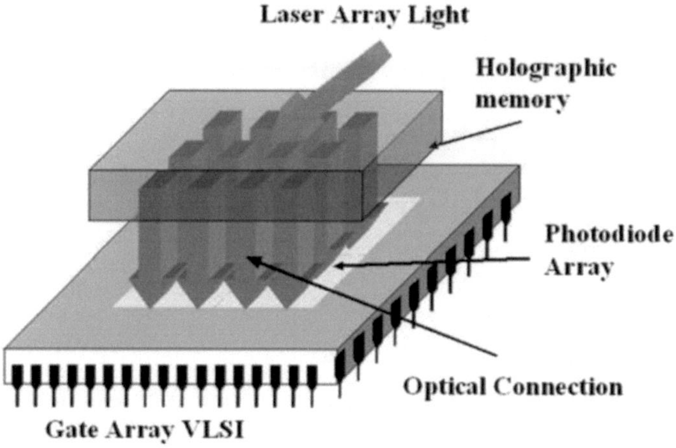

Fig. 4. Overview of an ORGA.

Technology	0.35μm double-poly triple-metal CMOS process
Chip size	4.9 mm × 4.9 mm
Photodiode size	25.5 μm × 25.5 μm
Distance between photodiodes	90 μm
Number of photodiodes	340
Gate count	68

Table 2. ORGA-VLSI Specifications.

described using this chip design as an example of ORGA-VLSI chips. The ORGA-VLSI chip consists of 4 optically reconfigurable logic blocks (ORLB), 5 optically reconfigurable switching matrices (ORSM), and 12 optically reconfigurable I/O bits (ORIOB) portrayed in Fig. 5(a). Each optically reconfigurable logic block is surrounded by wiring channels. One wiring channel has four connections. Switching matrices are located on the corners of optically reconfigurable logic blocks and are used as switches of wiring channels. In turn, the function of each block is described in the following sections.

3.3.1 Optically reconfigurable logic block

A block diagram of an optically reconfigurable logic block is presented in Fig. 5(b). Each optically reconfigurable logic block consists of a four-input one-output look-up table (LUT), six multiplexers, four transmission gates, and a delay flip-flop with a reset function. The input signals from the wiring channel, which are applied from optically reconfigurable I/O blocks, are transferred to a look-up table through four multiplexers. The look-up table is used for implementing Boolean functions such as AND circuits, OR circuits, and XOR circuits. The look-up table construction can be considered as a static memory. For example, a four-input one-output look-up table can be considered as a static memory with a four-bit address bus and a single-bit data bus. In this case, the address bus and data bus of the memory respectively represent signal inputs and signal output of a logic circuit. If some configuration information, "0001XXXXXXXXXXXX" is programmed to the static memory (X signifies a do not care state), then the look-up table can function as a two-input one-output

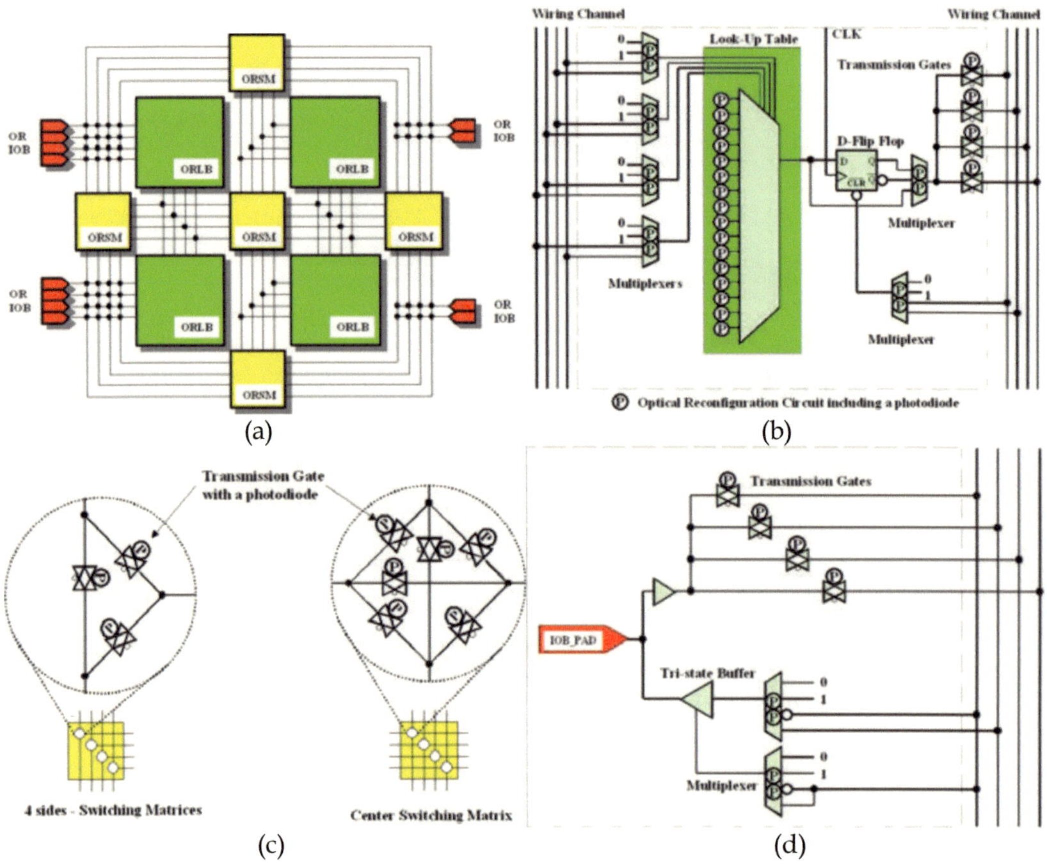

Fig. 5. Gate-array structure of a fabricated ORGA. Panels (a), (b), (c), and (d) respectively depict block diagram of a gate array, an optically reconfigurable logic block, an optically reconfigurable switching matrix, and an optically reconfigurable I/O bit.

Fig. 6. Photograph of an ORGA-VLSI board with a fabricated ORGA-VLSI chip. The ORGA-VLSI was fabricated using a 0.35 μm three-metal 4.9 × 4.9 mm^2 CMOS process chip. The gate count of a gate array on the chip is 68. In all, 340 photodiodes are used for optical configurations.

AND circuit. In this case, two multiplexers connected to upper two bits of the look-up table must be programmed to be *L*. The multiplexers connected to the inputs of look-up table can choose *L*, *H*, and two signals of the wiring channel for the look-up table inputs. In addition, the output of the look-up table is connected to a multiplexer through a delay type flip-flop. Therefore, a combinational circuit and sequential circuit can be chosen by changing the multiplexer. Finally, an output through the multiplexer from the look-up table is connected to the wiring channel through transmission gates, again. The last multiplexer controls the reset function of a delay-type flip-flop. Such a four-input one-output look-up table, each multiplexer, and each transmission gate respectively have 16 photodiodes, two photodiodes, and one photodiode. In all, 32 photodiodes are used for programming an optically reconfigurable logic block. In Fig. 5(b), the P mark shows a photodiode. In this optically reconfigurable logic block, although eight other photodiodes were implemented for special use of a differential reconfiguration strategy, that method is described in earlier reports [12], [21].

3.3.2 Optically reconfigurable switching matrix

Similarly, optically reconfigurable switching matrices are optically reconfigurable. A block diagram of the optically reconfigurable switching matrix is portrayed in Fig. 5(c). The basic construction is the same as that used by Xilinx Inc. One four-directional with 24 transmission gates and 4 three-directional switching matrices with 12 transmission gates were implemented in the gate array. In this construction, a connection in any direction can be realized using a transmission gate so that propagation delay can be decreased compared with the case using four transmission gate passes. Each transmission gate can be considered as a bi-directional switch. A photodiode is connected to a transmission gate and controls whether the transmission gate is closed or not. Based on that capability, four-direction and three-direction switching matrices can be programmed, respectively, as 24 and 12 optical connections.

3.3.3 Optically reconfigurable I/O block

Optically reconfigurable gate arrays are assumed to be reconfigured frequently. For that reason, an optical reconfiguration capability must be implemented for optically reconfigurable logic blocks and optically reconfigurable switching matrices. However, the I/O block might not always be reconfigured under such dynamic reconfiguration applications because such a dynamic reconfiguration arises inside the device and each state of Input, Output, or Input/Output, and each pin location of the I/O block must always be fixed because of limitations of an external environment. However, the ORGA-VLSI support optical reconfiguration for I/O blocks because optical reconfiguration information is provided optically from a holographic memory in ORGA. Consequently, electrically configurable I/O blocks are unsuitable for ORGAs. Here, each I/O block is also controlled using nine optical connections. Always, the configuration of the optically reconfigurable I/O block is executed only initially.

3.3.4 Physical implementation

Because the ORGA-VLSI has 340 photodiodes to program its gate array, the ORGA-VLSI can be reconfigured rapidly and perfectly in parallel. In this fabrication, the distance between each photodiode was designed as 90 μm. The photodiode size is set as 25.5 × 25.5 μm^2 to ease optical alignment. The photodiode was constructed between the N-well layer and P-substrate. The gate array's gate count is 68. It was confirmed experimentally that the ORGA-VLSI itself is reconfigurable within 10 ns.

4. Nine-configuration-context ORGA

This section presents an implementation example of a nine-configuration context ORGA.

4.1 Experimental system

An ORGA holographic memory system with nine configuration contexts using a liquid-crystal spatial light modulator (LC-SLM) as a holographic memory, nine 532 nm, 300 mW lasers (in the actual implementation, one laser emulated the nine lasers), and an ORGA-VLSI are depicted in Fig. 7. Each laser corresponds to a configuration context or a holographic recording area including the single configuration context and is used for addressing the configuration context. First, a nine-context holographic memory pattern is calculated using Eqs. 1 and 2. Here, distance L between a holographic memory and an ORGA-VLSI is 100 mm. The wavelength λ is 532 nm. The target LC-SLM is a projection TV panel (L3D07U-81G00; Seiko Epson Corp.). It is a 90° twisted nematic device with a thin-film transistor. The panel has 1,920 × 1,080 pixels, each of 8.5 × 8.5 μm^2, with 256 gradation levels. The calculated holographic pattern shown in Fig. 8(a) is displayed on the LC-SLM. The number of pixels of each recording area, including one reconfiguration context, is 450 × 250. Each interval between recording areas is 5 pixels. Therefore, the entire holographic memory pattern is 1,360 pixels × 760 pixels. Each laser beam is collimated: the beam is incident to its corresponding holographic recording area on the LC-SLM. By turning on a certain laser, one configuration context can be programmed onto the ORGA-VLSI. Optically parallel programming enables very high-speed configuration and reconfiguration.

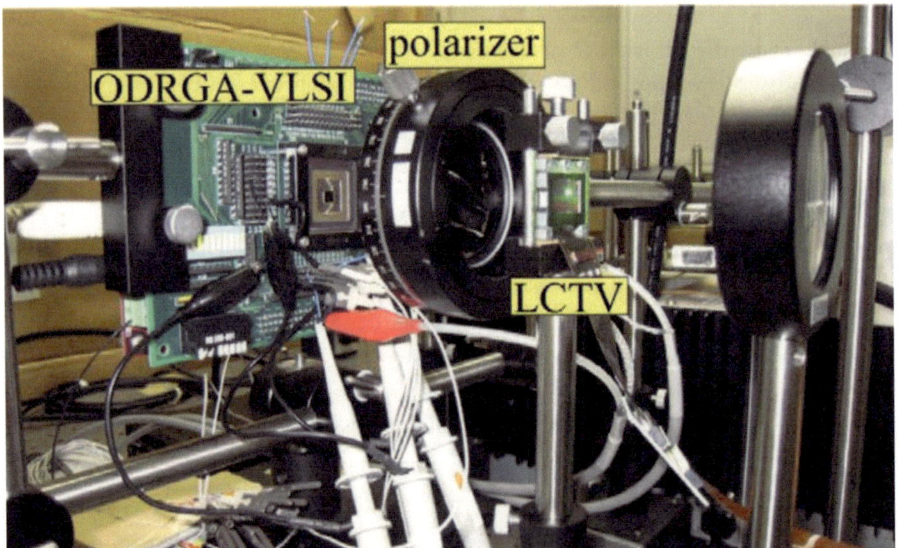

Fig. 7. ORGA system using a liquid crystal spatial light modulator as a holographic memory.

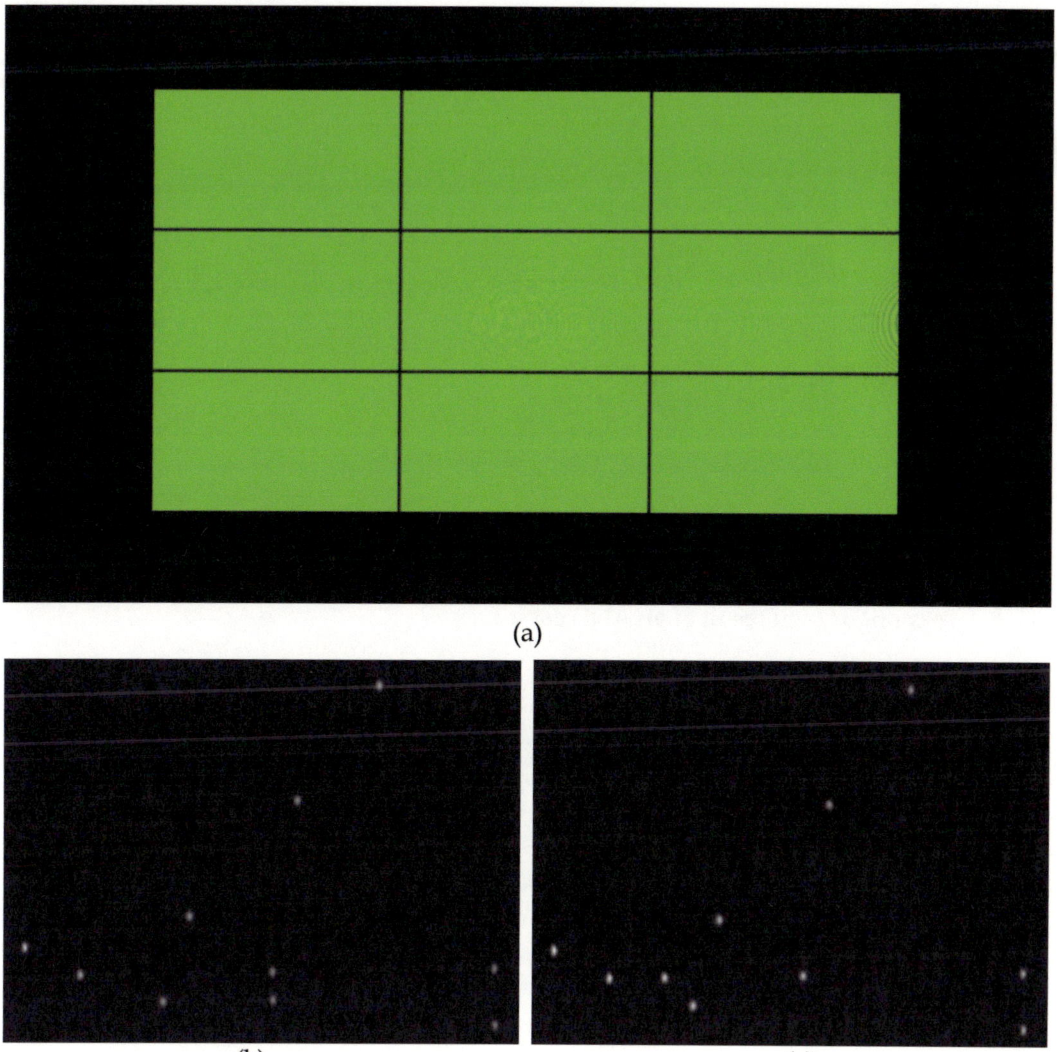

(a)

(b) (c)

Fig. 8. (a) Holographic memory pattern, and CCD-captured images of configuration contexts of (b) an AND and (c) a NOR circuits.

4.2 Configuration experiments

Here, among the nine configuration contexts, two configuration experiments of an AND circuit and a NOR circuit are introduced. A configuration context of an AND circuit was programmed at the top-left side of the holographic memory, while a configuration context of a NOR circuit was programmed at the bottom-left side of the holographic memory. Figures 8(b) and 8(c) depict CCD-captured images of configuration contexts of the AND circuit and the NOR circuit at the position of the ORGA-VLSI. Figure 9 shows that the AND circuit was programmed correctly onto ORGA-VLSI and that the AND circuit functioned correctly. Here, the configuration period of the AND circuit is 5 μs. Figure 10 shows that the NOR circuit was programmed correctly onto ORGA-VLSI and that the circuit functioned correctly. The configuration period of the NOR circuit was measured as 4 μs, thereby confirming the rapid configuration capability of the nine-configuration-context ORGA architecture.

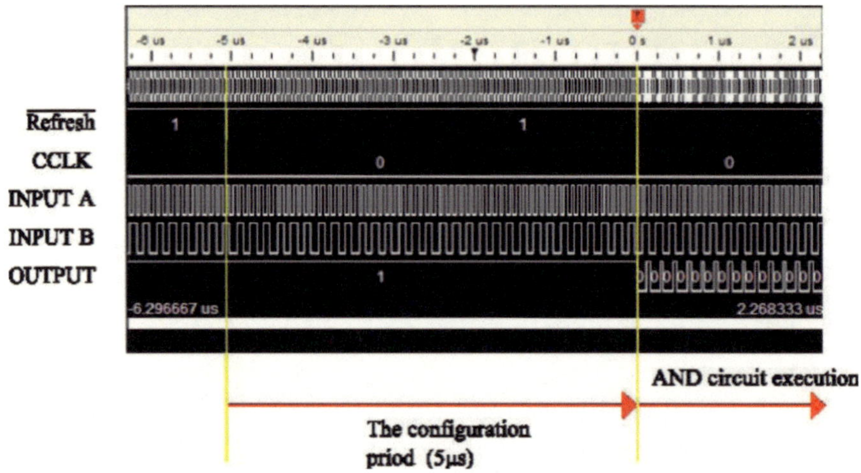

Fig. 9. 5 μs configuration result of an AND circuit.

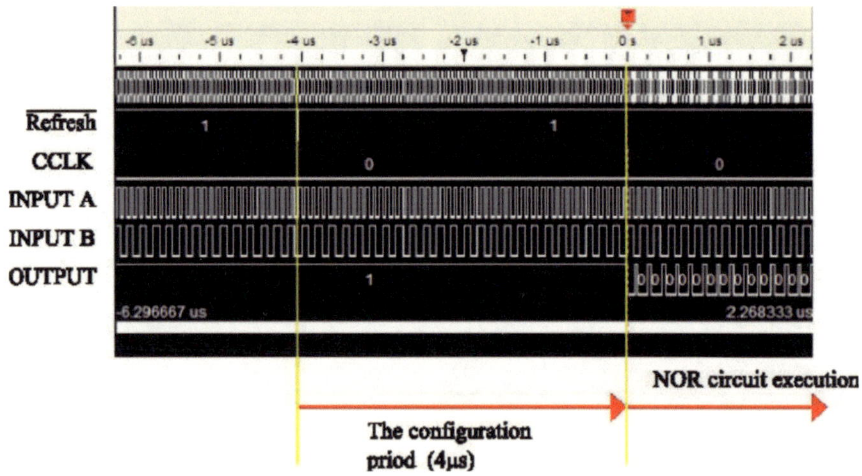

Fig. 10. 4 μs configuration result of an NOR circuit.

4.3 Response time of the liquid-crystal holographic memory

Next, the response time of a liquid-crystal holographic memory is estimated. The turn-on and turn-off times were measured experimentally using an L3D07U-81G00 panel provided by Seiko Epson Corp. The results show that the turn-on time is less than 12 ms. The turn-off time is less than 2 ms, as shown in Fig. 11.

5. Acceleration method

A liquid-crystal holographic memory is an easily rewritable material. For that reason, many reconfiguration contexts can be supplied dynamically to a gate array. Of course, a liquid-

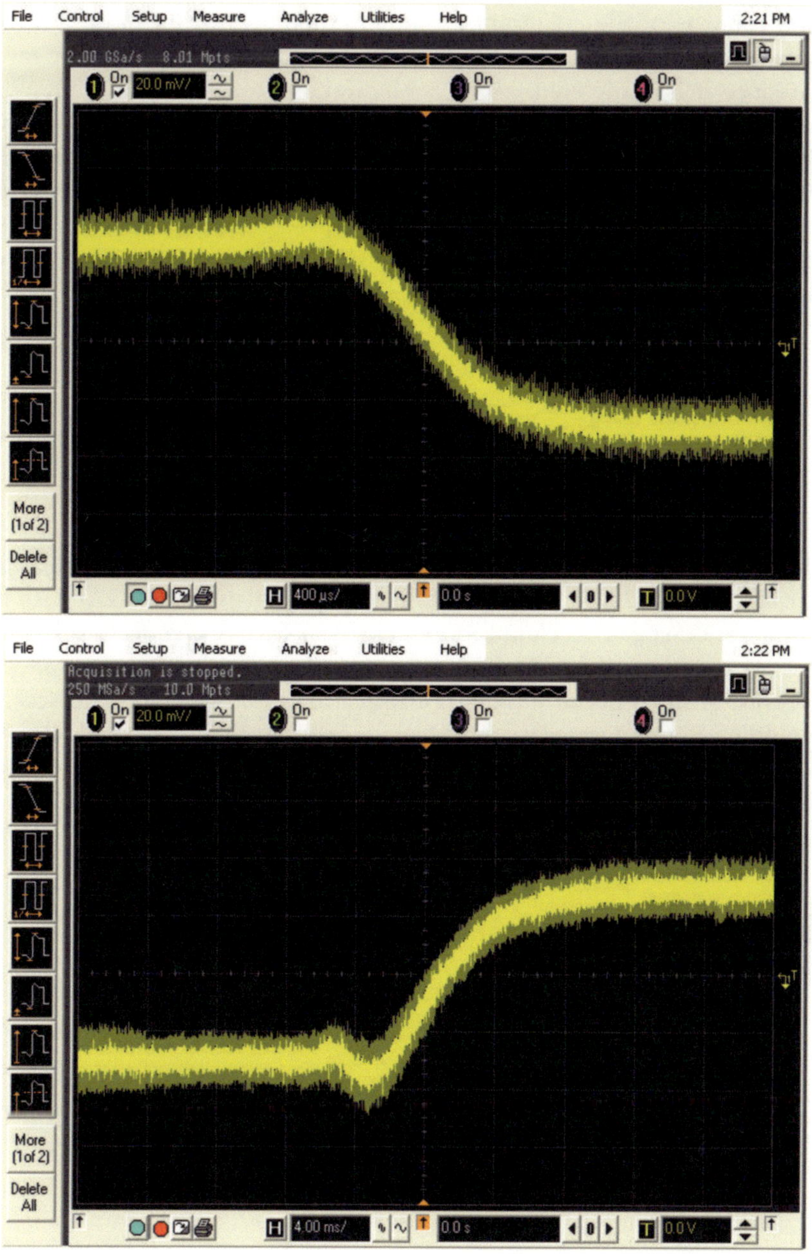

Fig. 11. Response time measurement of a liquid crystal holographic memory. The turn-on and turn-off times were measured experimentally using an LCD panel (L3D07U-81G00; Seiko Epson Corp.) Results show that the turn-on time is less than 12 ms; the turn-off time is less than 2 ms.

crystal holographic memory requires a period of a few milliseconds for changing holographic contexts. However, once a liquid-crystal holographic memory stores a holographic configuration context array, the holographic memory can successively write configuration contexts of the holographic configuration context array onto the gate array. Consequently, although a context preparation for a liquid-crystal holographic memory takes a long time, once it is completed, rapid configurations can be done easily. The average configuration time becomes the value of the response time of the liquid-crystal holographic memory divided by the array number of holographic configuration context array if it is assumed that the laser reconfiguration period is negligible compared to the response time of a liquid-crystal holographic memory. Therefore, when each configuration time $T_{Conf.}$ is much smaller than the switching time T_{LC} of a liquid-crystal holographic memory, the equation is estimated as the following.

$$T_{Virtualcycle} \simeq \frac{T_{LC}}{N}. \tag{13}$$

As configuration contexts become more numerous in the future, this architecture is expected to become increasingly useful.

6. Conclusion

This chapter has described an optically reconfigurable gate array (ORGA) with an LC-SLM and a reconfiguration-speed acceleration method. The ORGA architecture has enabled the achievement of microsecond-order reconfiguration and nine configuration contexts. Although the LC-SLM response time is not faster than that of silicon devices, the use of multiple configurations decreases the average configuration period. Based on that improvement, this easily programmable LC-SLM was demonstrated as useful for ORGA applications.

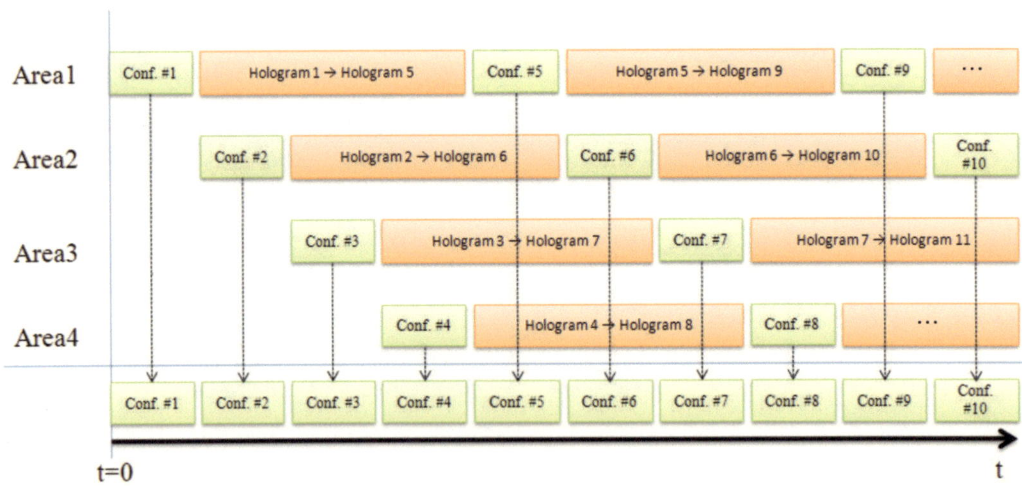

Fig. 12. Successive configuration method for a liquid crystal holographic memory. The figure presents an example of a four-context liquid crystal holographic memory. In this case, successive configurations can be executed as three or four times faster than the switching speed of the liquid crystal holographic memory.

7. Acknowledgments

This research was supported by the Ministry of Education, Science, Sports and Culture, Grant-in-Aid for Scientific Research on Innovative Areas, No. 20200027. The VLSI chip in this study was fabricated in the chip fabrication program of VLSI Design and Education Center (VDEC), the University of Tokyo in collaboration with Rohm Co. Ltd. and Toppan Printing Co. Ltd. Also, some experiments were supported by Mr. M. Nakajima and Mr. T. Mabuchi.

8. References

[1] Takuya Otani, "Seiko Epson Develops New Liquid Crystal Panel for Rear Projection TV," Nikkei Electronics, 2006.

[2] SEIKO EPSON CORPORATION, "Epson Develops Next-Generation HTPS Panel for LCD Projection TVs," News Release, 2006.

[3] Sony Corporation, "Sony develops "SXRD", a display device capable of generating high resolution, high contrast images of film quality smoothness," Press Releases, 2003.

[4] Canon Incorporated, "CANON ANNOUNCES DEVELOPMENT OF OWN LCOS REFLECTIVE LCD PANEL," Press Release, 2008.

[5] J. Mumbru, G. Panotopoulos, D. Psaltis, X. An, F. Mok, S. Ay, S. Barna, E. Fossum, "Optically Programmable Gate Array," SPIE of Optics in Computing 2000, Vol. 4089, pp. 763-771, 2000.

[6] J. Mumbru, G. Zhou, X. An, W. Liu, G. Panotopoulos, F. Mok, and D. Psaltis, "Optical memory for computing and information processing," SPIE on Algorithms, Devices, and Systems for Optical Information Processing III, Vol. 3804, pp. 14-24, 1999.

[7] J. Mumbru, G. Zhou, S. Ay, X. An, G. Panotopoulos, F. Mok, and D. Psaltis, "Optically Reconfigurable Processors," SPIE Critical Review 1999 Euro-American Workshop on Optoelectronic Information Processing, Vol. 74, pp. 265-288, 1999.

[8] Altera Corporation, "Altera Devices," http://www. altera.com.

[9] Xilinx Inc., "Xilinx Product Data Sheets," http://www. xilinx.com.

[10] Lattice Semiconductor Corporation, "LatticeECP and EC Family Data Sheet," http://www. latticesemi.co.jp/products, 2005.

[11] M. Watanabe, F. Kobayashi, "Dynamic Optically Reconfigurable Gate Array," Japanese Journal of Applied Physics, Vol. 45, No. 4B, pp. 3510-3515, 2006.

[12] M. Miyano, M. Watanabe, F. Kobayashi, "Optically Differential Reconfigurable Gate Array," Electronics and Computers in Japan, Part II, Issue 11, vol. 90, pp. 132-139, 2007.

[13] N. Yamaguchi, M. Watanabe, "Liquid crystal holographic configurations for ORGAs," Applied Optics, Vol. 47, No. 28, pp. 4692-4700, 2008.

[14] D. Seto, M. Watanabe, "A dynamic optically reconfigurable gate array - perfect emulation," IEEE Journal of Quantum Electronics, Vol. 44, Issue 5, pp. 493-500, 2008.

[15] http://www.ipflex.co.jp

[16] H. Nakano, T. Shindo, T. Kazami, M. Motomura, "Development of dynamically reconfigurable processor LSI," NEC Tech. J. (Japan), vol. 56, no. 4, pp. 99-102, 2003.

[17] A. Dehon, "Dynamically Programmable Gate Arrays: A Step Toward Increased Computational Density," Fourth Canadian Workshop on Field Programmable Devices, pp. 47-54, 1996.

[18] S.M.Scalera and J.R.Vazquez, "The design and implementation of a context switching FPGA," IEEE symposium on FPGAs for Custom Computing Machines, pp. 78-85, 1998.

[19] S.Trimberger, et al. "A Time--Multiplexed FPGA," FCCM, pp. 22-28, 1997.

[20] D. Jones, D.M.Lewis, "A time--multiplexed FPGA architecture for logic emulation," Custom Integrated Circuits Conference, pp. 495 - 498, 1995.

[21] M. Nakajima, M. Watanabe, "Fast reconfiguration experiments of an optically differential reconfigurable gate array with 9 configuration contexts, IEEE International Symposium on Circuits and Systems, pp. 2013-2016, 2009.

[22] M. Nakajima, M. Watanabe, "Fast optical reconfigurations of a nine-context DORGA," International Workshop on Applied Reconfigurable Computing, Vol. 5453, pp. 123-132, 2009.

[23] T. Mabuchi, M. Watanabe, "A 9-context Optically Reconfigurable Gate Array," International SoC Design Conference, pp. 1-4, 2008.